Windows 10 for the Internet of Things

Controlling Internet-Connected Devices from Raspberry Pi

Second Edition

Charles Bell

Apress®

Windows 10 for the Internet of Things: Controlling Internet-Connected Devices from Raspberry Pi

Charles Bell
Warsaw, VA, USA

ISBN-13 (pbk): 978-1-4842-6608-3   ISBN-13 (electronic): 978-1-4842-6609-0
https://doi.org/10.1007/978-1-4842-6609-0

Managing Director, Apress Media LLC: Welmoed Spahr
Acquisitions Editor: Jonathan Gennick
Development Editor: Laura Berendson
Coordinating Editor: Jill Balzano

Cover image designed by Freepik (www.freepik.com)

Distributed to the book trade worldwide by Springer Science+Business Media LLC, 1 New York Plaza, Suite 4600, New York, NY 10004. Phone 1-800-SPRINGER, fax (201) 348-4505, e-mail orders-ny@springer-sbm.com, or visit www.springeronline.com. Apress Media, LLC is a California LLC and the sole member (owner) is Springer Science + Business Media Finance Inc (SSBM Finance Inc). SSBM Finance Inc is a **Delaware** corporation.

For information on translations, please e-mail booktranslations@springernature.com; for reprint, paperback, or audio rights, please e-mail bookpermissions@springernature.com.

Apress titles may be purchased in bulk for academic, corporate, or promotional use. eBook versions and licenses are also available for most titles. For more information, reference our Print and eBook Bulk Sales web page at http://www.apress.com/bulk-sales.

Any source code or other supplementary material referenced by the author in this book is available to readers on GitHub via the book's product page, located at www.apress.com/9781484266083. For more detailed information, please visit http://www.apress.com/source-code.

Printed on acid-free paper

*This book is dedicated to the medical professionals
who have gone above and beyond what is humanly possible
during this tragic moment in history.
God bless you all.*

Table of Contents

About the Author

 **Charles Bell** conducts research in emerging technologies. He is a member of the Oracle MySQL Development team working on the MySQL Database Service project. He lives in a small town in rural Virginia with his loving wife. He received his Doctor of Philosophy in Engineering from Virginia Commonwealth University in 2005. Dr. Bell is an expert in the database field and has extensive knowledge and experience in software development and systems engineering. His research interests include 3D printers, microcontrollers, three-dimensional printing, database systems, software engineering, and sensor networks. He spends his limited free time as a practicing maker focusing on microcontroller projects and refinement of three-dimensional printers. Dr. Bell maintains a blog on his research projects and many other interests.

About the Technical Reviewer

Reggie Burnett is currently employed as senior software development manager for Oracle Corp., where he is in charge of development projects spanning many different platforms and architectures. Specializing in Windows and .NET technologies, Reggie has written articles for publications such as the *.NET Developers Journal.*

Reggie is married and has four children. He lives in central Tennessee where he plays golf and pool and works on his next geeky project.

Acknowledgments

I would like to thank all of the many talented and energetic professionals at Apress. I appreciate the understanding and patience of my editor, Jonathan Gennick, and managing editor, Jill Balzano. They were instrumental in the success of this project. I would also like to thank the army of publishing professionals at Apress for making me look so good in print. Thank you all very much!

I'd like to especially thank the technical reviewer, Reggie Burnett, for his often-profound insights, constructive criticism, and encouragement. I'd also like to thank my friends for their encouragement and suggestions for things to include in the book.

I'd like to also thank the nice folks at AAEON for their expertise and generosity in the loan of one of their enterprise IoT boards and a special thanks to Michelle Tseng for her patience answering and brokering my questions.

Most importantly, I want to thank my wife, Annette, for her unending patience and understanding while I spent so much time with my laptop.

Introduction

Internet of Things (IoT) solutions are not nearly as complicated as the name may seem to indicate. Indeed, the IoT is largely another name for what we have already been doing. You may have heard of connected devices or Internet-ready or even cloud-enabled. All of these refer to the same thing—be it a single device such as a toaster or a plant monitor or a complex, multidevice product like home automation solutions. They all share one thing in common: they can be accessed via the Internet to either display data or interact with the devices directly. The trick is applying knowledge of technologies to leverage them to the best advantages for your IoT solution. Until the release of Windows 10 IoT Core, those who use Windows wanting to experiment with IoT solutions and in particular hardware like the Raspberry Pi had to learn a new operating system in order to get started. That is no longer true! In this book, we explore how to leverage Windows 10 in your IoT solutions.

Intended Audience

I wrote this book to share my passion for IoT solutions and Windows 10. I especially wanted to show how anyone could use Windows 10 along with a low-cost computing board to create cool IoT projects—all without having to learn a new operating system!

The intended audience therefore includes anyone interested in learning how to use Windows 10 for IoT projects, such as hobbyists and enthusiasts, and even designers and engineers building commercial Windows 10–based IoT solutions.

How This Book Is Structured

The book was written to guide the reader from a general knowledge of IoT to expertise in developing Windows 10 solutions for the IoT. The first several chapters cover general topics, which include a short introduction to the Internet of Things, the Windows 10 IoT Core technologies, and some of the available hardware for IoT. Additional chapters are primers on how to write IoT solutions in a variety of programming languages. Rather than focusing on a single language, which often forces readers unfamiliar with the language to learn new skills just to read the book, I've included tutorials in a number of languages to make the book usable by more readers. Throughout the book are examples of how to implement IoT solutions in the various languages. As you will see, some languages are better suited for certain projects. The book contains five detailed and increasingly complex projects for you to explore and enjoy as you develop IoT solutions with Windows 10. The book also provides two chapters that show you how to work with Microsoft Azure to store and visualize your IoT data via a bonus sixth project. The book concludes with a look at how to grow beyond the material presented. An appendix listing the hardware components for each chapter is included for your convenience. The following is a brief overview of each chapter in this book:

- *Chapter 1, What Is the Internet of Things?*: This chapter answers general questions about the IoT and how IoT solutions are constructed. You are introduced to some terminology describing the architecture of IoT solutions, and you are provided examples of well-known IoT solutions. The chapter concludes with a brief introduction to Windows 10.

- *Chapter 2, Introducing the Windows 10 IoT Core*: This chapter presents a version of Windows 10 called the Windows 10 IoT Core that runs on low-cost computers, such as the Raspberry Pi. You discover the basic features of Windows 10, including how to prepare your PC and get started with Windows 10 on your device. You will also see how to boot up the Raspberry Pi with Windows 10!

- *Chapter 3, Introducing the Raspberry Pi*: This chapter explores the Raspberry Pi and how to set up and configure it using the Linux operating system in order to understand the platform and supporting technologies. You'll also discover a few key concepts of how to work with Linux and get a brief look at writing Python scripts, which are used to write Windows 10 IoT applications in later chapters.

- *Chapter 4, Developing IoT Solutions with Windows 10*: This chapter presents a demonstration on how to get started using Visual Studio 2019. The chapter introduces several Windows 10 IoT Core–compatible hardware boards, including the layout of the GPIO headers. The chapter demonstrates how to build, deploy, and test your first Windows 10 IoT Core application.

- *Chapter 5, Windows 10 IoT Development with C++*: This chapter provides a crash course on the basics of C++ programming in Visual Studio, including an explanation of some of the most commonly used language features. As such, this chapter provides you with the skills that you need to understand the growing number of IoT project examples available on the Internet. The chapter concludes by walking through a C++ example project that shows you how to interact with hardware.

- *Chapter 6, Windows 10 IoT Development with C#*: This chapter offers a crash course on the basics of C# programming in Visual Studio, including an explanation of some of the most commonly used language features. As such, this chapter provides you with the skills that you need to understand the growing number of IoT project examples available on the Internet. The chapter concludes by walking through a C# example project that shows you how to interact with hardware.

- *Chapter 7, Windows 10 IoT Development with Visual Basic*: This chapter is a crash course on the basics of Visual Basic programming in Visual Studio, including an explanation of some of the most commonly used language features. As such, this chapter provides you with the skills that you need to understand the growing number of IoT project examples available on the Internet. The chapter concludes by walking through a Visual Basic example project that shows you how to interact with hardware.

- *Chapter 8, Electronics for Beginners*: This chapter presents an overview of electronics for those who want to work with the types of electronic components commonly found in IoT projects. The chapter includes an overview of some of the basics, descriptions of common components, and a look at sensors. If you are new to electronics, this chapter gives you the extra boost that you need to understand the components used in the projects in this book.

- *Chapter 9, The Adafruit Microsoft IoT Pack for Raspberry Pi*: This chapter explores the Adafruit Microsoft IoT Pack for Raspberry Pi 3 and demonstrates a small project that uses the components in the kit (well, mostly) to read data from a simple sensor.

- *Chapter 10, Project 1: Building an LED Power Meter*: This chapter walks through a project using LEDs to display power (volts). You see how to use a potentiometer as a variable input device, read from an analog-to-digital converter (ADC), learn how to set up and use a Serial Peripheral Interface (SPI), discover a powerful debugging technique, and learn how to create a class to encapsulate functionality.

- *Chapter 11, Project 2: Measuring Light*: This chapter explores a solution that demonstrates how to measure light using a sensor. The project measures the ambient light in the room and then calculates how much power to send to the LED using a technique called *pulse-width modulation* (PWM).

- *Chapter 12, Project 3: Using Weather Sensors*: This chapter demonstrates a very common type of IoT solution—a weather station. In this case, the project uses sensors from the Adafruit kit and implements the code by mixing C# and C++ in the same solution, reusing existing code, and combining it with new code in another language.

- *Chapter 13, Project 4: Using MySQL to Store Data*: This chapter revisits the project from Chapter 12 and modifies it to store the IoT data collected in a MySQL database. Thus, you see an example of how to complete the data storage element of your IoT solutions.

- *Chapter 14, Project 5: Remote Control Hardware*: This chapter presents one method for building IoT solutions that control hardware remotely using a web page.

- *Chapter 15, Azure IoT Solutions: Cloud Services*: This chapter introduces the Microsoft Azure IoT products including a tutorial on how Azure IoT applications work.

- *Chapter 16, Azure IoT Solutions: Building an Azure IoT Solution*: This chapter presents a bonus project that builds on the Azure tutorial from Chapter 15.

- *Chapter 17, Where to Go from Here?:* This chapter explores what you can do to continue your craft of building IoT solutions. Most people want to simply continue to develop projects for themselves, either for fun or to solve problems around the home or office. However, some want to take their skills to the next level. Others may want to know how to take their solutions into the enterprise. This chapter shows you how to do just that.

- *Appendix*: The appendix contains a list of the required hardware components for each chapter.

How to Use This Book

This book is designed to guide you through learning more about what the Internet of Things is, discovering the power of Windows 10 IoT Core, and seeing how to build your IoT solutions using the best language suited for the task.

If you are familiar with some of the topics early in the book, I recommend you skim them so that you are familiar with the context presented so that the later chapters—especially the examples—are easy to understand and implement on your own. You may also want to read some of the chapters out of order so that you can get your project moving, but I recommend going back to the chapters you skip to ensure that you get all of the data presented.

If you are just getting started with Windows 10 or are not well versed in using Visual Studio, I recommend reading Chapters 1–9 in their entirety before developing your own IoT solution or jumping to the example projects. That said, many of the examples permit you to build small examples that you can use to learn the concepts.

Downloading the Code

The code for the examples shown in this book is available on the Apress website, www.apress.com. You can find a link on the book's information page on the Source Code/Downloads tab. This tab is located in the Related Titles section of the page.

Note   Source code can also be found at https://github.com/Apress/win-10-for-internet-of-things.

Contacting the Author

Should you have any questions or comments—or even spot a mistake you think I should know about—you can contact me, the author, at drcharlesbell@gmail.com.

CHAPTER 1

What Is the Internet of Things?

Much has been written about the Internet of Things (hence IoT). Some sources are more about promoting IoT as their next big feature; other sources seem to suggest IoT is something everyone needs to learn or be left behind. Fortunately, books and similar media avoid the sales pitch to expand on the science and technology for implementing and managing the data for IoT, while other texts concentrate on the future or the inevitable evolution of our society as we become more connected to the world around us each and every day. However, you need not dive into such tomes or be able to recite rhetoric to get started with the IoT. In fact, through the efforts of many companies, including Microsoft, you can explore the IoT without intensive training or expensive hardware and software.

This book is intended to be a guide to help you understand the IoT and to begin building solutions that you can use to learn more about it. We will explore many aspects—from understanding what the IoT is to basic knowledge of electronics and even how to write custom software for building solutions for the Internet of Things. Best of all, we do so by using one of the most popular platforms for personal computers—Windows 10—and one of the bestselling hardware for IoT development—the Raspberry Pi.

So, what is this IoT?[1] I'll begin by explaining what it isn't. The IoT is not a new device or proprietary software, or some new piece of hardware. It is not a new marketing scheme to sell you more of what you already have by renaming it and pronouncing it "new and improved."[2] While it is true that the IoT employs technology and techniques

---

[1] https://en.wikipedia.org/wiki/Internet_of_Things

[2] For example, everything seems to be cloud-this, cloud-that when in reality nothing was changed.

© Charles Bell 2021
C. Bell, *Windows 10 for the Internet of Things,* https://doi.org/10.1007/978-1-4842-6609-0_1

that already exist, the way they are employed, coupled with the ability to access the solution from anywhere in the world, makes the IoT an exciting concept to explore. Now let's discuss what the IoT is.

The essence of the IoT is simply interconnected devices that generate and exchange data from observations, facts, and other data, making it available to anyone. While there seem to be some marketing efforts attempting to make anything connected to the Internet an IoT solution or device (not unlike the shameless labeling of everything "cloud"), IoT solutions are designed to make our knowledge of the world around us more timely and relevant by making it possible to get data about anything from anywhere at any time.

As you can imagine, if we were to connect every device around us to the Internet and make sensory data available for those devices, it is clear there is potential for the number of IoT devices to exceed the human population of the planet and for the data generated to rapidly exceed the capabilities of all but the most sophisticated database systems. These concepts are commonly known as addressability and big data, which are two of the most active and debated topics in IoT.

However, the IoT is all about understanding the world around us. That is, we can leverage the data to make our world and our understanding of it better.

The Internet of Things and You

The human body is a marvel of ingenious sensory apparatus that allow us to see, hear, taste, and even feel through touch anything we encounter or get near. Even our brains can store visual and auditory events recalling them at will. IoT solutions mimic many of these sensory capabilities and therefore can become an extension of our own abilities.

While that may sound a bit grandiose (and it is), IoT solutions can record observations in the form of data from one or more sensors. Sensors are devices that produce either analog or digital values. We can then use the data collected to draw conclusions about the subject matter.

This could be as simple as a sensor to detect when a mailbox is opened. In this case, the knowledge we gain from a simple switch opening or closing (depending on how it is implemented and interpreted) may be used to predict when incoming mail has arrived or when outgoing mail has been picked up. I use the term predict because the sensor (switch) only tells us the door was opened or closed, not that anything was placed in or removed from the mailbox itself—that would require additional sensors.

When working with IoT projects that include sensors, you should always think about what conclusions you can draw from the data. Sometimes, like the switch in the mailbox, it can be only a few things, which is most often the case. By defining what we can perceive (learn) from the sensor data, we can better understand what our IoT project and its data can do for us.

A more sophisticated example is using a series of sensors to record atmospheric data such as temperature, humidity, barometric pressure, wind speed, ambient light, rainfall, and so forth, to monitor the weather and perform analysis on the data to predict trends in weather. That is, we can predict within a reasonable certainty that precipitation is in the area and to some extent its severity.

Now, add the ability to see this data not only in real time (as it occurs) but also remotely from anywhere in the world, and the solution becomes more than a simple weather station. It becomes a way to observe the weather about one place from anywhere in the world.

This example may be a bit commonplace since you can tune into any number of television, Web, and radio broadcasts to hear the weather from anywhere in the world. But consider the implications of building such a solution in your home. Now you can see data about the weather at your own home from anywhere!

In the same way, but perhaps on a smaller scale, we can build solutions to monitor plants to help us understand how often they need water and other nutrients. Or perhaps we can monitor our pets while we are away at work. Further, we can record data about wildlife in our area to better understand our effect on nature.

IoT Is More Than Just Connected to the Internet

So, if a device is connected to the Internet, does that make it an IoT solution? That depends on whom you ask. Some believe the answer is yes. However, others (like me) contend that the answer is not unless there is some benefit from doing so.

For example, if you could connect your toaster to the Internet, what would be the benefit of doing so? It would be pointless (or at least extremely eccentric) to get a text on your phone from your toaster stating that your toast is ready given that it only takes a couple of minutes to complete. In this case, the answer is no. However, if you have someone—such as a child or perhaps an older adult—whom you would like to monitor, it may be helpful to be able to check to see how often and when they use a device like a toaster so that you can check on them.[3] That is, you can use the data to help you make decisions about their care and safety.

Allow me to illustrate with another example. I was fortunate to participate in a design workshop held on the Microsoft campus in the late 1990s. During our tour of the campus, we were introduced to the world's first Internet-enabled refrigerator (also called a smart refrigerator). There were sensors in the shelves to detect the weight of food. It was suggested that, with a little ingenuity, you could use the sensors to notify your grocer when your milk supply ran low, which would enable people to have their grocery shopping not only online but also automatic. This would have been great if you lived in a location where your grocer delivers, but not very helpful for those of us who live in rural areas.[4] While it wasn't touted an IoT device (the term was coined later), many felt the device illustrated what could be possible if devices were connected to the Internet.

Thus, being connected to the Internet doesn't make something IoT. Rather, IoT solutions must be those things that provide some meaning—however small that benefit is to someone or some other device or service. More importantly, IoT solutions allow us to sense the world around us and learn from those observations. The real tricky part is in how the data is collected, stored, and presented. We will see all of these in practice through examples in later chapters.

IoT solutions can also take advantage of companies that provide services that can help enhance or provide features that you can use in your IoT solutions. These features are commonly called IoT services and range from storage and presentation to infrastructure services, such as hosting.

IoT Services

Sadly, there are companies that tout having IoT products and services that are nothing more than marketing hype—much like what some companies have done by prepending "cloud" or appending "for the cloud" to the name. Fortunately, there are some good products and services being built especially for IoT. These range from data storage and hosting to specialized hardware.

---

[3]Toasters and toaster ovens have appeared in the top 5 most dangerous appliances in the home. Scary.

[4]However, given the COVID-19 stay-at-home orders in many places, this idea may have come back into practicality.

Indeed, businesses are adding IoT services to their product offerings, and it isn't the usual suspects, such as the Internet giants. I have seen IoT solutions and services being offered by Cisco, AT&T, HP, and countless startups and smaller businesses. I use the term IoT vendor to describe those businesses that provide services for IoT solutions.

You may be wondering what these services and products are and why someone would consider using them. That is, what is an IoT service and why would you decide to buy it? The biggest concerns in the decision to buy a service are cost and time to market.

If your developers do not have the resources or expertise and obtaining them will require more than the cost of the service, it may be more economical to purchase the service. However, you should also consider any additional software or hardware changes (sometimes called retooling) necessary in the decision. I once encountered a well-meaning and well-documented contracted service that permitted a product to go to market sooner than projected at a massive savings. Sadly, while the champions of that contract won awards for technical achievement, they failed to consider the fact that the systems had to be retooled to use the new service. More specifically, it took longer to adopt the new service than it would to write one from scratch. So instead of saving money, the organization spent nearly twice the original budget and was late to market. Clearly, you must consider all factors.

Similarly, if your time is short or you have hard deadlines to meet to make your solution production-ready, it may be quicker to purchase an IoT service rather than create or adapt your own. This may require spending a bit more, but in this case, the motivation is time and not (necessarily) cost. Of course, project planning is a balance of cost, time, and features.

So, what are some of the IoT services available? The following lists a few that have emerged in the last few years. It is likely more will be offered as IoT solutions and services mature.

- *Enterprise IoT data hosting and presentation*: Services that allow your users to develop enterprise IoT solutions from connecting to, managing, and customizing data presentation in a friendly form, such as graphs, charts, and so forth

- *IoT data storage*: Services that permit you to store your IoT data and get simple reports

- *Networking*: Services that provide networking and similar communication protocols or platforms for IoT. Most specialize in machine-to-machine (M2M) services

- *IoT hardware platforms*: Vendors that permit you to rapidly develop and prototype IoT devices using a hardware platform and a host of supported modules and tools for building devices ranging from a simple component to a complete device

Now that you know more about what IoT is, let's look at a few examples of IoT solutions to get a better idea of what IoT solutions can do and how they are employed.

A Brief Look at IoT Solutions

An IoT solution is simply a set of devices designed to produce, consume, or present data about some event or series of events or observations. This can include devices that generate data, such as a sensor, devices that combine data to deduce something, devices or services designed to tabulate and store the data, and devices or systems designed to present the data. Any or all of these may be connected to the Internet.

IoT solutions may include one or all of these qualities, whether it is combined into a single device such as a web camera; used a sensor package and monitoring unit, such as a weather station; or used a complex system of dedicated sensors, aggregators, data storage, and presentation, such as complete home automation system. Figure 1-1 shows a futuristic picture of all devices—everywhere—connected to the Internet through databases, data collectors or integrators, display services, or other devices.

Figure 1-1. *The future of IoT—all devices, everywhere[5]*

Let's take a look at some example IoT solutions. The IoT solutions described in this section are a mix of solutions that should give you an idea of the ranges of sizes and complexities of IoT solutions. I also point out how some of these solutions leverage services from IoT vendors.

Sensor Networks

Sensor networks are one of the most common forms of IoT solutions. Simply stated, sensor networks allow you to observe the world around you and make sense of it. Sensor networks could take the form of a pond monitoring system that alerts you to water level, water purity (contamination), or water temperature; or detects predators; or even turns on features automatically, such as lighting or fish feeders.

If you, or someone you know, have spent any time in a medical facility, it's likely that a sensor network was employed to monitor body functions, such as temperature, cardiac and respiratory rates, and even movement. Modern automobiles also contain sensor networks dedicated to monitoring the engine, climate, and, even in some cars, road conditions. For example, the lane-warning feature uses sensors (typically a camera, microprocessor, and software) to detect when you drift too far toward lane or road demarcations.

---

[5] https://pixabay.com/en/network-iot-internet-of-things-782707/

Thus, sensor networks employ one or more sensors that take measurements (observations) about an event or state and communicate that data to another component or node in the network, which is then presented, in some form or another, for analysis. Let's take a look at an example of an important medical IoT solution.

Medical Applications

Medical applications—including health monitoring and fitness—are gaining a lot of attention as consumer products. These solutions cover a wide range of capabilities, such as the fitness features built into the Apple Watch to fitness bands that keep track of your workout, and even medical applications that help you control life-threatening conditions. For example, there are solutions that can help you manage diabetes.

Diabetes is a disease that affects millions of people worldwide (www.diabetes.org). There are several forms: the most serious being type 1 (www.diabetes.org/diabetes-basics/type-1/?loc=db-slabnav). Those afflicted with type 1 diabetes do not produce enough (or any) insulin due to genetic deficiencies, birth defects, or injuries to the pancreas. Insulin is a hormone that the body uses to extract a simple sugar called glucose, which is created from sugars and starches, from blood for use in cells.

Thus, type 1 diabetics must monitor their blood glucose to ensure that they are using their medications (primarily insulin) properly and balanced with a healthy lifestyle and diet. If their blood glucose levels become too low or too high, they can suffer from a host of symptoms. Worse, extremely low blood glucose levels are very dangerous and can be fatal.

One of the newest versions of a blood glucose tester consists of a small sensor that is inserted in the body along with a monitor that connects to the sensor via Bluetooth. You wear the monitor on your body (or keep it within 20 feet at all times). The solution is marketed by Dexcom (www.dexcom.com) and is called a continuous glucose monitor (CGM) that permits the patient to share their data to others via their phone. Thus, the patient pairs their CGM with their phone and then shares the data over the Internet to others. This could be loved ones, those that help with their care, or even medical professionals. Figure 1-2 shows an example of the Dexcom CGM monitor and sensor. The monitor is on the left, and the sensor and transmitter are on the right. The sensor is the size of a small syringe needle and remains inserted in the body for up to a week.

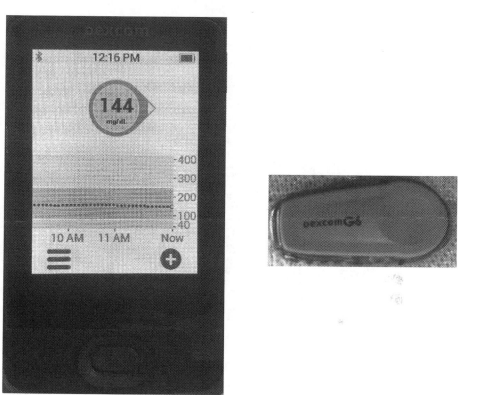

Figure 1-2. *Dexcom continuous glucose monitor with sensor*

WHAT ABOUT BLOOD GLUCOSE TESTERS (GLUCOMETERS)?

Until solutions like the Dexcom CGM came about, diabetics had to use a manual tester. Traditional blood glucose testers are single-use events that require the patient to prick their finger or arm and draw a small amount of blood onto a test strip. While this device has been used for many years, it is only recently that manufacturers have started making blood glucose testers with memory features and even connectivity to other devices, such as laptops or phones. The ultimate evolution of these devices is a solution like Dexcom, which has become a medical IoT device that improves the quality of life for diabetics.

Dexcom also provides a free web-based reporting software called Clarity that is accessed from a special uploading application called the Clarity Uploader (see `http://dexcom.com/clarity` for more details)[6] to allow patients to see the data collected and generate a host of reports they can use to see their glucose levels over time. Reports include averages, patterns, daily trends, and more. They can even share their data with their doctor. Figure 1-3 shows an example of the Dexcom Clarity with typical data loaded.

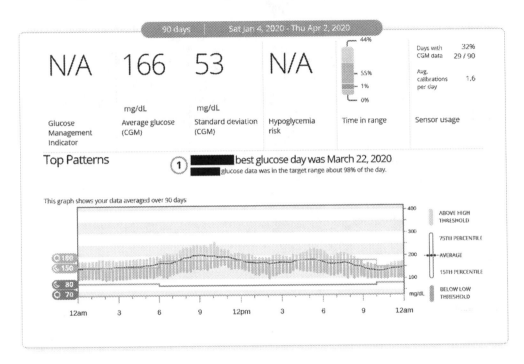

Figure 1-3. *Dexcom Clarity*

A feature called Dexcom Share permits the patient to make their data available to others via an app on their phone. That is, the patient's phone transmits data to the Dexcom cloud servers, which is then sent to anyone who has the Dexcom Share app and has been given permission to see the data. Figure 1-4 shows an example of the Dexcom Share CGM report from the Dexcom Share iOS app, which allows you to check the blood glucose of a friend easily and quickly or loved one.

---

[6]Dexcom also provides a mobile version of Clarity for iOS or Android.

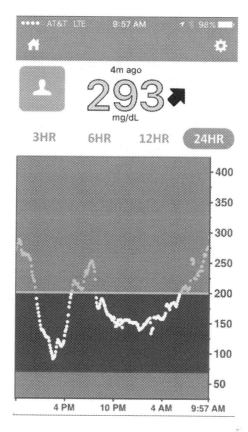

Figure 1-4. *Dexcom Share app report*

Not only does the app allow the visualization of the data, it can also relay alerts for low or high blood glucose levels, which has profound implications for patients who suffer from additional ailments or complications from diabetes. For example, if the patient's blood glucose level drops while they are alone, incapacitated, or unable to get treatment, loved ones with the Dexcom Share app can respond by checking on the patient and potentially avoiding a critical diabetic event.

While this solution is a single sensor connected to the Internet via a proprietary application, it is an excellent example of a medical IoT device that can enhance the lives of not only the patient but everyone who cares for them.

Combined with the programmable alerts, you and your loved ones can help manage the effects of diabetes. If you have a loved one who suffers with diabetes, a CGM is worth every penny for peace of mind alone. This is the true power of IoT materialized in a potentially life-saving solution.

Automotive IoT Solutions

Another personal IoT solution is the use of Internet-connected automotive features. One of the oldest products is called OnStar (www.onstar.com), which is available on most late-model and new General Motors (GM) vehicles. While OnStar predates the IoT evolution, it is a satellite-based service that has several levels and many fee-based options; it incorporates the Internet to permit communication with vehicle owners. Indeed, the newest GM vehicles come with a Wi-Fi access point built into the car! Better still, there are some basic features that are free to GM owners that, in my opinion, are very valuable.

The free, basic features include regular maintenance reports sent to you via email and the ability to use an app on your phone to remotely unlock, lock, and start the car—all the features on your key fob. This is a really cool feature if you have ever locked your keys in your car! Figure 1-5 shows an example of the remote key fob app on iOS. Of course, there are even more features available for a fee, including navigation, telephone, Wi-Fi, and on-call support.

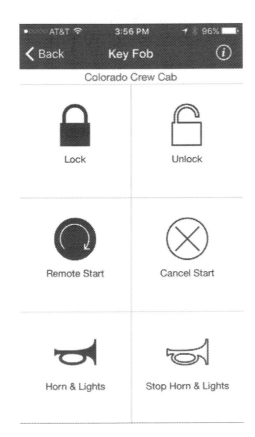

Figure 1-5. *OnStar app key fob feature*

The OnStar app works by connecting to the OnStar services in the cloud, requesting the feature (e.g., unlock) that is sent to the vehicle via the OnStar satellite network. So, it is an excellent example of how IoT solutions use multiple communication protocols.

The feature I like most is the maintenance reports. You will receive an email with an overview of the maintenance status of your vehicle. The report includes such things as oil life, tire pressure, engine and transmission warnings, emissions, air bag, and more. Figure 1-6 shows an excerpt of a typical email that you receive.

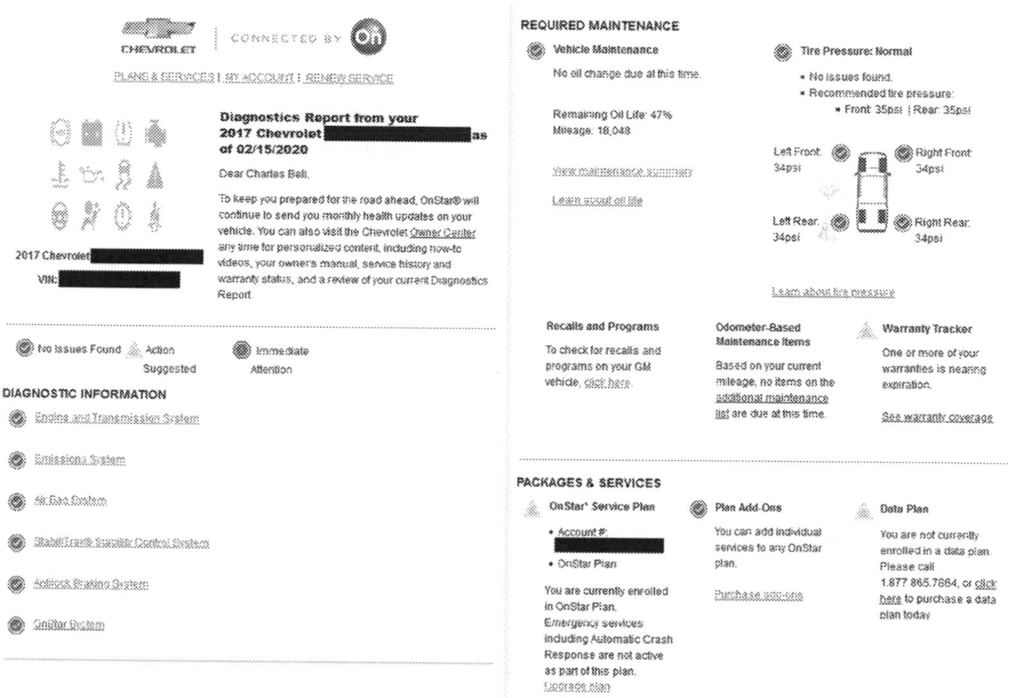

Figure 1-6. *OnStar maintenance report*

Notice the information displayed. This is no mere indicator light. Actual data is transmitted to OnStar from your vehicle. For example, the odometer reading and tire pressure data is taken directly from the vehicle's onboard data storage. Data from the sensors is read and interpreted, and the report is generated for you. This feature demonstrates how automatic compilation of data in an IoT solution can help us keep our vehicles in good mechanical condition with early warning of needed maintenance. This serves us best by helping us keep our vehicles in prime condition and thus in a state of high resell value.

I should note that GM is not the only automotive manufacturer offering such services. Many others are working on their own solutions, ranging from an OnStar-like feature set to solutions that focus on entertainment and connectivity.

Fleet Management

Another example of an IoT solution is a fleet management system.[7] While developed and deployed well before the coining of the phrase Internet of Things, fleet management systems allow businesses to monitor their cars, trucks, and ships—just about any mobile unit—to not only track their current location but also to use the location data (GPS coordinates taken over time) to plan more efficient routes, thereby reducing the cost of shipment.

Fleet management systems are not just for routing. Indeed, fleet management systems also allow businesses to monitor each unit to conduct diagnostics. For example, it is possible to know the amount of fuel in each truck, when its last maintenance was performed—or more importantly, when the next maintenance is due—and much more. The combination of vehicle geographic tracking and diagnostics is called telematics. Figure 1-7 shows a drawing of a fleet management system.

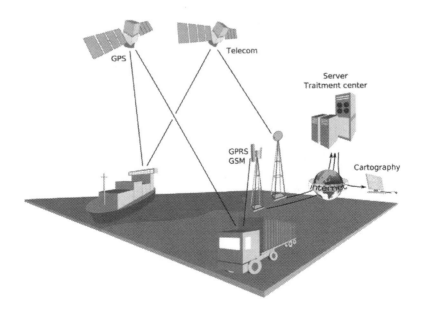

Figure 1-7. *Fleet management example[8]*

---

[7]https://en.wikipedia.org/wiki/Fleet_management

[8]Éric Chassaing—via CC BY-SA 3.0 (http://creativecommons.org/licenses/by-sa/3.0/).

In Figure 1-7, you see the application of GPS systems to track location as well as satellite communication to transmit additional data, such as diagnostics, payload states, and more. All these ultimately traverse the Internet, and the data becomes accessible by the business analysts.

You may think fleet management systems are only for large shipping companies, but with the proliferation of GPS modules and even the microcontroller market, anyone can create a fleet management system. That is, they do not cost millions of dollars to develop.

For example, if you owned a bicycle delivery company, you could easily incorporate GPS modules with either cellular or wireless connectivity on each delivery person to track their location, average travel time, and more. More specifically, you can use such a solution to minimize delivery times by allowing packages to be handed off from one delivery person to another, rather than having them return to the depot each time they complete a set of deliveries.

CAMERA DRONES AND THE IOT

One possible use of the IoT is making data that drones generate available over the Internet. Some people feel that drones are an invasion of privacy. I agree in situations where they are misused, or established laws are violated. Fortunately, most drone owners obey local laws, regulations, and property owners' wishes.[9]

However, there are many legitimate uses of drones, be they land, air, or sea based. For example, I can imagine home monitoring solutions where you can check on your home remotely by viewing data from fixed cameras as well as data from mobile drones. I for one would love to see a solution that allowed me to program a predetermined sentry flight path to monitor my properties with a flying camera drone.

While some vendors have Wi-Fi-enabled drones, there are not many consumer-grade options available that stream data in real time over the Internet. However, it is just a matter of time before we see solutions that include drones. Of course, the current controversy and the movement of the US government to register and track drones, along with increasing restrictions on their use, may limit the expansion of drones and IoT solutions that include drone-acquired data.

---

[9]Drones are increasingly under scrutiny, and the rules change often. If you have a drone and operate in the United States, be sure to check the following website for the latest rules: https://registermyuas.faa.gov/.

IoT and Security

The recent rash of massive data breaches proves that basic security simply is not good enough. We have seen everything from outright theft to exploitation of the data stolen from very well-known businesses, like popular brick-and-mortar retailers, convenience stores, and even some government agencies!

IoT solutions are not immune to security threats. Indeed, as IoT solutions become more and more integrated into our lives, so too will our personal data. Thus, security must be taken extremely seriously and built into the solution from the start.

This includes solutions that we develop ourselves. More specifically, if you design a weather station for your own use, you should take reasonable steps to ensure that the data is protected from both accidental and deliberate exploitation. You may think weather data is not a high risk, but consider the case where you include GPS coordinates for your sensors (a reasonable feature) so that people can see where this weather is being observed. If someone could see that information and determine the solution uses an Internet connection, it is possible they could gain physical access to the Internet device and possibly use it to further penetrate and exploit your systems. Thus, security is not just about the data; it should encompass all aspects of the solution—from data to software to hardware to physical access.

There are four areas where you may want to consider spending extra care ensuring that your IoT solution is protected with good security. As you will see, this includes several things you should consider for your existing infrastructure, computers, and even safe computing habits. By leveraging all these areas, you will be building a layered approach to security, often called a defense-in-depth method.

Security Begins at Home

Before introducing an IoT solution to your home network, you should consider taking precautions to ensure that the machines on your home network are protected. Some of the best practices for securing your home networking include the following:

- *Passwords*: This may seem like a simple thing, but always make sure that you use passwords on all your computers and devices. Also, adopt good password habits, such as requiring longer strings, mixed case, numbers, and symbols to ensure that the passwords are not easily guessed.[10]

- *Secure your Wi-Fi*: If you have a Wi-Fi network, make sure that you add a password and use the latest security protocols, such as WPA2, or even better the built-in secure setup features of some wireless routers.

- *Use a firewall*: You should also use a firewall to block all unused ports (TCP or UDP). For example, lock down all ports except those your solution uses, such as port 80 for HTML.

- *Restrict physical access*: Lock your doors! Just because your network has a great password and your computers use super world-espionage spy-encrypted biometric access, these things are meaningless if someone can gain access to your networking hardware directly. For IoT solutions, this means any external components should be installed in tamper-proof enclosures or locked away so that they cannot be discovered. This also includes any network wiring.

Secure Your Devices

As I mentioned, your IoT devices also need to be secured. The following are some practices to consider:

- *Use passwords*: Always add passwords to the user accounts you use on your IoT devices. This includes making sure that you rename any default passwords. For example, you may be tempted to consider a wee Raspberry Pi or BeagleBone Black too small of a device to be a security concern, but if you consider that these devices run one of the most powerful operating systems available (forms of Linux), a Raspberry Pi can be a very powerful hacking tool.

---

[10]You also need to balance complexity of passwords with your ability to remember them. If you have to write it down, you've just defeated your own security!

- *Keep your software up to date*: You should try to use the latest versions of any software that you use. This includes the operating system as well as any firmware that you may be running. Newer versions often have improved security or fewer security vulnerabilities.

- *If your software offers security features, use them*: If you have servers or services running on your devices, and they offer features such as automatic lockout for missed passwords, turn them on. Not all software has these features, but if they are available, they can be a great way to defeat repeated attacks.

Use Encryption

This is one area that is often overlooked. You can further protect yourself and your data if you encrypt the data as it is stored and the communication mechanism as it is transmitted. If you encrypt your data, it cannot be easily deciphered, even if someone were to gain physical access to the storage device. Use the same care with your encryption keys and passcodes as you do your computer passwords.

Security Doesn't End at the Cloud

There are many considerations for connecting IoT devices to cloud services. Indeed, Microsoft has made it very easy to use cloud services with your IoT solutions. However, there are two important considerations for security and your IoT data:

- *Do you need the cloud?* The first thing you should consider is whether you need to put any of your data in the cloud. It is often the case that cloud services make it very easy to store and view your data, but is it really necessary to do so? For example, you may be eager to view logistical data for where your dog spends his time while you are at work, but who else would really care to view this data? In this case, storing the data in the cloud to make it available to everyone is not necessary.

- *Do not relax!* Many people seem to let their guard down when working with cloud services. For whatever reason, they consider the cloud more secure. The fact is—it is not! In fact, you must apply the very same security best practices when working in the cloud that you do for your own network, computers, and security policies. Indeed, if anything, you need to be even more vigilant because cloud services are not in your control with respect to protecting against physical access (however remote and unlikely) nor are you guaranteed your data isn't on the same devices as tens, hundreds, or even thousands of other users' data.

Now that you have an idea of how you should include security in your projects, let's look at how Windows 10 has evolved into a modern platform that supports not only the usual productivity and gaming tasks but also help us build IoT solutions.

Introducing Windows 10

Microsoft has not been idle in recent years. In fact, the latest release of the Windows operating system, Windows 10, has shown Microsoft listens to its customers and delivers features that people want. More than any release in the past, Windows 10 is both familiar and capable on desktop, laptop, and tablets. In fact, Windows 10 has become the most stable Windows release in history.

As a long-term platform-independent user, I have had my favorites over the years, but some versions of Windows have not been high on the list and at times not on the list at all. This was mostly due to the changing face of the PC from beige boxes[11] to personal, tactile, sensitive devices through the proliferation of tablets and other smart devices.

However, with the release of Windows 10 and their line of Surface computers (desktop, laptop, and tablet varieties), I consider my Surface computers a platform that I am very comfortable using and wouldn't hesitate to use for almost any task. It just works the way a Windows 10 computer should. In fact, I used my Surface Laptop for all the examples in this book.

In case you are new to Windows 10, the following sections introduce a number of the newest features of Windows 10, including some familiar behavior that has been missing for some time, and some new things that make using Windows 10 across several platforms seamless. I have included this information for those that have yet

---

[11]Yes, I was using PCs when IBM put the PC in personal computer. My first PC had an Intel 8088 processor running at 8MHz with a modest 512KB of memory. Most phones exceed these capabilities by orders of magnitude.

to experience Windows 10 or those that have delayed upgrading. To use this book, you need a machine running Windows 10. If you have not upgraded yet, the following sections will be helpful.

However, if you are already using Windows 10 or have been for some time, you may want to skim this section so that you are familiar with the newest features. I have found that it is always helpful to read the impressions of others because I often discover features that I was not aware of or had yet to encounter. Plus, it gives me a greater depth of knowledge on the subject.

Overview of Windows 10 Features

This section explores the major advances and new features of Windows 10. If you are thinking about upgrading to Windows 10, this section should convince you to do so, because it covers what the latest Windows operating system has become. I cover the most important features related to developing IoT applications. Thus, this is not a complete list of the many features of the new version. For a complete list, see the Microsoft Windows 10 site (`www.microsoft.com/en-us/windows/features`).

The Return of the Desktop

One of the evolutions I found to be most unappealing was the shift from a desktop with a Start menu to that of a panel of small application interfaces. While I understand the reason for the evolution (the rise of the touch screen and tablets), I found the dual interface of Windows 8 confusing. It was as if the operating system had two heads: one for "legacy" users, complete with an abbreviated Start menu, and another for "new" touch-enabled applications (which seemed to only include the latest Office applications). Switching from one to another—particularly on a typical desktop without a touch screen—was awkward and often frustrating.

Fortunately, Windows 10 brings back the desktop concept with an all-new design that incorporates the best of the Windows 8 Start screen with a much improved Start menu. That is, we once again have the familiar menu, floating windows, tray, and more. Figure 1-8 shows a snapshot of the new desktop. Does this look familiar?

If you use a tablet or a machine that can switch from laptop to tablet (also called a 2-in-1), Windows allows you to use the new desktop when in desktop mode (a keyboard is attached) and the more tablet-friendly Start screen when in tablet mode (the keyboard is removed). Of course, you can configure this behavior to your liking.

Tip   You can access the power user menu by pressing *Windows key + X.*

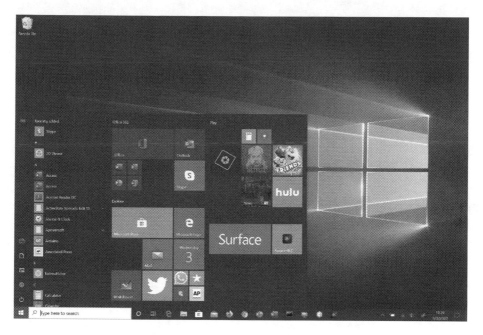

Figure 1-8. Windows 10 desktop

After having used the new desktop for some time, I must say it just works the way a PC should. That may seem like false praise, but it is not. There are many things about Windows 10 that work like it should—as it should have in previous versions.

One feature of the desktop I welcome more than any other is the use of virtual desktops. The other platforms I use have had this feature for some time. Having it in Windows 10 allows me to use my PCs in a very familiar manner: by placing my most frequently used (and running) applications maximized in their own desktops.

To create a new virtual desktop, click the task view icon on the taskbar (see the red arrow in Figure 1-9). You will then see a pop-up pane that shows a thumbnail of all the virtual desktops that are active. To add a new virtual desktop, click the plus sign to the right. You can close a virtual desktop by clicking the *X* icon on the thumbnail. Figure 1-9 shows how to access the virtual desktop feature depicting an excerpt of a full screen snapshot. You may find using virtual desktops to be very helpful when developing applications or working on productivity applications alongside your mail and other communication applications.

Figure 1-9. *Virtual desktops in Windows 10*

The task view is also used in tablet mode. In fact, the virtual desktop is very similar to the tablet mode of the desktop. You can switch from one to the other by using the task view icon.

Compatibility

With so many changes and new features, it is reasonable to expect Windows 10 to have problems running older applications. However, Microsoft has worked very hard to make the new version run the older applications. In fact, I have several rather old (Windows XP era) applications that I have installed on Windows 10, and they all work well. There are a few things that you can do to adjust compatibility, but most applications should run unaltered and without jumping through menus to customize. Thus, if you are concerned about being able to run your older applications, you should not have to worry.

Notifications and Action Center

The Action Center is an interesting feature that allows application developers to display notices. The Action Center is accessed by a right edge swipe or by clicking the Action Center icon on the system tray.

When applications trigger a notification, the Action Center displays the notification briefly as a small fly-in dialog. I like this new feature, especially since the other platforms I have used have their own implementation. Not only is it convenient to know what is going on—such as getting new emails or receiving a bid on your auction—it is also a great way to take a look at the events of the day. Just swipe and view all your notifications in a list (see Figure 1-10).

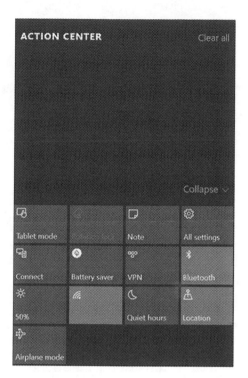

Figure 1-10. *Windows 10 Action Center*

At the bottom of the Action Center are a few shortcut buttons for many common tasks, which put the action in Action Center. Here we find buttons for switching to tablet mode, brightness, Bluetooth, network connections, and more. The Action Center, coupled with the new desktop and Start menu, completes the user experience.

Fortunately, you can develop your IoT solutions to include notification to users. I encourage you to do so in your own Windows applications.

Windows 10 and the IoT

There are several versions of Windows 10—ranging from those that run on phones to those that run on desktop computers and laptop replacement tablets. Indeed, Microsoft Windows 10 runs on more devices than any previous version of Windows, but that is not the end of the Windows 10 proliferation. Windows 10 also comes in a version designed to run on low-cost computing hardware, such as single-board computers designed for integration with hardware and embedding in other solutions, which makes it an excellent choice for use in IoT solutions.

The Windows 10 version designed to run on low-cost computing boards is called the Windows 10 IoT Core. We use Windows 10 IoT Core in this book. The Windows 10 IoT Core is designed to run on smaller devices, such as the Raspberry Pi, MinnowBoard Max, and other small computing boards. It is optimized to run in smaller memory without the need for advanced processors or even a graphical user interface. Thus, it supports only console or background applications.

Windows 10 IoT Core supports the Universal Windows Platform (UWP), allowing you to create your applications and deploy them. As mentioned previously, we will use Visual Studio 2019 to build these applications. As you will see in Chapter 4, Visual Studio 2019 includes all the tools you need to build UWP applications for deployment to the Raspberry Pi. This includes APIs and drivers for accessing the general-purpose input/output (GPIO) pins, as well as interfaces such as I2C and SPI. Best of all, there are a host of example code that you can use for your own projects.

This really is an exciting element to Windows 10. Indeed, except for some rather limited exploration of Windows Embedded Compact (Windows CE), Windows 10 represents the first time that you can use the Windows 10 IoT Core to leverage the power of Windows on your smaller devices. You explore the Windows 10 IoT Core in greater detail in the next chapter.

Summary

The Internet of Things is an exciting new world for us all. Those of us young at heart but old enough to remember *The Jetsons* TV series recall seeing a taste of what is possible in the land of make believe. Talking toasters, flying cars that spring from briefcases, and robots with attitude notwithstanding, television fantasy of decades ago is now coming true. We have wristwatches that double as phones and video players. We can unlock our cars from around the world, find out if our dog has gone outside, and even answer the door from across the city. All of this is possible and working today with the advent of the IoT.

In this chapter, we discovered what the IoT is and saw some examples of well-known IoT solutions. We also discovered how Microsoft is opening doors for Windows users by expanding its Windows 10 operating system to the IoT via the Raspberry Pi and similar low-cost computing hardware. This is a very exciting opportunity for people who do not want to learn the nuances of a Linux-based operating system to explore the world of hardware and IoT from a familiar and well-understood platform.

In the next chapter, we will discover more about the Windows 10 IoT Core including what hardware it runs on and how to get started running Windows 10 on a Raspberry Pi. As you will see, it is not difficult. We will then explore the Raspberry Pi in more detail in Chapter 3 to complete our tour of getting started with Windows 10 IoT Core.

## CHAPTER 2

Introducing the Windows 10 IoT Core

Windows 10 represents an exciting entry in the IoT arena, giving Windows users a native toolset to experiment with building IoT solutions. While some platforms such as the Arduino are very Windows-friendly, other platforms have forced users to learn about new, sometimes very different, operating systems or tools that are, in contrast to Visual Studio, very challenging to learn. In fact, I've heard of some people giving up altogether or not even trying because the operating system and tools seemed too intimidating.[1] All of these became roadblocks for those wanting a familiar and easy-to-use platform to develop IoT solutions.

Microsoft has risen to the occasion, creating a unique way to develop applications for and deploy solutions to hardware that traditionally has been off limits for many Windows users who did not want to learn a new operating system, such as Linux, which is the most popular choice for embedded hardware development. Regardless, I recommend you understand the basics of these other platforms. In fact, there is a short primer on the Raspberry Pi in the next chapter including a look at the base or preferred operating system. But don't worry; you need not become a Linux expert to use Windows 10 with the Raspberry Pi.

In this chapter, you discover a version of Windows 10 called the Windows 10 IoT Core that runs on low-cost computers, such as the Raspberry Pi. You will discover the basic features of Windows 10, including how to prepare your PC and get started with the Windows 10 on your device. You will even see how to boot up the Raspberry Pi with Windows 10! Let's start with what you will need to get started.

---

[1]Learning Linux isn't really so terrible, as you shall see, but if you've never used such an operating system, it can be frustratingly difficult to learn how to do even simple tasks.

© Charles Bell 2021
C. Bell, *Windows 10 for the Internet of Things*, https://doi.org/10.1007/978-1-4842-6609-0_2

Things You Will Need

Working with Windows 10 IoT Core is easy, but you will need a few things to get started. First and foremost, you will need an IoT board that you can use to run Windows 10 IoT Core. You will also need to download the correct image for your IoT board from Microsoft, install it on the correct medium, then use that to boot your IoT board. Finally, to create IoT solutions, you will need to install and configure Visual Studio 2019 on your Windows 10 machine. This may sound like a lot, but most of the steps need only run once.

In this section, we will examine each of these topics and introduce the components you need, starting with Windows 10 IoT Core.

Windows 10 IoT Core

Windows IoT Core gives the ability for Windows users to leverage their experience and knowledge of developing applications for Windows on smaller devices. While Microsoft has offered several products designated as "embedded" or "compact" or "embedded compact" in the past, which were scaled-down versions of the operating system, there were many differences and a few bridges that had to be crossed to use them. While highly touted, the offerings never really lived up to the "write the code once and deploy everywhere" mantra. That is, until now.

WHAT MAKES IOT CORE SO SPECIAL?

Unlike the previous limited featured Windows products meant for smaller platforms, Windows 10 IoT Core shares many of the same components as the flagship operating system for PCs. That is, it has the same core components and kernel, and even some of the middleware is based on the same core code. In fact, the code generated can be binary compatible with the other platforms, which means you can write code that can run on either the IoT device or your PC. It should be noted that this capability is highly dependent on what the code does. For example, if your code accesses the general-purpose input/output (GPIO) hardware pins on the low-cost computing board, you cannot run the application on the PC (there are no GPIO pins on the PC).[2]

---

[2]Well, most PCs. Some low-cost computing boards are simply a fancy case away from a fully functional PC.

Interacting with Hardware

The ability to access hardware directly—such as the GPIO pins—is what makes Windows 10 IoT Core so attractive to hobbyists and IoT enthusiasts who want to build custom hardware solutions using small, inexpensive hardware.

For example, if you wanted to build a simple device to signal you when someone opens your screen door, you would likely not use a PC costing several hundreds of dollars. Not only would that be expensive and bulky, there isn't an easy way to connect a simple switch (sensor) to your PC, much less to a PC located elsewhere. It would be much more cost effective to use a simple switch connected to a small, inexpensive set of electronic components using a simple application to turn on an LED or ring a buzzer. What makes the Windows 10 IoT Core even more appealing is you don't have to relearn how to write software—you can write a Windows application to run on the small device.

Video Support

Since most IoT devices do not include a monitor (some may), Windows 10 IoT Core is designed to run headless (without a monitor) or headed (with a monitor). Headless solutions require less memory since they do not load any video libraries or subsystems. Headed solutions are possible if the hardware chosen supports video (all current hardware options have HDMI video capabilities).

Thus, you can create IoT solutions with visual components or interactive applications, such as those for kiosks, or even interactive help systems. You choose whether the application is headless or headed by the configuration of the device. In fact, the configuration is accessible from the device or remotely through a set of tools running on your PC. You'll see more about these features later in this chapter.

One Platform, Many Devices

For developers of Windows 10 applications, including IoT solutions, Microsoft has adopted a "one Windows" philosophy where developers can develop their code once and run it on any installation of Windows. This is accomplished with a technology called the Universal Windows Platform API (sometimes called UWP or universal applications or UWP apps).

Thus, developers can create an application that runs on phones, tablets, desktop, and even servers without having to change their code or exchange different libraries. As you will see once you start with the projects in this book, you are developing your applications (apps, scripts, etc.) on your Windows 10 desktop (tablet) and deploying them to the Raspberry Pi—all without having to move the code to the Raspberry Pi, alter it, compile it, and so on. This is a huge improvement for IoT developers over other development choices.

For example, if you chose to use a Raspberry Pi with its default operating system, you would have to learn how to develop Linux applications—complete with learning new development tools (if not a new code editor). With Windows 10, you use an old favorite— Visual Studio—to build and deploy the application. How cool is that?

The real power of the UWP API is discussed in Chapter 4 as you explore how a single application (code) can be compiled on your PC and deployed to the Raspberry Pi. Indeed, the UWP API allows you to write one solution (source code) and deploy it to any Windows 10 device from a phone, low-cost computing board, PC, tablet, and so forth. This opens the possibility of using any of the Windows 10 supported devices in your projects.

Windows 10 IoT Hardware

Windows 10 IoT Core is designed and optimized to run on smaller devices, such as low-cost computing boards. Furthermore, Windows 10 IoT Core can run headless[3] (without a display), thereby removing the need for sophisticated graphics (but still supports graphic applications with special libraries). All this is possible with the extensible Universal Windows Platform (UWP) API, as described earlier.

Currently, the Windows 10 IoT Core runs on the Raspberry Pi 2B or 3B (3B is recommended), MinnowBoard Max–compatible boards, and the Arrow DragonBoard 410c. All of these boards are considered low-cost computing platforms. I describe each of these briefly in the upcoming sections.

Note   At the time of this writing, the Microsoft documentation still includes the MinnowBoard, but the links to the MinnowBoard site and Wiki no longer exist. Thus, we omit this board from our survey of low-cost IoT boards for Windows 10 IoT Core. However, Microsoft is still providing a download for the binary to run on

---

[3]Kind of like the headless horseman, but more like the horseless headsman.

these boards at `https://go.microsoft.com/fwlink/?linkid=846057`. If you want to use this board, you can follow the directions from the first edition of this book.

If you are considering a commercial or enterprise-level project, Windows 10 IoT Core is also supported on several high-performance (and thus more costly) boards including the AAEON UP Squared (compact Intel Atom PC for serious makers), Keith & Koep i-PAN M7/T7 CoverLens (touch panel computers), and NXP i.MX 6/7/8 (multi-core ARM processor boards for advanced graphics, imaging, machine vision, audio, voice, video, and safety-critical applications). We will discuss using an enterprise board in Chapter 17 but will concentrate on the low-cost boards to illustrate the concepts and to experiment with sample projects.

Tip For more information about each of the high-performance boards including links to the manufacturer sites, see `https://docs.microsoft.com/en-us/windows/iot-core/tutorials/quickstarter/prototypeboards`.

WHY ARE THEY CALLED LOW-POWER COMPUTING PLATFORMS?

Low-powered computing platforms, sometimes called low-cost computer boards or mini-computers, are built from inexpensive components designed to run a low resource–intensive operating system. Most boards have all the normal features you would expect from a low-cost computer, including video, USB, and networking features. However, not all boards have all of these features.

The reason they are sometimes called low power isn't because of their smaller CPUs or memory capabilities; rather, it is because of their power requirements, which are typically between 5V and 24V. Since they do not require a massive, PC-like power supply, these boards can be used in projects that need the capabilities of a computer with a real operating system but do not have space for a full-sized computer, cannot devote the cost of a computer, or must run on a lower voltage.

There are many varieties of low-cost computing boards. Some support the full features of a typical computer (and can be used as a pretty decent laptop alternative), while others have the bare essentials to make them usable as embedded computers. For example, some boards permit you to connect a network cable, keyboard, mouse, and monitor for use as a normal laptop or desktop computer, while others have only networking and USB interfaces, requiring you to remotely access them in order to use them. Fortunately, all the low-cost computing boards available for Windows 10 have support for networking, video, and USB peripherals.

Raspberry Pi

The Raspberry Pi 3 Model B is one of the older iterations of the Raspberry Pi (www. raspberrypi.org/products/raspberry-pi-3-model-b/). It has all the features of the original Raspberry Pi 2B but with a faster 64-bit quad core processor and onboard Wi-Fi (a first for the Raspberry Pi). However, the Raspberry Pi 2B is more than capable of running Windows 10 IoT Core solutions. Since the 2B and 3B are older boards, you can find them on various websites for less than the current models. In fact, if you know someone who has been working with the Raspberry Pi, they may have a few older boards laying around you could borrow.

So, what about the Raspberry Pi Zero, 3B+, and 4B? It turns out, Microsoft recommends using the 2B or 3B boards (not Zero, 3B+, or 4B) for Windows 10 IoT Core, citing the newer boards are not supported. Indeed, there are no images of Windows 10 IoT Core to download for these boards. Thus, we will use the Raspberry Pi 3B for this book and hence refer to it as the Raspberry Pi.

BUT WAIT, THERE IS HOPE FOR THE RASPBERRY PI 3B+!

If you have a Raspberry Pi 3B+, you can use the technology preview build that Microsoft provides (unofficially). Download the image from https://software-download. microsoft.com/pr/Windows10_InsiderPreview_IoTCore_RPi3B_en-us_17661. iso?t=04fac461-a78b-4e26-a432-e0547db8ebad&e=1592249349&h=c8faa3509876 86ea4881cbfffacb5aab and use that to build a disk image for your 3B+. If that link expires, search for InsiderPreview_IoTCore_RPi3B on the Microsoft website.

The Raspberry Pi is a popular board with IoT developers mainly because of its low cost and ease of use. Given the popularity of the Raspberry Pi, I cover it in greater detail in Chapter 3, including a short tutorial on how to get started using it with its native operating system. Thus, I briefly cover the highlights here and reserve a more detailed discussion on using the board for Chapter 3.

Note   I describe the Raspberry Pi 3B here, but you can use the Raspberry Pi 2B.

The Raspberry Pi 3B hardware includes a 1.2GHz 64-bit ARM CPU, 1GB RAM, video graphics with HDMI output, four USB ports, Ethernet, a camera interface (CSI), a display interface (DSI), a micro-SD card, and 40 GPIO pins. Figure 2-1 shows the Raspberry Pi 3B board.

Figure 2-1. *Raspberry Pi model 3B (courtesy of www.raspberrypi.org)*

The camera interface is interesting. You can buy one of several camera modules like the ones at Adafruit (`http://adafruit.com/categories/177`) including low-light and high-precision versions. You then connect it to the board with a flat ribbon cable. You can use it for all manner of things like video conferencing or for use as a remote video-monitoring component. I've used this feature extensively by turning a couple of my Raspberry Pi boards into 3D printing hubs where I can send print jobs over the network to print and check the progress of the prints remotely or as low-cost video surveillance devices.

The LCD interface is also interesting because there is now a 7-inch LCD touch panel that connects to the DSI port (`http://element14.com/community/docs/DOC-78156/l/raspberry-pi-7-touchscreen-display`). I have also seen a number of interesting Raspberry Pi tablets built using the new LCD touch panel. You can learn about one promising example (made by Adafruit, so I expect it to be excellent) at `http://thingiverse.com/thing:1082431`.

To date, the Raspberry Pi has been my go-to board for all manner of small projects due to its low cost and availability. There are also many examples from the community on how to employ the Raspberry Pi in your projects. For more information about the Raspberry Pi, see Chapter 3.

One of the things I like about the Raspberry Pi is you can run a number of operating systems on it by installing the operating system on a micro-SD card. This allows me to use a single Raspberry Pi for a host of projects, each with its own micro-SD card. In fact, the basic setup at `www.raspberrypi.org` includes a special boot loader that permits you to install the operating system of your choice.

Arrow DragonBoard 410c

The Arrow DragonBoard 410c is a low-cost computing board that incorporates the Qualcomm quad core Snapdragon 410 processor. This processor is an ARM Cortex–based single-chip system supporting a wide variety of hardware from USB to networking. The processor runs up to 1.2GHz per core in either 32- or 64-bit mode, which is a bit more powerful than the Raspberry Pi.

The board is a fully featured low-cost computing platform complete with 1GB of RAM, 8GB onboard storage (eMMC), an HDMI 1080p display (with audio over HDMI), Wi-Fi, Bluetooth, GPS (yes, GPS!), USB ports, and even a micro-SD card. Figure 2-2 shows the DragonBoard 410c.

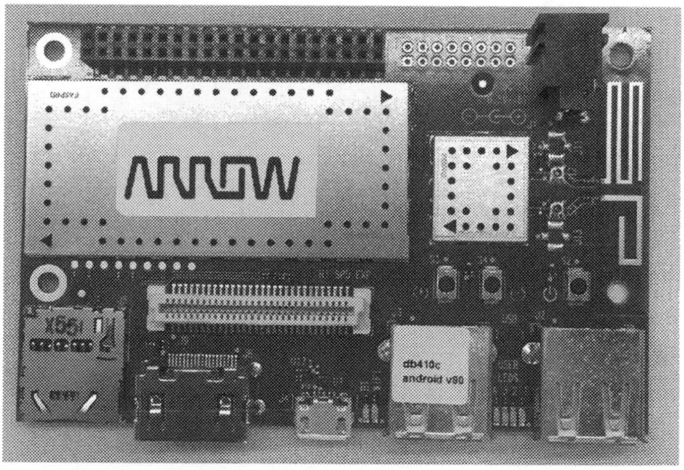

Figure 2-2. *Arrow DragonBoard 410c*

Interestingly, the DragonBoard 410c can be booted from the onboard memory using the Android 5.1 operating system, provided you haven't loaded Windows 10 IoT Core because you will overwrite the base operating system. However, you can recover the factory settings by following the procedure at `https://github.com/96boards/documentation/wiki/Dragonboard-410c-Installation-Guide-for-Linux-and-Android`. Figure 2-3 shows the default operating system (Android-based) of the DragonBoard 410c. Thus, you could use the DragonBoard 410c as an ultra-compact desktop or laptop computer.

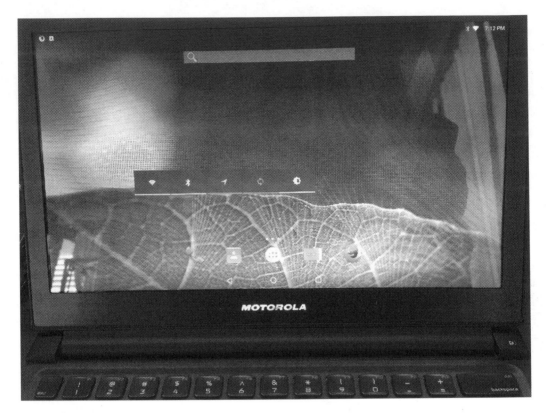

Figure 2-3. *Onboard Android OS on the DragonBoard 410c*

Note   The DragonBoard 410c does not use an SD card to boot Windows 10 IoT Core. I discuss these differences in a later section.

Given its small size, onboard Wi-Fi, USB, GPIO header, and more, the DragonBoard 410c is a good alternative to the Raspberry Pi. Yes, it does cost more (about $75 vs. about $35 for the Raspberry Pi), but if you need the more powerful processor and convenience of onboard Wi-Fi, you may want to consider it for solutions that need a bit more processing power.

Tip   For more details on the DragonBoard 410c, visit the Arrow data sheet at www.arrow.com/en/products/dragonboard410c/arrow-development-tools#partsDetailsDatasheet.

The best source for purchasing an Arrow DragonBoard 410c is from the manufacturer directly; go to `www.arrow.com` for details on ordering a board to complete your low-cost computer board arsenal. Note that the manufacturer stocks a host of additional electronic components, making them another source for gathering components for your IoT project. You can also find it on Maker Shed at `www.makershed.com/products/dragonboard`.

Which One Should I Choose?

The three boards are those that are currently supported for use with Windows 10 IoT Core. Which you choose is largely up to you as each has their merits. Perhaps the most compelling reasons to choose one over the others for most hobbyists and enthusiasts are cost and availability.

At the time of this writing, the Raspberry Pi costs less than the other boards and is much easier to find. The Raspberry Pi costs about $35, the DragonBoard 410c about $75, making the Raspberry Pi the most economical for initial cost.

Since most readers want to limit their investment (hardware can get expensive quickly once you start buying sensors and other bits and bobs you need), I focus on the Raspberry Pi in this book. Even if you plan to use one of the other boards, following along with the Raspberry Pi helps you learn the skills that you need without spending considerably more for the base component. However, if you need to use one of the other boards, the examples in the rest of this book can be adapted without much fuss.

Consider another possibility. If you are new to electronics, or you make a mistake with powering your board or components, you could damage the board.[4] Yikes! I have a small, sad box of components I've managed to destroy over the years. Fortunately, it is a small box with only a few inexpensive (but quite dead) components. I keep it around to remind me what a simple mistake reading a wiring diagram can do to your wallet. Wouldn't a $35 board that you can get from a host of vendors be a bit easier to accept if you kill it?

Although this book favors the Raspberry Pi for its economy and availability, the other boards are strong alternatives that you may want to consider if you need more powerful hardware. If cost were not an option, I would likely use the other boards more often.

---

[4]Hey, it happens to everyone.

Additional Hardware

If you are just getting started with these boards, there are a number of things that you need, including some additional hardware (e.g., cables) to connect to and use the boards. You also need some software to write and deploy your software to the board.

To use these boards, you need, at a minimum, a power supply, network connection, and a micro-SD card. There is some optional hardware you may want to have on hand as well. I explain some of these in more detail.

Power Supply

The Raspberry Pi can be powered by a 5V USB power supply via a USB type A male to a micro-A male cable (a commonly used cable for small electronic devices). Be careful with this cable, as the smaller end is rather fragile and easy to damage. Be sure to use the newer 2.0A or 2.5A power supply for best results.

The DragonBoard 410c must be powered by a dedicated power supply capable of delivering 6.5–18V, whereas the Raspberry Pi boards require a 5V 2.5A power supply. You can buy the correct power supply for each of these boards from the supplier, but you can use any power supply rated for the correct voltage and amperage.

I like to use universal power supplies with a variety of connectors that can be switched to different voltages. Figure 2-4 shows a typical universal power supply with several tips. If you want to minimize the gear knocking around on your workbench, get a universal power supply like this one. However, be sure to test the adapter at the proper setting. Some inexpensive universal power supplies are quite inaccurate and may produce more or less voltage or amperage than what is advertised. Thus, you should buy one that has been reviewed by others and has good reviews from buyers.

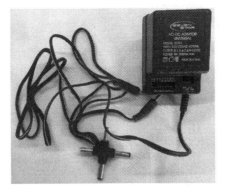

Figure 2-4. *Universal power supply*

Be sure to get one that can generate at least 3.3V, has a variety of tips (sizes), and the polarity of the plug can be selected. That is, some devices require the center pin to be positive. Having the ability to switch the polarity of the center pin makes the power supply usable on a wider variety of boards. I should note that most of the power supplies I've seen that have replaceable tips permit selecting polarity by plugging in the tip in one of two ways. However, this isn't always easy to tell which way the tip is oriented, so check it twice before using it.

Caution   Some universal power supplies may not generate the required amperage for your board. Be sure to check the power rating of the power supply to be sure it matches your board before you buy or use it to power the board.

Networking

The Raspberry Pi has an Ethernet port. While you can use wireless connections with both boards (e.g., via a USB dongle), it is easier to simply use an Ethernet cable and plug it into your network. The DragonBoard 410c, on the other hand, has wireless networking onboard and can be used to connect to your wireless network. If you choose to use a Raspberry Pi 3, you can use the onboard Wi-Fi.

Optional Hardware

There are a number of optional hardware components you may want to have on hand. In fact, they can be quite convenient for getting started with the board. Fortunately, all the boards have built-in video and USB host capabilities, making it easy to set them up with interactive peripherals.

You need an HDMI-compatible monitor. The monitor doesn't have to be an expensive, 30-inch 4K display, since none of these boards has that sort of video capability. I recommend a small HDMI monitor of about 7 inches or larger. Naturally, you also need the appropriate HDMI cable. The Raspberry Pi and DragonBoard 410c use a standard HDMI cable. When buying HDMI cables, be sure to purchase high-quality cables because not all cables are fully wired, and they may not work. I've found the best are those designed to support audio and Ethernet, but these do cost a bit more. The cables at www.mycablemart.com are of sufficient high quality, and you can get whatever configuration of connectors you need including an assortment of adapters if you already have some HDMI cables.

If you want to interact with the device for setup or configuration, you should consider a small USB keyboard and mouse. Only the Raspberry Pi has a surplus of USB ports, so you may want to consider a keyboard that has a USB hub or the mouse built in. Figure 2-5 shows an example of a compact wireless keyboard that I use for my low-cost computing boards (see www.adafruit.com/products/922). I like it for its small size, built-in mouse, and even a small speaker for audio.[5]

Figure 2-5. *Mini wireless keyboard*

The keyboard is only about 6 inches long, making it easy to pack away in your kit bag, but typing on one won't earn you any speed typing merit badges and can be a bit tedious. The keyboard comes with its own USB dongle that is compatible with Windows 10 IoT Core. You can find these under various vendor names on Amazon and other popular online computer vendors.

However, since these small keyboards are sold in many slight variations, you may want to buy one from a vendor that is willing to accept it as a return if it doesn't work. That said, a wired USB keyboard and mouse are the safer alternatives.

Tip   For a complete list of hardware supported by Windows 10 IoT Core including USB devices, see https://docs.microsoft.com/en-us/windows/ iot-core/learn-about-hardware/hardwarecompatlist.

---

[5]Not supported on all platforms but works great with Android OS.

Software Development Tools

The software development tool of choice for Windows 10 IoT Core is Visual Studio 2019 (www.visualstudio.com). You can use any version of Visual Studio 2019, including the free community version. Yes, this means developing applications for the Windows 10 IoT Core uses a very familiar tool with a robust feature set. As you will see, you can leverage nearly all the features of Visual Studio when developing and deploying your IoT solutions.

Better still, there is a growing list of add-ons, sample applications, and more resources available for Visual Studio 2019 and Windows 10 IoT Core. You can develop your UWP applications in a variety of languages, including C++, C#, Visual Basic, and more. However, most examples are written in C++, which is the more popular choice among the examples from Microsoft and the community.

If you have never used Visual Studio before, do not despair—I include a step-by-step description of how to use the tools in each of the proceeding project chapters with a quick-start walk-through in Chapter 4. The following section presents an overview of how to get started with the Windows 10 IoT Core and your low-cost computing board. That is, the section demonstrates how easy it is to set up your PC and your board to begin developing an IoT.

Windows 10 IoT Core Editions

Before we take a deeper look at Windows 10 IoT Core, let's discuss the various versions of that release. Yes, there's more than one! The following sections describe each briefly at a high level. We will concentrate on Windows 10 IoT Core (base) for most of this book but will discuss the enterprise and server editions in Chapter 17.

Windows 10 IoT Core

This is the base version most will use to work with low-cost hardware to create IoT solutions. In fact, the next section gives a walk-through on its features and use.

Windows 10 IoT Enterprise

This version isn't run on your low-cost hardware; rather, it is used to provide manageability and security for an enterprise of IoT solutions. It is a full version of Windows 10. Microsoft states it is a binary "equivalent to Windows 10 Enterprise," so you can expect the same experience and use all of the same tools as you would on your PC. The difference is in licensing (it is a paid product), and it is intended for use by original equipment manufacturers (OEMs).

Tip   See `https://docs.microsoft.com/en-us/windows/iot-core/` `windows-iot-enterprise` for more information about this version.

Windows Server IoT 2019

This version is like the IoT Enterprise version in that it runs on your PC server hardware. However, this version is a variant of the Windows Server 2019 and also uses the same tools and has the same experience as the Windows Server 2019. Once again, the difference is in licensing. Windows Server IoT 2019 is only licensed through the OEM channel under special dedicated use rights.

Tip   See `https://docs.microsoft.com/en-us/windows/iot-core/` `windows-server` for more information about this version.

Getting Started with Windows 10 IoT Core

Now that you know more about the Windows 10 IoT Core and the hardware it runs on and the accessories you need to hook things up, let's get your hands into the hardware and boot up Windows 10 IoT Core for the first time. As you may imagine, there are a few things that you need to do to get things going.

In this section, you see how to get all the prerequisites settled in order for you to start using the Windows 10 IoT Core. As you will see, this requires configuring your computer as well as setting up your board. I walk you through all of these steps for each of the boards available. Although this book focuses on the Raspberry Pi, I include all three boards so that when you want to work with one of the other boards, you will have everything that you need to get started.

Let's begin with setting up your computer.

Setting Up Your Computer

While most would expect this, the first thing you must know about using the Windows 10 IoT Core is that you need to have a Windows 10 PC. Moreover, you must be running Windows 10 version 10.0.1.10240 or greater. To check your Windows version, go to the search box next to the *Start* button and enter System Information. Click the menu item shown. You see the *System Information* dialog, as shown in Figure 2-6. I have Windows 10 version 10.0.18363, which is newer than the minimum requirement.

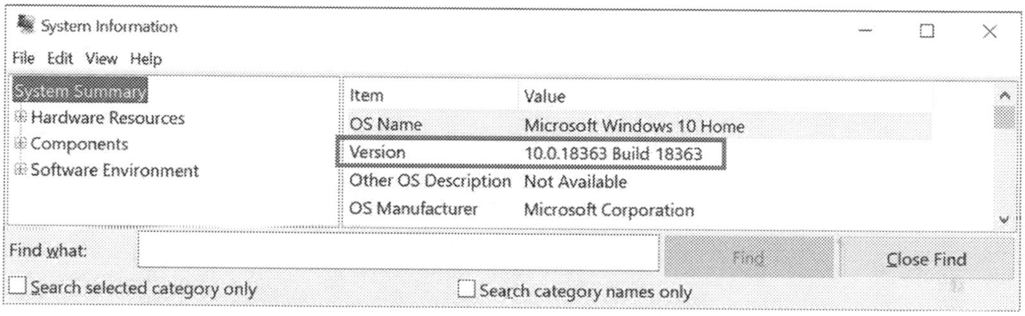

Figure 2-6. *System Information*

I've read some criticism about requiring a Windows 10 PC to use Windows 10 IoT Core, but again for Windows 10 users wanting to explore the IoT, it's a non-issue. However, if you normally use another platform (such as Linux or Mac), you will need a Windows 10 machine going forward.

But there is more. You need to configure your PC for use with the Windows 10 IoT Core tools. Briefly, this includes the following. I explain each of these steps in more detail. Once all of these steps are complete, your PC is ready to set up and use Windows 10 IoT Core on your low-cost computing board.

- Enable developer mode.

- Install Visual Studio 2019 and the sample templates for the IoT.

- Install the Windows 10 IoT Core Dashboard.

Enable Developer Mode

This step is one that is often overlooked and easily forgotten, especially if you have more than one Windows 10 PC. Windows 10 has initiated a new licensing mechanism for developing applications. Rather than require a special developer license to develop and deploy your applications, you simply enable your Windows 10 PC to turn on developer mode, which allows you to compile, deploy, and test applications for Windows 10 IoT Core.

To enable developer mode, enter use developer features in the search box next to the *Start* button. Choose the *Settings* menu item by the same name. Alternatively, you can open the settings application, click *Update & Security*, and then click *For developers*. Once you have the dialog open, select the *Developer mode* radio button. Once you click the button, you are asked for confirmation to turn on developer mode. The message explains that using developer mode may increase your security risk. Be sure to take appropriate actions to ensure that you are protected while online. Click *Yes* in that dialog. Figure 2-7 shows the developer mode dialog with the correct settings selected.

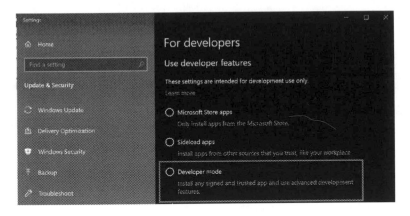

Figure 2-7. *Enabling developer mode*

Next, you need to install the software development tools. In this case, you want to install Visual Studio 2019. You also need to install the IoT templates and add-ons. You won't use Visual Studio in this chapter, but since it is required, you'll discover how to install it so that you can complete the process to prepare your PC for developing IoT solutions.

Install Visual Studio Community 2019

If you do not have Visual Studio 2019, you can go to `https://visualstudio.microsoft.com/` and hover over the `Download Visual Studio` drop-down box and click `Community 2019`. This version has a license, making it free to use for individuals, open source projects, academic institutions, students, and small project teams. Visual Studio 2019 Community has everything that you need to develop Windows 10 IoT solutions.

The download consists of a small executable named `vs_Community.exe` or similar, which you can execute once the download is complete. This is the Internet installation version, which downloads the components needed during installation. You use this version in this chapter since it is the easiest to do and requires less download time.[6]

Once Visual Studio Community Edition is downloaded, you can double-click the executable to begin the installation. Once the installer launches, you see the splash screen followed shortly by the installation type page. You also may need to authorize changes to the system via a pop-up dialog. Figure 2-8 shows the downloader installing its prerequisites.

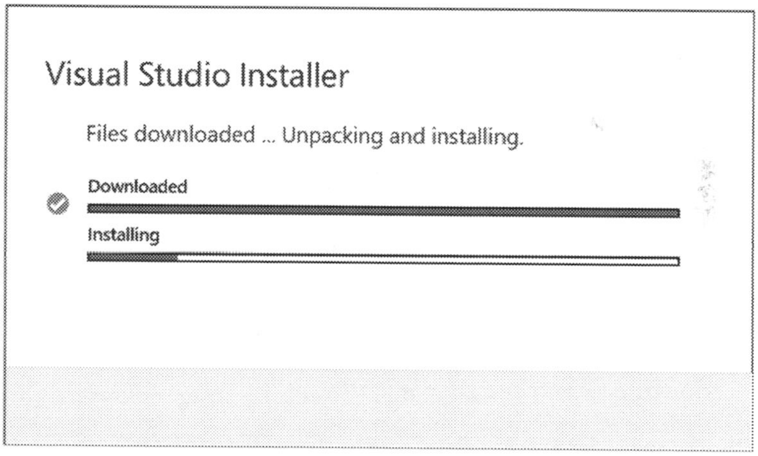

Figure 2-8. *Visual Studio 2019 downloader prerequisite installation*

---

[6]This could be a big deal if your data plan is limited to a fixed amount of data per month. The full download with all options is approximately 5.8GB or more.

Once the prerequisites are installed, the installer will open a new window where you can choose what components (called workloads) you want. This is vastly improved over previous installers because it allows you to choose what you want at a glance rather than wade through long lists of selections (but that option is still there!).

When it opens, you will see a page with the *Workloads* tab open that permits you what workloads you want to install. You simply scroll through the list and click the ones you want. You can also click the *Individual components* tab to select each module separately (a dizzying list it is indeed). The *Language Packs* allow you to optionally choose one or more languages to support in your tools and projects. You can also change the location of the installation on the *Installation locations* tab, but I recommend accepting the default locations. Figure 2-9 shows the installation page with the needed workloads selected.

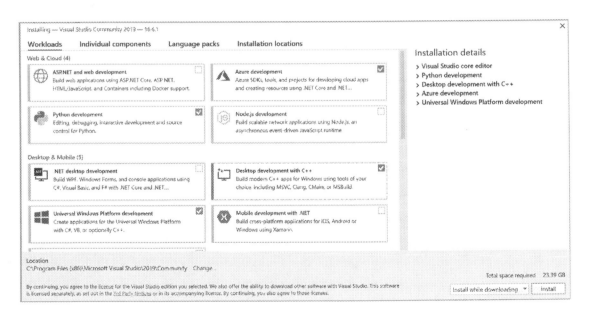

Figure 2-9. *Visual Studio 2019 Community: installation type*

The basic components that you need for this book include the UWP and Visual C++ features. You can select these by clicking the workload, which places a check in the upper-right corner. The components that you need are listed next. Once you have the workloads you want checked, click *Install* to start the installation. If your Internet download speed is very slow, you may want to choose a custom installation and uncheck everything to install the basic components and install any additional components later.

- Development with C++

- Python development (optional)

- Universal Windows Platform development

- Azure development (optional)

Be advised, if you check select all features, the installation could exceed 15GB and require over 6GB of download data.

Tip   If you would like to use Python to develop IoT applications to run on your device, you can, but it requires setting up Python on your device and manually copying the files and dependencies. This is more work than a beginner's book should contain. For more information, see `https://docs.microsoft.com/en-us/windows/iot-core/developer-tools/python`.

There is one more option you may be interested in. At the bottom of the dialog, you can click the *Install while downloading* drop-down and change it to *Download all, then install* to permit you to install Visual Studio without being connected to the Internet (or network). Microsoft added this to help those who want an offline installation. Interestingly, you can copy those files to another PC to avoid downloading everything twice.

Should you find you need to add other workloads (features), you can do so by simply launching the Visual Studio Installer application. To launch the application, use the search tool to search for `Visual Studio Installer` and then open the application. Once open, simply click *Modify* and select the workloads you want and click *Install*. Figure 2-10 shows how to modify the installation.

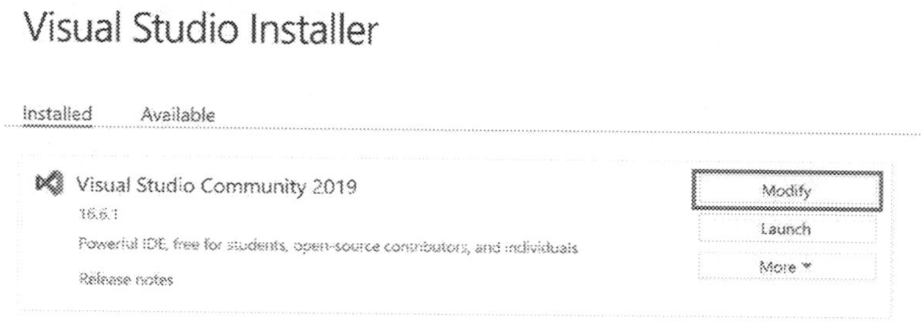

Figure 2-10. *Visual Studio 2019 modify installation*

Now comes the stage where the installer begins downloading the components you selected from the Internet and installs them. Depending on what options you chose, this could be a long list. Also, depending on your Internet download speed, downloads can take some time to complete.

Further, the installation of the components can also take a long time. It is not unusual to take several hours to complete the installation. Again, if this is a concern, you can use a custom installation and choose one component at a time. Once underway, you see a progress page like the excerpt shown in Figure 2-11, which shows an example of the progress dialog.

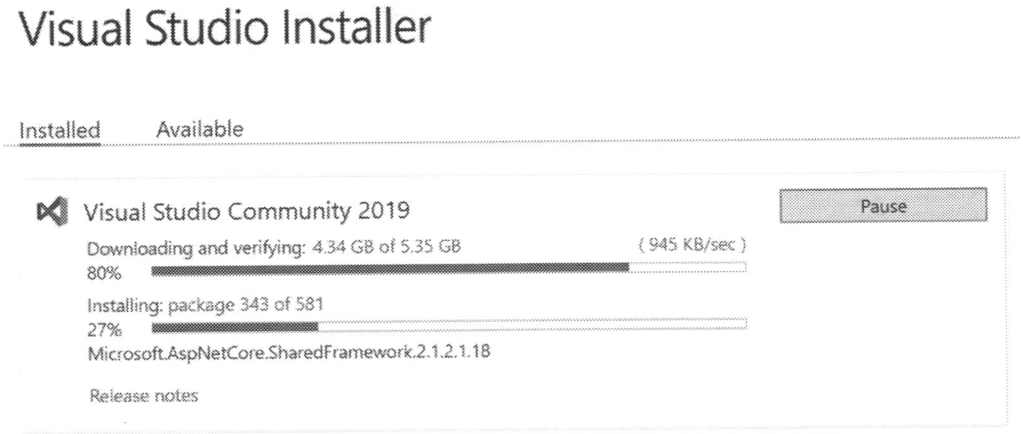

Visual Studio Installer

Installed   Available

Visual Studio Community 2019   Pause
Downloading and verifying: 4.34 GB of 5.35 GB   (945 KB/sec)
80%
Installing: package 343 of 581
27%
Microsoft.AspNetCore.SharedFramework.2.1.2.1.18

Release notes

Figure 2-11. *Visual Studio 2019 installation progress*

Tip   If you have a slow Internet connection, you may want to start the installation before you go to bed and let it run overnight. To do so, you can click the *Pause* button and then relaunch the installer from the *Start* menu later when your bandwidth improves.

When the installation completes, you see a dialog page that informs you that you need to restart your PC to complete the installation. Figure 2-12 shows an example of what you should see.

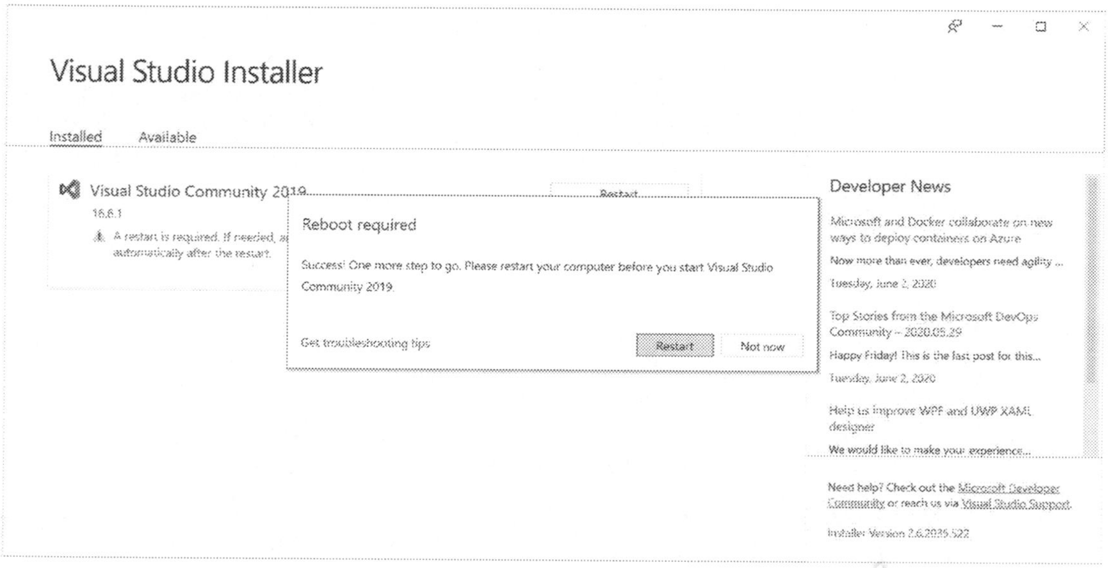

Figure 2-12. *Visual Studio 2019 Community: installation complete*

Simply click Restart and your computer will shut down. Note that it is possible the shutdown and restart may take a while to complete. So, don't be surprised if you see the blue screen with the twirling dots for a longer than normal period.

Once the restart completes, you can launch Visual Studio 2019. Since this is the first time launching and if you choose to install the Azure components, you may be prompted to sign in to complete the Azure setup. But you can do that later by clicking *Not now, maybe later*. Figure 2-13 shows the welcome dialog.

Figure 2-13. *Welcome dialog*

Once you sign in or bypass the welcome dialog, you may experience a delay as the application configures your system and the options you chose during installation.

The next step is to select the default environment and color scheme. I chose C++ and the default blue color scheme. Figure 2-14 shows the environment. When you are happy with your selection, click *Start Visual Studio*. You may see another delay as the environment is set up.

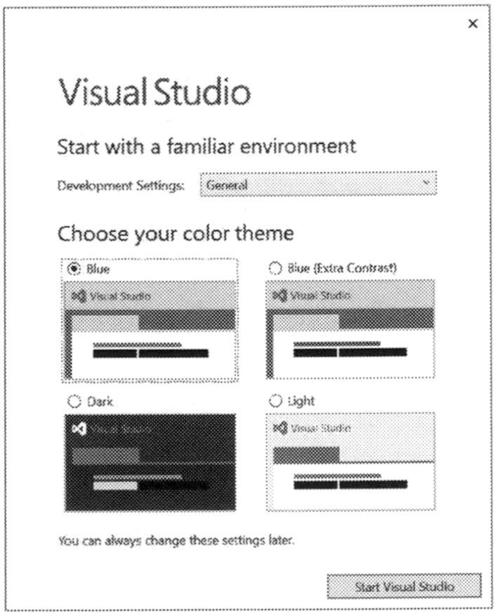

Figure 2-14. *Select environment*

Next, you will see the open project dialog where you can open an existing project or create a new one as shown in Figure 2-15. For now, choose Continue without code.

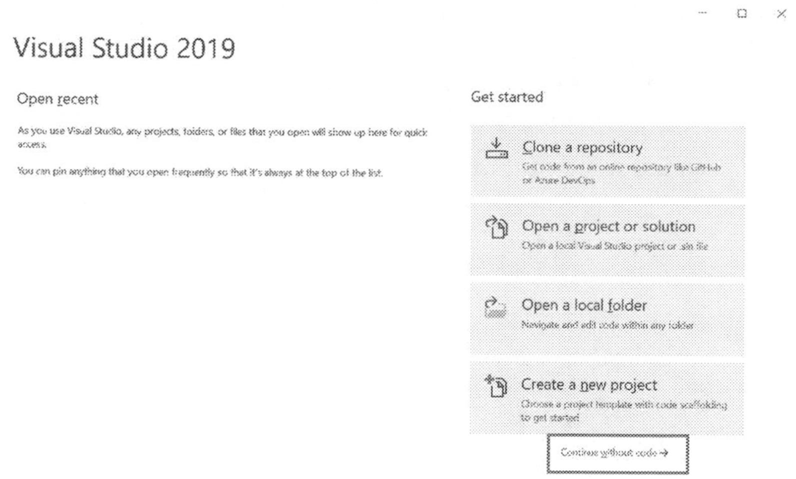

Visual Studio 2019

Open recent

As you use Visual Studio, any projects, folders, or files that you open will show up here for quick access.

You can pin anything that you open frequently so that it's always at the top of the list.

Get started

Clone a repository
Get code from an online repository like GitHub or Azure DevOps

Open a project or solution
Open a local Visual Studio project or .sln file

Open a local folder
Navigate and edit code within any folder

Create a new project
Choose a project template with code scaffolding to get started

Continue without code →

Figure 2-15. *Open project dialog*

Figure 2-16 shows a typical layout of Visual Studio. Don't worry about what all the panels, menus, and hundreds of options are at this time. You learn the essentials of what you need to know in Chapters 4 through 6.

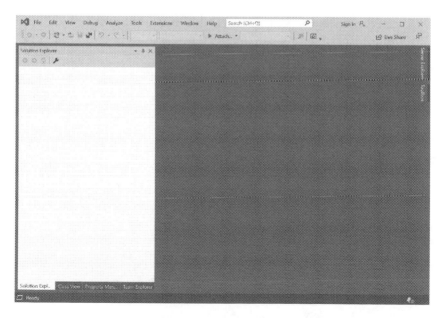

Figure 2-16. *Visual Studio 2019 Community user interface*

Install the Windows 10 IoT Core Templates

There is one more thing you need to install: the Windows 10 IoT Core templates. To install the templates, click *Extensions* ➤ *Manage Extensions*. Then in the dialog that opens, type IoT in the search box and press enter. Figure 2-17 shows the results you should see.

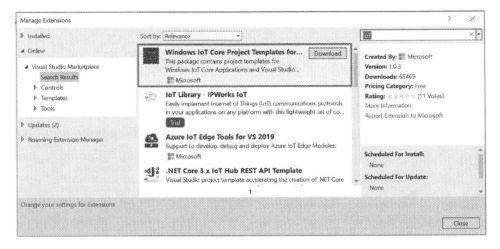

Figure 2-17. *Searching for the Windows IoT Core Templates*

When the list is populated, select the *Windows IoT Core Templates* entry in the list and click *Download*. Once downloaded, you will need to restart Visual Studio. When you do so, you will see the templates and supporting components installed. You may need to click *Modify* in the dialog to accept the license and continue. This process may take a few moments to complete, and when it does, you can close the dialog.

Validate Your Visual Studio 2019 Community Installation

OK, now you have Visual Studio installed and added the IoT templates. You can validate the installation by opening Visual Studio and then selecting Help ➤ About Microsoft Visual Studio. Observe the version information, as shown in Figure 2-18.

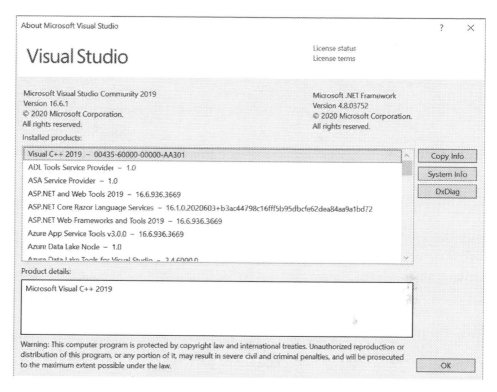

Figure 2-18. *Validating Visual Studio: About dialog*

Troubleshooting Visual Studio Installation Problems

While the installation of Visual Studio can take a very long time to complete, there are sometimes cases where the installation fails. This is most often caused by loss of connectivity to the Internet or simply failed download of one or more components. You can recover from this form of failure by simply restarting the installation. Indeed, most installation failures can be fixed in this manner.

However, if something goes really wonky, you may need to repair the installation. Fortunately, you can restart the installation or open the Program and Features application. Select the Visual Studio 2019 entry and then click *Repair*. You are presented with a dialog that allows you to modify or repair the installation. Click repair to recover from the installation failure.

On very rare occasions, the installation could fail in such a way that some components will not install, resulting in a failed installation. In these rare occasions, you can view the log of the failed installation (there is a link on the final dialog page for the installation) and try to determine the cause. Since there are so many things that could go wrong, it isn't possible to list them. Thus, you must examine the log and fix each error as described. I recommend doing a search for the error on the Internet and read the suggested solutions. Be sure to read several solutions thoroughly before you attempt them. Also, make a restore point before continuing.

Tip   Always make a restore point when attempting to fix installation problems. If the solution fails or makes it worse, you can restore the system to the last checkpoint and try another solution.

When you find yourself spending a lot of time trying to fix really odd errors for which you can find no solutions on the Internet, you should first attempt to uninstall using the command line with the `/uninstall /force` options with `vs_community.exe / uninstall /force` and then restart the installation. If this does not work or gives the same errors, you can try deleting the contents of the `ProgramData/Program Cache` folder. Use this as a last resort because not only will it force installation to download all the packages again, but it also removes cached packages from other applications. However, I've found that this trick works very well.

Caution   Deleting cached packages from `ProgramData/Package Cache` may require downloading the packages for new applications. Use with care and only as a last resort.

Now that you have Visual Studio installed, you need only one more thing on your PC: the Windows 10 IoT tools.

Tip   The latest downloads for Windows 10 IoT Core can be found at `https:// docs.microsoft.com/en-us/windows/iot-core/downloads`. This includes the tools as well as the binary images for the boards.

Install the Windows 10 IoT Core Dashboard

The last step is to install the Windows 10 IoT Core development tools. You need these tools to complete the installation of Windows 10 IoT Core on your low-cost computing board. To download the installation, go to `https://go.microsoft.com/ fwlink/?LinkID=708576`. This link downloads an installation named setup.exe. Once downloaded, launch the executable and follow the prompts. The installation begins with a small download of the tools, as shown in Figure 2-19.

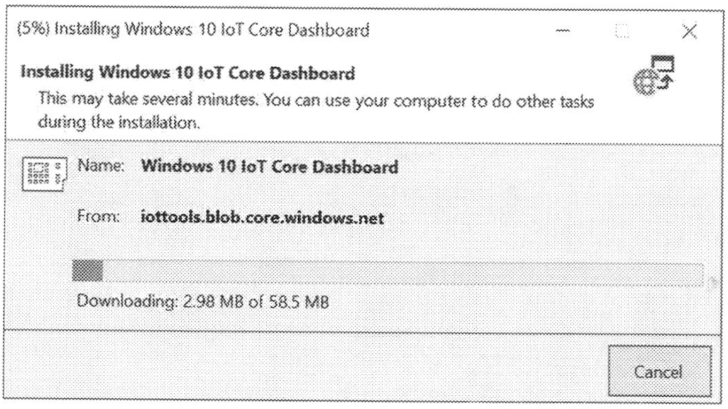

Figure 2-19. *Windows 10 IoT Core Dashboard installation*

Once the installation is complete, the dashboard launches. However, you may see a permissions dialog like the one shown in Figure 2-20. Click *Allow access* to allow the dashboard to run.

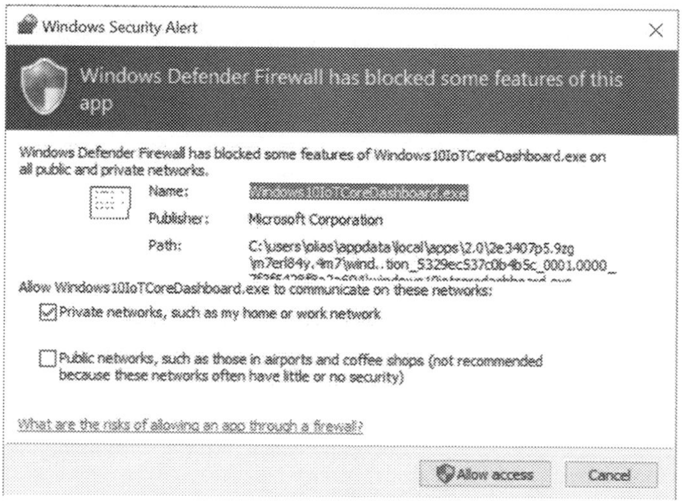

Figure 2-20. *Allow access to Windows 10 IoT Core Dashboard*

Once that dialog is closed, the dashboard will launch. Figure 2-21 shows the Windows 10 IoT Core Dashboard, which launches the *Set up a new device page* by default. This page is used to configure a bootable SD card for your device. You see how to use this in the next section. On subsequent launches, the application checks for updates and automatically downloads and installs them.

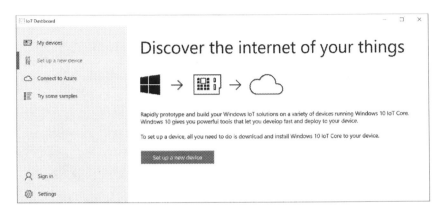

Figure 2-21.   *Windows 10 IoT Core Dashboard*

There are six tags or links on the left side of the window. The *My devices* page lists the devices running Windows 10 IoT Core on your network. The *Set up a new device* link lets you set up a new device. The *Connect to Azure* link lets you log in to Azure for hosting your data. There is also a *Try some samples* link to a set of sample applications that you can use to get started. You see one of these in action in Chapter 4. The *Sign in* link lets you log in to Microsoft using your Microsoft account. Finally, the Settings link displays the version and builds information for the application.[7] You explore this tool in more detail in the following sections as you set up your hardware for use with the Windows 10 IoT Core.

Now that your PC is ready to go, let's see how each of the low-cost computing boards is set up to install Windows 10 IoT Core and boot for the first time.

---

[7]Oddly, there are no settings that you can change.

Getting Started with Your Board

Windows 10 IoT Core is very easy to set up and get your board running. The general process is as follows. I describe each in more detail through examples of each board. I recommend reading through the setup of all the boards, especially if you haven't decided which board (or SD card create process) you want to use. But first, I present some tips that may be helpful in setting up your board:

1. Use the Windows 10 IoT Core Dashboard to download and install the Windows 10 board-specific installation package on your SD card.

2. Connect your hardware to power, monitor, keyboard, and mouse.

3. Power on the device.

4. Configure basic settings and connect to the network.

5. Connect to your device with the IoT Core Dashboard.

Note   There is an alternative method where you download the disk image (`.iso`) manually and install it with a special flashing application that comes with the image (or from the manufacturer). I detail this process in the following sections.

Tip: Be Patient and Thorough

I should encourage you to exercise patience and perseverance should you encounter problems. Although I will explain the steps you need to perform in detail, there are some things that could go wrong. I have included as many pitfalls as I can, but my experience has shown troubleshooting problems with the hardware may still arise.

For example, I spent quite some time trying to boot one of my boards only to discover one of the cables was defective. I neglected to consider the cable as the culprit because it was new. Thus, you should approach problems with an iterative mindset where you check each component (SD card, cable, power supply, etc.) one at a time for correct working order and, more importantly, change only one thing at a time. If it doesn't solve the problem, return it to the original setting. That is, if you swap the HDMI cable and it doesn't fix the problem, go back to the original one.

Tip: Downloading the .iso Image

Downloading a bootable image can be accomplished in two ways: you can download the boot image (as an .iso file) or you can use the Windows 10 IoT Core Dashboard to download the image. There are also two ways to write the image to the SD card (for the Raspberry Pi). You can either use the Windows 10 IoT Core Dashboard or a helper application that comes with the .iso boot image. You'll see how to do both in the following sections.

Tip: Use Class 10 SD Cards

Microsoft recommends using class 10 SD cards or higher.[8] Currently, only two SD cards have been tested—Ultra cards from SanDisk and the EVO cards from Samsung. I've found other class 10 SD cards work well, and some of the slower classes may work but results are mixed. If you observe your device booting slowly or the startup sequence seems jumpy (the screen flickers), you may need to use a faster SD card. Incompatible SD cards can result in unacceptable performance or failure to boot. If you are having problems with your device after initial setup, try a faster SD card. I prefer the SanDisk Ultra SD cards since they are more plentiful and thus can be a bit cheaper.

Tip: Double-Check Your Power Supply

Be sure to double-check your power supply to ensure that it has the proper rating for your board. The Raspberry Pi needs a 5V 2.0A–2.5A power supply with a micro-USB connector, and the DragonBoard 410c works best with a 12V 2A power supply with a 4mm tip. If your board does not boot or powers off while running, it is possible the power supply is faulty or insufficient for the board and its peripherals.

Caution   Using the wrong tip can damage the power connector on the board.[9]

Now that you understand the basic process and have foreknowledge of some of the pitfalls, let's see how to set up each of the boards in turn starting with the Raspberry Pi. You see how to connect to the board after you learn how to configure all the boards.

---

[8]https://en.wikipedia.org/wiki/Secure_Digital#Speed_class_rating

[9]Can you guess how I know this? It is far too easy to bend the tabs inside the connector. This is more likely when using universal power supplies.

Raspberry Pi Configuration

This section demonstrates how to install and boot Windows 10 IoT Core on the Raspberry Pi. You see the specific steps needed to get your board ready for Windows 10 IoT Core, including the hardware that you need, how to connect to the board once Windows 10 IoT Core boots, and how to configure the board for your network.

Let's begin with the prerequisite hardware that you need.

Prerequisites

The following are the miscellaneous hardware that you need to use Windows 10 IoT Core with your board. You will see where these cables are plugged into your board later in this section. For now, just gather the items you need:

- Raspberry Pi

- 5V 2.5A micro-USB power adapter, like the one at `www.adafruit.com/products/1995`

- USB wired or wireless (not Bluetooth) keyboard and mouse

- HDMI monitor with HDMI cable or a suitable adapter for use with a DVI monitor

- Ethernet cable

- Micro-SD card 8GB or larger: class 10 or better

- SD card reader (if your computer doesn't have one)

Download and Install the Image with the Windows 10 IoT Core Dashboard

Microsoft has provided an easy process to build SD cards for your board, which includes downloading the image, formatting the card, and writing the image to the SD card. Fortunately, we can do all of the steps needed through the Windows IoT Core Dashboard.

Simply insert an SD card in your card reader and then open the application, select *Set up a new device*, and then select the options from the dialog excerpt shown in Figure 2-22. Here, we select the device type (Raspberry Pi), the latest build, the SD card (if more than one), device name, and password. Be sure to tick the software license checkbox.

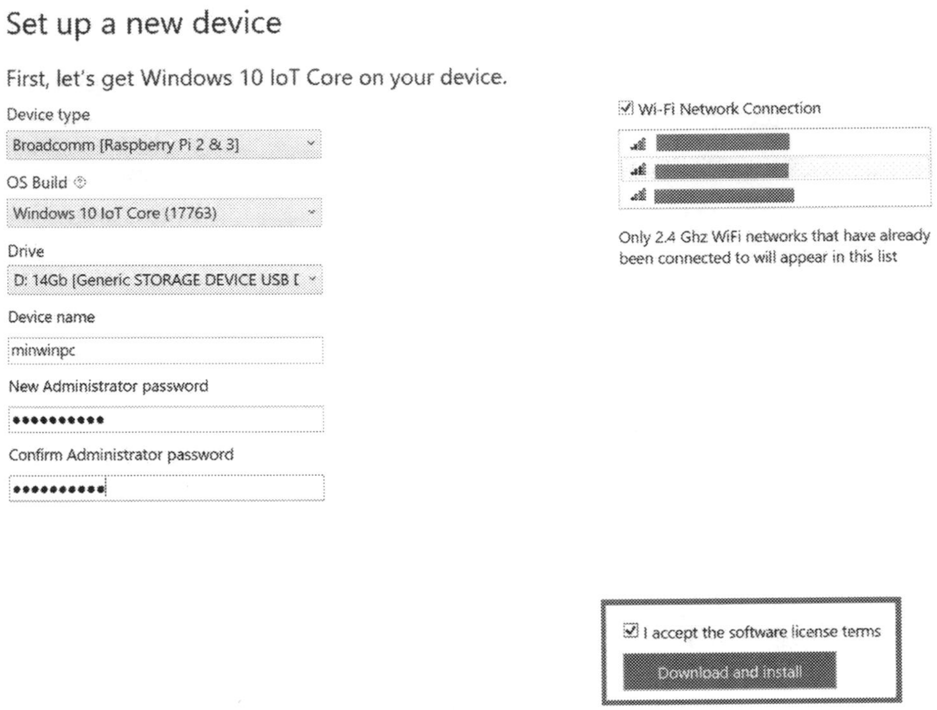

Set up a new device

First, let's get Windows 10 IoT Core on your device.

Device type

Broadcomm [Raspberry Pi 2 & 3]

OS Build ⊕

Windows 10 IoT Core (17763)

Drive

D: 14Gb [Generic STORAGE DEVICE USB [

Device name

minwinpc

New Administrator password

●●●●●●●●●●

Confirm Administrator password

●●●●●●●●●●

☑ Wi-Fi Network Connection

Only 2.4 Ghz WiFi networks that have already been connected to will appear in this list

☑ I accept the software license terms

Download and install

Figure 2-22. *Set up a new device in IoT Dashboard*

Once you have made your selections, you can click *Download and install* to begin the process. This will download the Raspberry Pi image and write it to the SD card for you. The process can take a few minutes depending on the speed of your Internet connection. Figure 2-23 shows an excerpt of the progress.

Downloading Windows 10 IoT Core

443 MB downloading - 50%

Cancel

Flashing your SD card/Device

Pending

☑ I accept the software license terms

Download and install

Figure 2-23. *Dashboard download progress*

When the download is complete, the flash process will begin. Figure 2-24 shows an excerpt of the progress.

Downloading Windows 10 IoT Core

Download complete

Flashing your SD card/Device

Unpacking the installer

☑ I accept the software license terms

Download and install

Figure 2-24. *Dashboard flash progress*

During the process, you may see one or more permissions dialog terminal windows open, such as formatting or erasing the SD card like the one shown in Figure 2-25.

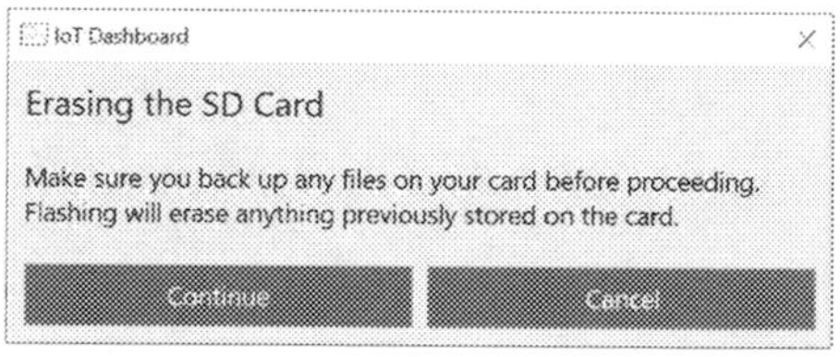

IoT Dashboard   ×

Erasing the SD Card

Make sure you back up any files on your card before proceeding.
Flashing will erase anything previously stored on the card.

Continue   Cancel

Figure 2-25. *Erase SD card dialog*

Eventually, you will see a terminal open that tracks the progress of the SD card write operation as shown in Figure 2-26. This terminal will close on its own, so don't close it.

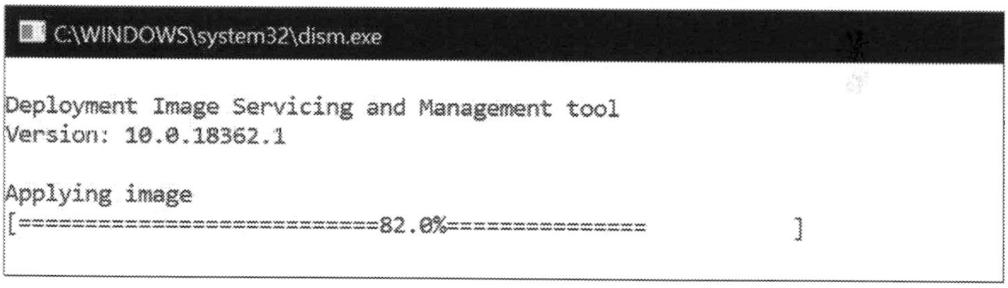

■ C:\WINDOWS\system32\dism.exe

Deployment Image Servicing and Management tool
Version: 10.0.18362.1

Applying image
[==========================82.0%================   ]

Figure 2-26. *Write disk image terminal window*

Once that terminal closes, you will see a completion and welcome dialog from the dashboard as shown in Figure 2-27.

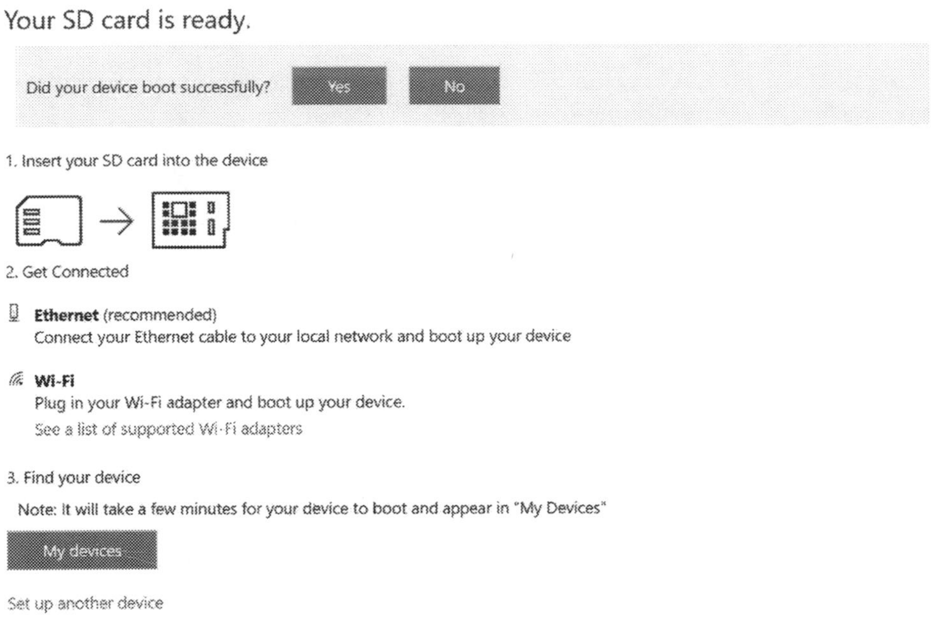

Your SD card is ready.

Did your device boot successfully?   Yes   No

1. Insert your SD card into the device

2. Get Connected

 Ethernet (recommended)
Connect your Ethernet cable to your local network and boot up your device

 Wi-Fi
Plug in your Wi-Fi adapter and boot up your device.
See a list of supported Wi-Fi adapters

3. Find your device
Note: It will take a few minutes for your device to boot and appear in "My Devices"

My devices

Set up another device

Figure 2-27. *Operation complete dialog*

Notice the dialog asks you to confirm that your device booted correctly, but that is a bit premature since we haven't even taken the SD card out of the computer! Regardless, you can wait to click *Yes* later or click *No* to satisfy the dialog (won't hurt anything to do either). Clicking *No* results in some helpful links to troubleshooting.

Download and Install the Image Manually

In this section, we will examine how to create the SD card image manually by downloading the disk image, extracting it, and installing the tools that come with the disk image. Use this process if you want to use an image that is not included in the dashboard. The key file we want in this process is the `flash.ffu` file, which we will use with the dashboard to create the SD card image.

To download the disk image for the Raspberry Pi, visit `https://docs.microsoft.com/en-us/windows/iot-core/downloads`. Scroll down to the April 2018 section and click the link for the Raspberry Pi. This will download a file named `Windows10_IoTCore_RPi_ARM32_en-us_17763Oct .iso` or similar. Double-click or right-click and choose *Mount* to mount the drive. In that drive, you will see an installer named

`Windows_10_IoT_Core_for_RPi.msi` or similar. Double-click that file to install this application. You'll need to do the usual steps for most installations, including accepting the license, permitting the change on your computer, and so on.

The binary image for the Raspberry Pi is named `flash.ffu` and installed in the `c:\ Program Files (x86)\Microsoft IoT\FFU\RaspberryPi2`. Once the installation is complete, you can unmount the virtual drive. If you changed the installation folder, be sure to note the correct path—you need it in the next step.

Now you are ready to write the Windows 10 IoT Core bootable files to the micro-SD card. Open the Windows 10 IoT Core Dashboard and insert a suitable SD card into your SD card reader. Make sure that you've backed up the data on the card before you proceed because the next step overwrites the contents. Like we did in the previous section, click *Set up a new device* and fill in the device name and password. Choose the device type (Raspberry Pi), but instead of selecting the latest build, choose Custom and then browse for the *flash.ffu* file as shown in Figure 2-28. Be sure to tick the software license checkbox.

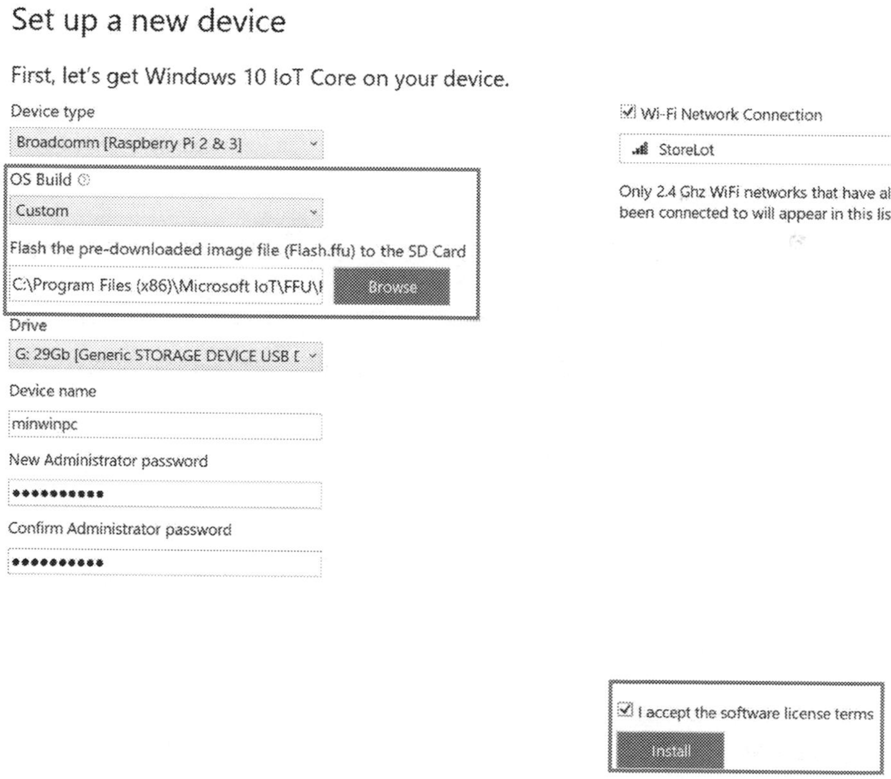

Figure 2-28. *Windows 10 IoT Core Dashboard: Set up a new device with custom image*

When you are ready, click the *Install* button. Since you choose the custom installation, it copies the file from your computer and does not download anything. Once the process begins, you see a new terminal window to copy the image as before.

When the process is complete, you will see the complete/welcome screen, and you can now take the SD card out of your computer. Be sure to eject it properly like you would any other USB or removable drive. You're now ready to connect the hardware and boot the image for the first time. Before you do that, click the *My devices* tab (or button) in the Windows 10 IoT Core Dashboard. You'll be using this screen to connect to the Raspberry Pi in a later step.

Connecting the Hardware

If this is your first time using a Raspberry Pi, orient the board on the table with the Raspberry Pi logo facing you. Figure 2-29 helps you locate the connection points. Insert the micro-SD card into the Raspberry Pi SD card reader located on the bottom of the left side. Connect the HDMI monitor to the HDMI port located on the bottom. You can connect your Ethernet cable and USB mouse and keyboard to the ports on the right side of the board.

Figure 2-29. *Connections for the Raspberry Pi*

OK, you're now ready to power the board and boot up Windows 10 IoT Core!

Booting Windows 10 IoT Core for the First Time

Now you can power on the board and boot from the SD card. Plug your AC adapter into a power source and then insert the micro-USB power located in the bottom-left corner of the board, as shown. You should see the power LED illuminate and the SD card activity LEDs blink. These LEDs are located on the left side of the board.

You see the Windows logo and an activity cursor appear on the monitor connected to the Raspberry Pi. The first boot may take some time, but eventually you will be asked to choose the default language. Use the mouse or keyboard to select your language.

Next, you see the bootup screen, as shown in Figure 2-30. You can configure your board using this screen, as well as shut down or reboot the board. The *Device Settings* button is the small gear located in the upper-right portion of the screen. If you click it, you see the *Device Settings* screen that allows you to change the default language or connect to a Wi-Fi network if you have a wireless network adapter plugged into the Raspberry Pi. Finally, you can shut down or restart the board by clicking the power button in the upper right of the screen.

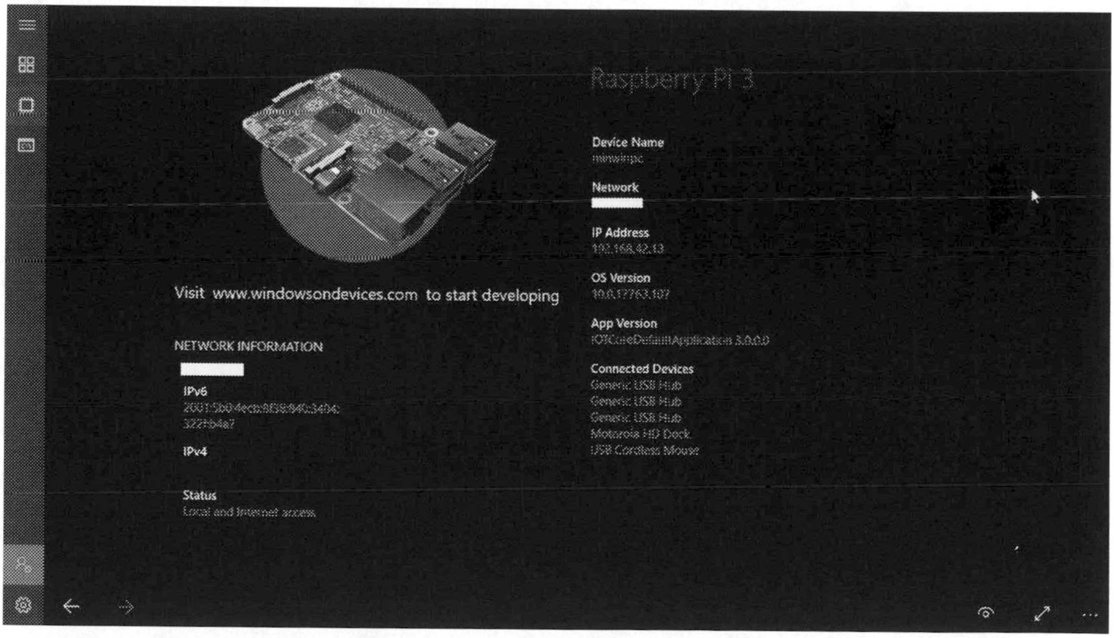

Figure 2-30. *Windows 10 IoT Core boot screen: Raspberry Pi*

Take note of the IP address as you may need this if you want to connect to the Raspberry Pi from your computer. Since the methods for connecting are the same for all the boards, you will see how to connect to the board after we see how to configure the DragonBoard 410c. After that section, you learn how you can connect to your board from your PC.

DragonBoard 410c Configuration

This section demonstrates how to install and boot Windows 10 IoT Core on the DragonBoard 410c. You see the specific steps needed to get your board ready for Windows 10 IoT Core, including the hardware that you need, how to connect to the board once Windows 10 IoT Core boots, and how to configure the board for your network.

Many of the steps are similar for this board as the Raspberry Pi. However, you do not use an SD card to boot the DragonBoard 410c. Instead, you will use a special application to download the boot image into a special memory drive (rewritable, nonvolatile memory) on the DragonBoard 410c. Thus, that step is completely different than either of the other boards, but as you will see, you still download and install the firmware tools.

Tip   If you do not have a DragonBoard 410c, you should still read the following sections to learn some of the features of using the dashboard and later connecting to your board.

Let's begin with the prerequisite hardware that you need.

Prerequisites

The following are the miscellaneous hardware that you need to use Windows 10 IoT Core with your board. You see where these cables are plugged into your board later in this section. For now, just gather the items you need.

DragonBoard 410c

- +6.5V to +18V power adapter with a 4.75 x 1.75mm barrel male plug

- USB wired or wireless (not Bluetooth) keyboard and mouse

- HDMI monitor with HDMI cable or a suitable adapter for use with a DVI monitor

- Micro-SD card 8GB or larger: class 10 or better

- SD card reader (if your computer doesn't have one)

The process to download the image to the DragonBoard 410c is a little different than the Raspberry Pi because it requires us to put the board into a special USB mode to write the image to the onboard memory. We will also use the manual method described previously. The following section describes the process in more detail.

Download and Install the Image Manually

The first step is to download the .iso file for the DragonBoard 410c, mount the .iso file, and then install the board-specific setup using the `.msi` file. The installation installs the `flash.ffu` image which you will need.

Once the .iso file is downloaded, simply double-click it or right-click and choose *Mount*. This mounts as a virtual drive that opens automatically. You see a file named `Windows_10_IoT_Core_QCDB410C.msi` or similar. Double-click to start the installation. You'll need to do the usual steps for most installations, including accepting the license, permitting the change on your computer, and so on.

The binary image for the DragonBoard is named `flash.ffu` and installed in the `c:\ Program Files (x86)\Microsoft IoT\FFU\QCDB410C`. Once the installation is complete, you can unmount the virtual drive. If you changed the installation folder, be sure to note the correct path—you need it in the next step.

Connecting the Hardware

If this is your first time using a DragonBoard 410c, orient the board on the table with the Arrow logo facing you. Figure 2-31 helps you locate the connection points.

Figure 2-31. *Connections for the DragonBoard 410c*

Connect the HDMI monitor to the HDMI port located on the bottom of the board. Connect the USB mouse and keyboard to the USB ports on the bottom of the board to the right of the HDMI connector. Now that all the cables are connected, you can connect the power and let the board boot up.

Notice the circle drawn on the right side of the board. This indicates the + button, which you will use to put the board into the USB boot mode in the next step.

Writing the Image to the DragonBoard 410c

We can use the dashboard to write the image to the board. Go ahead and launch the dashboard now. In the dashboard, recall we click *Set up a new device* and then make the selections we need. In this case, we want to change the device type to Qualcomm [DragonBoard 410c] and then the *OS Build* to *Custom* and browse for the `flash.ffu` file we downloaded earlier. Recall, this is located in the `c:\Program Files (x86)\ Microsoft IoT\FFU\QCDB410C` folder.

At this point, you must place the DragonBoard 410c in USB boot mode. But first, connect the board to a monitor and keyboard (so you can see the boot screen or errors if something goes wrong), use a USB to micro-USB cable to connect the board to your PC, then power it off (if not already), press and hold the + button on top of the board, and then power it back on. Figure 2-32 shows an excerpt of the dashboard state waiting for the board to enter USB boot mode.

Set up a new device

First, let's get Windows 10 IoT Core on your device.

Device type

Qualcomm [DragonBoard 410c]

OS Build ⓘ

Custom

Flash the pre-downloaded image file (Flash.ffu) to the SD Card

C:\Program Files (x86)\Microsoft IoT\FFU\(Browse

To turn on Flash Mode :

1. Connect the DragonBoard and the host PC with a microUSB cable.

2. While holding down the volume up (+) button, power on the device.

If your device is not showing up in the list, click here for more information.

Figure 2-32. *Windows 10 IoT Core Dashboard waiting for DragonBoard 410c USB boot mode*

Once your board enters USB boot mode, the dashboard will switch to indicate it has connected to the board as shown in the excerpt of the dashboard in Figure 2-33.

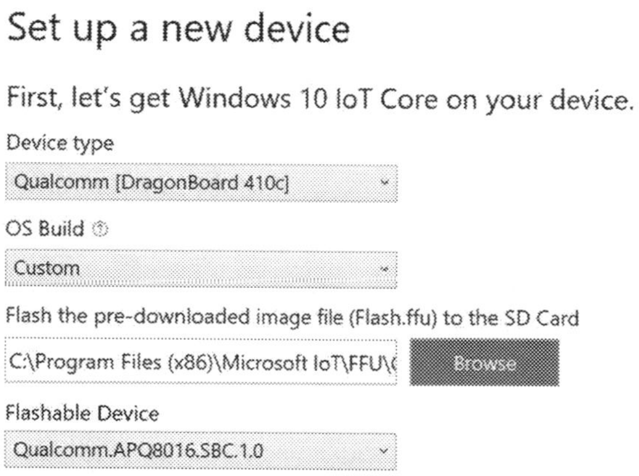

Set up a new device

First, let's get Windows 10 IoT Core on your device.

Device type

Qualcomm [DragonBoard 410c]

OS Build ⓘ

Custom

Flash the pre-downloaded image file (Flash.ffu) to the SD Card

C:\Program Files (x86)\Microsoft IoT\FFU\(Browse

Flashable Device

Qualcomm.APQ8016.SBC.1.0

Figure 2-33. *Windows IoT Core Dashboard DragonBoard in USB boot mode*

Tip   If the DragonBoard Update Tool does not connect, try using a different USB cable and check that the board is powered on. If that doesn't work, reinstall the DragonBoard 410c tools and reboot your computer. If you are still having problems, check the Device Manager for issues with the USB driver.

At this point, you're ready to write the file. Simply check the software license tick box and then click the Install button. This process will take a few moments, and you should see a progress similar to Figure 2-34.

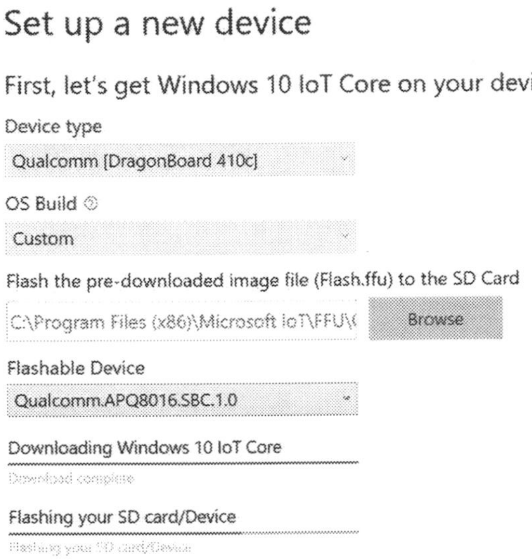

Set up a new device

First, let's get Windows 10 IoT Core on your device.

Device type

Qualcomm [DragonBoard 410c]

OS Build ⊙

Custom

Flash the pre-downloaded image file (Flash.ffu) to the SD Card

C:\Program Files (x86)\Microsoft IoT\FFU\(Browse

Flashable Device

Qualcomm.APQ8016.SBC.1.0

Downloading Windows 10 IoT Core

Download complete

Flashing your SD card/Device

Flashing your SD card/Device

Figure 2-34. *Windows IoT Core Dashboard writing image to DragonBoard 410c*

When the process is complete, you will see the welcome/complete screen like we saw previously. At this point, it is OK to power off the board and disconnect the USB cable from your computer. You will not need the USB cable for the next step.

Booting Windows 10 IoT Core for the First Time

Now it's time to boot the board with Windows 10 IoT Core. If you have not already, connect the HDMI monitor, USB keyboard and mouse (optional), and then the power adapter. Power on the device. It takes a couple of minutes for Windows to boot the first time, but subsequent boots are a bit faster. You see the Windows logo and an activity cursor appear on the monitor. The first boot may take some time, but eventually you will be asked to choose the default language. Use the mouse or keyboard to select your language.

Once the language and communication settings are selected, you see the bootup screen, as shown in Figure 2-35. Recall from the Raspberry Pi section, you can configure your board using this screen, as well as shut down or reboot the board. The *Device*

Settings button is the small gear located in the upper-right portion of the screen. If you click it, you see the *Device Settings* screen that allows you to change the default language and connect to a Wi-Fi network.

Figure 2-35. *Windows 10 IoT Core boot screen excerpt: DragonBoard 410c*

OK, now you're ready to connect to your board and explore its capabilities. I encourage you to do so as there are many things you can do to configure your board. What follows is a brief overview. You can find much more detail in Microsoft's online documentation at `https://docs.microsoft.com/en-us/windows/iot-core/`.

Connecting to Your Board

Now that your board is booted and connected to your network, you can log in remotely from your computer. Indeed, once you've set up the board, you normally would not connect it to a monitor and keyboard. That is, it is more common that you would deploy your board in your solution and run it headless.

Connecting to your board can be done in a number of ways, including the Windows 10 IoT Core Dashboard, a secure shell (SSH) connection via the command line, or using Windows PowerShell. There are other ways, but these are the most common and most useful. I'll discuss each of these in the following sections. These methods are the same for all the boards—there are no special steps for the DragonBoard 410c.

Connect with the Windows 10 IoT Core Dashboard

There are several ways to connect to your board. You can use the Windows 10 IoT Core Dashboard and external tools, or you can use the dashboard to launch the external tools. The easiest method is to use the dashboard. Once the board is booted, click the *My Devices* tab to see the list of all of your Windows 10 IoT Core devices. It could take a few minutes for your board to appear in the list. If it does not, make sure that the board is connected to the same network as your computer. Then right-click the board you want to connect to. This will display a connection menu as shown in Figure 2-36.

Note   Once the board is booted, it could take a few minutes for the board to show up in the My Devices list.

Figure 2-36. Dashboard connection menu

Here, we see we can connect to the board with the Device Portal, launch the Microsoft PowerShell and the Remote Client, and get information about the device. Notice we can also shut down and restart the board from this menu, which is really nice.

Let's examine a couple of these options starting with the Device Portal.

Connecting with the Device Portal

The Device Portal is a website hosted on your board that allows you to do all manner of activities from configuring the board to managing it to deploying applications. It is your best, one-stop tool. You are likely to spend a lot of time here.

To connect to the Device Portal, simply right-click the board in the My Devices tab of the dashboard and choose *Device Portal* or double-click it and choose the *Open Windows Device Portal in browser* link. You may see a login dialog similar to Figure 2-37. If you are using a Raspberry Pi, use the user account Administrator and the password you set when you created the SD card image. If you are using the DragonBoard 410c or another, similar board, you may need to use the default password (p@ssw0rd). Be sure to change the password later if it is the default.

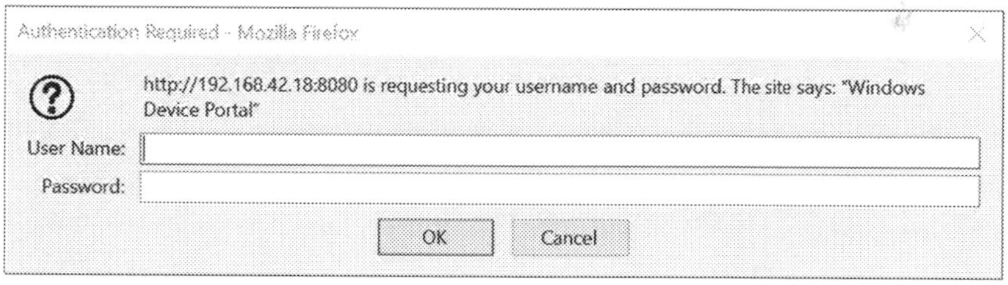

Figure 2-37.   Device Portal login

Enter the user and password and then click *OK*. When the page opens, you will see a new page open in your default browser similar to Figure 2-38.

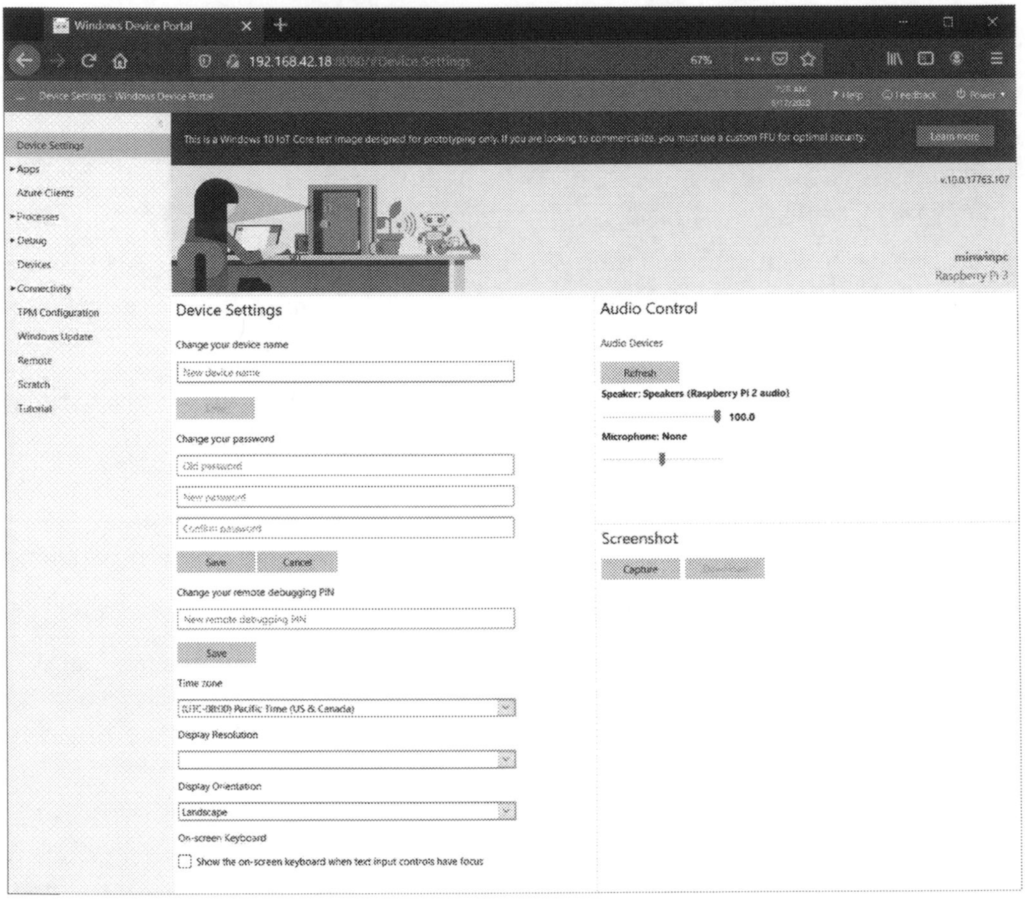

Figure 2-38. *Device Portal*

Wow, that's a lot of information! As you can see, you can change all manner of settings from the default page, but there are many more things you can do here. On the left side of the page are a number of tabs. Figure 2-39 shows the list in more detail.

Figure 2-39. *Tabs in Device Portal*

The following briefly describes each of the tabs. I encourage you to poke around with these and learn more about each. You may find using these tools will make your IoT projects more powerful and easier to set up and maintain.

- *Device Settings*: Changes basic settings about the device (default page).

- *Apps*: Allows you to install or uninstall packages on your device via the Apps manager or explore files with the File explorer or launch one of several sample applications.

- *Azure Clients*: Connects to and works with Microsoft Azure. Note that TPM must be configured first.

- *Processes*: Allows you to see which processes are currently running as well as the memory used.

- *Debug*: Provides tools to help you diagnose problems with your application including real-time debugging, tracing, and performance.

- *Devices*: Lists and controls the devices connected to the board.

- *TPM Configuration*: Configures the Trusted Platform Module (TPM) cryptographic coprocessor settings.

- *Connectivity*: Manages network and Bluetooth connections as well as connection sharing.

- *Windows Update*: Allows you to update the system files on the board (just like your computer).

- *Remote*: Enables or disables the Windows IoT Remote Server, which permits computers to connect to this device remotely using an application (disabled by default).

- *Scratch*: A placeholder for you to add your own extensions. Cool.

- *Tutorial*: A stepwise tutorial on the Device Portal. Highly recommended for those new to Windows IoT Core and the Device Portal.

Don't worry about learning what each and every one of these do; you'll discover the most frequently used throughout this book. As you can see, some of these are advanced features for performance testing, debugging, and event tracing. Most hobbyists and enthusiasts do not use these advanced features, but they're there if you need them. Sadly, there isn't a lot of documentation for how to use some of them.

Tip   See `https://docs.microsoft.com/en-us/windows/iot-core/ manage-your-device/deviceportal` for more information about the Device Portal.

Connecting with SSH

Another, popular method for connecting to your board and indeed a method I use often is to use secure shell (SSH). There are several varieties available as third-party applications if you don't want to use the standard SSH application. One of the ones I like is called PuTTY, which is actually a general terminal session application. The SSH feature in PuTTY is very easy to use. What I like most about PuTTY is it is open source software and therefore free to download and use.

You can download PuTTY from `www.putty.org`. Simply download the installer file (`.msi`) and install it. There are other applications available from the download site, but you only need the PuTTY application for connecting to your boards.

Tip   For more information about PuTTY, visit `www.putty.org`.

To connect to your board, you need the IP address from the Windows 10 IoT Core boot screen on the board. You captured this when you booted the board for the first time, or you can find it in the dashboard under *My Devices* by right-clicking the board and choosing *Copy IPv4 address*, which copies it to the clipboard. When you launch the putty.exe executable, you see the initial dialog, as shown in Figure 2-40. Select the SSH radio button and enter the IP address of your board in the Host Name (or IP address) box. Leave the port as 22.

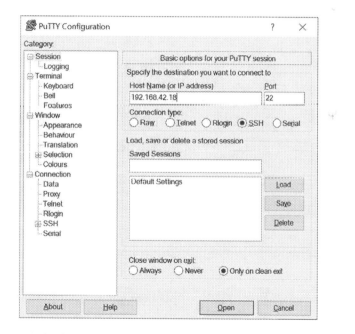

Figure 2-40. *PuTTY dialog*

When you are ready, click the *Open* button. You may be asked to put the host IP address in the hosts file. Confirm that request to proceed with the connection. Once the connection completes, you see a new command window open and are asked to log in.

Enter Administrator as the user and press Enter. Then enter the password that you set in a previous step (or the default p@ssw0rd, if you have not changed it) and press Enter. Figure 2-41 shows an example of the SSH session.

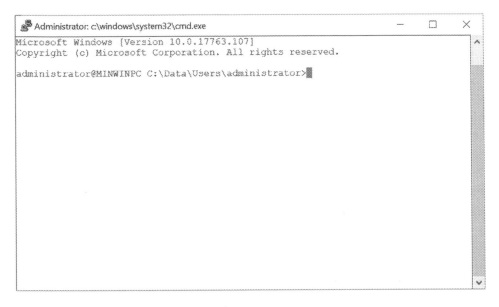

Figure 2-41. *SSH connection example*

There are a variety of commands you can use. You can perform nearly all the operations on the home screen via command-line utilities, such as boot configuration, startup applications, shutdown, restart, and more.

Thus, you can do everything you need to do remotely via the command line rather than use the Windows Device Portal (but I've found the portal more convenient). For more details about the command utilities available, see https://docs.microsoft.com/ en-us/windows/iot-core/manage-your-device/commandlineutils.

Connecting with the Windows PowerShell

You can also connect to your board with Windows PowerShell. However, it isn't quite as obvious or intuitive unless you've used PowerShell previously. To connect to your board with the PowerShell, use the dashboard on the *My Devices* page and right-click your board and then select *Launch PowerShell*. This launches the PowerShell beginning with a prompt for logging in as shown in Figure 2-42. Once again, use the user and password you defined earlier.

Figure 2-42. *Login prompt (PowerShell)*

Once you log in, you will see the PowerShell is connected to your board, and you can run the IoT utility commands such as those found at `https://docs.microsoft.com/en-us/windows/iot-core/manage-your-device/commandlineutils`. Figure 2-43 shows an example of the PowerShell connected to a Raspberry Pi.

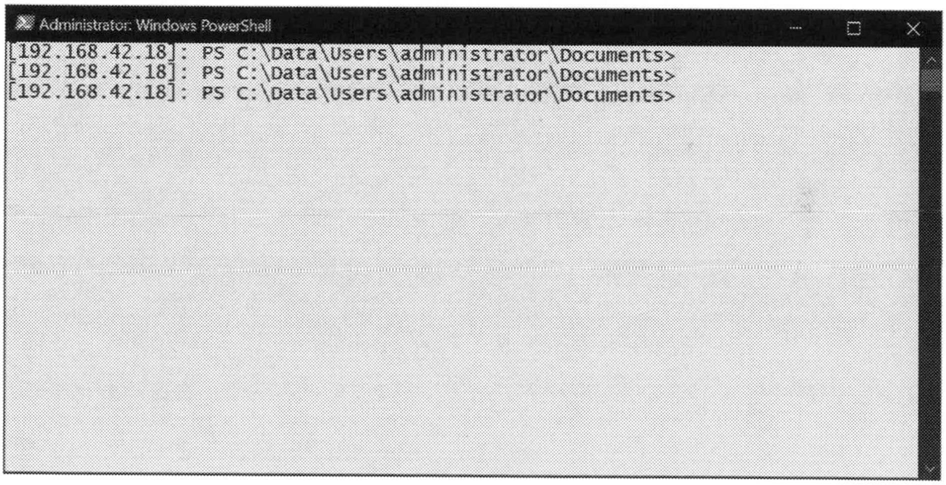

```
Administrator: Windows PowerShell                                              —   □   ×
[192.168.42.18]: PS C:\Data\Users\administrator\Documents>
[192.168.42.18]: PS C:\Data\Users\administrator\Documents>
[192.168.42.18]: PS C:\Data\Users\administrator\Documents>
```

Figure 2-43. *Connection with PowerShell*

You can also launch the PowerShell manually and connect, but you should do so as the administrator since some commands require elevated privileges. Simply search for `powershell`. When the application appears in the list, right-click it and select the *Run as administrator* option.

The steps you need to perform require starting the Windows remote management service, creating a trust relationship between your PC and the board, and then opening a session. See, not so intuitive, is it? Let's see how to do this.

Note   Connecting to your board may take several seconds from the PowerShell.

Once the PowerShell launches, start the Windows remote management service with the following command. Since this is a network operation, you use the net command. You see messages stating that the service has started.

```
> net start WinRM
```

Next, you create the trust relationship with the set-item command. Type the following command substituting the IP address from the board as you recorded earlier when you booted Windows 10 IoT Core for the first time. You are asked to confirm the change. Enter *Y* to continue.

```
> Set-Item WSMan:\localhost\Client\TrustedHosts -Value 192.168.42.18
```

Tip? If you want to connect to more than one board, you can place the list of IP addresses inside double quotes separated by commas, such as `"192.168.42.18,192.168.42.22"`.

Finally, you can start a session with your Windows IoT Core device using the `Enter-PSSession` command. Yes, another unintuitive command.[10] Here, you provide the computer name (IP address is fine) and username (Administrator). You are prompted for the password in a pop-up dialog. The following shows the command I used to open a session to my board at `192.168.42.18`. Figure 2-44 shows the PowerShell session once login succeeds.

```
> Enter-PSSession -ComputerName 192.168.42.18 -Credential 192.168.42.18\
Administrator
```

---

[10]For muggles.

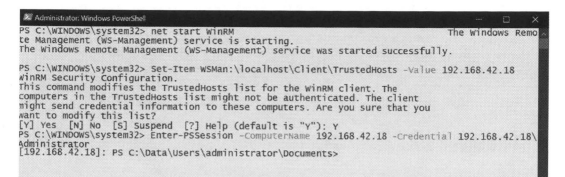

Figure 2-44. Connecting with the Windows PowerShell: session established

The IP address of the board is in square brackets. This lets you know which board you are currently connected to, should you connect to more than one at the same time. From here, you can use the command-line utilities described in the previous section.

Shutting Down the Device

There is just one more thing to discuss—shutting down the board! There are various ways to do this. If you have the board connected to a monitor with at least a keyboard, you can choose the small gear icon on the bottom left and then choose *Power Options* and *Shutdown*. Or, if you are using the Device Portal, simply click *Power* and then *Shutdown* in the upper-right corner as shown in Figure 2-45.

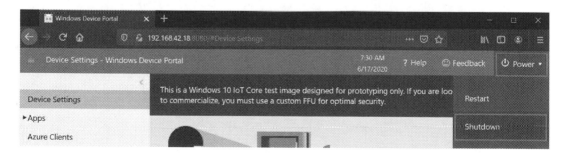

Figure 2-45. Shut down the board in Device Portal

Finally, you can also use the dashboard *My Devices* tab to right-click the board and choose *Shutdown*. Let the board shut down normally before unplugging the power. It takes a couple of minutes.

Now that you've had a crash course on getting your board up and running, we can talk more about how we can configure and write IoT applications. Go ahead and spend some time working with your board in the Device Portal and try out some of the example applications.

Summary

The Windows 10 IoT Core offers an option to explore electronics and the IoT for the rest of us. While there have been and continue to be options for exploring electronics and IoT for users of other platforms, Windows users have had to learn alternative operating systems and (seemingly) arcane commands in order to explore even the most basic of hobbyist electronics. With Windows 10, those days are gone!

This chapter explored the Windows 10 IoT Core and the hardware it runs on. You discovered some of the key features of the platform along with the details of the three low-cost computing boards that support Windows 10 IoT Core. You also discovered how to get your PC and your hardware configured to run Windows 10 IoT Core along with a walk-through for each of the available boards, from creating the boot image to booting and configuring the board to connecting to the board from your PC.

In the next chapter, you take a short break to examine the Raspberry Pi in more detail. You will learn more about the features and specifications of the Raspberry Pi, including a quick overview of how to boot it in its native operating system as well as how to write a simple program to illuminate an LED from code written in Python. While the chapter does not use Windows 10, you learn a great deal about the Raspberry Pi and a little bit about Python, which you can use to write scripts for the Raspberry Pi to work with the GPIO. How cool is that?

CHAPTER 3

Introducing the Raspberry Pi

The Raspberry Pi is one of the latest disruptive devices in recent years that has changed the way that we think about and design embedded solutions and the IoT. In fact, the Raspberry Pi has had tremendous success among hobbyists and enthusiasts. This is partly due to its low cost but also because it is a full-fledged computer running an open source operating system that has a wide audience: Linux.

Fortunately, for us, Windows 10 runs on the Raspberry Pi! However, given the popularity of the Raspberry Pi, it is likely that you will encounter examples and resources that are written for or only work with Linux. Thus, learning more about the Raspberry Pi and its native environment allows you to leverage the plethora of data for the Raspberry Pi and adapt, or helps you develop similar solutions for Windows 10. Plus, it gives you a brief insight into the non-Windows world of Raspberry Pi.[1]

This chapter explains how to set up and configure the Raspberry Pi using the Linux operating system. You'll also discover a few key concepts of how to work with Linux and even a brief look at writing Python scripts, which you can use to write IoT applications. Let us begin with an in-depth look at the Raspberry Pi.

Note While there are several versions of the Raspberry Pi, including the compute module, Zero, Model 1 A+, 2 Model B, 3 Model B/B+, and Model 4B, I discuss only those versions that run Windows 10 IoT Core: the Raspberry Pi 2 Model B and Raspberry Pi 3 Model B. For more details about all of the Raspberry Pi boards, see `www.raspberrypi.org`.

---

[1]This should reinforce how well Windows 10 IoT Core works!

© Charles Bell 2021

C. Bell, *Windows 10 for the Internet of Things*, https://doi.org/10.1007/978-1-4842-6609-0_3

Getting Started with the Raspberry Pi

The Raspberry Pi is a small, inexpensive personal computer, also called a low-cost computing board or simply low-cost computer. Although it lacks the capacity for memory expansion and can't accommodate onboard devices such as CD, DVD, and hard drives, it has everything a simple personal computer requires. That is, it has USB ports, an Ethernet port, HDMI, and even an audio connector for sound. The Raspberry Pi 3 Model B even has Bluetooth and Wi-Fi!

The Raspberry Pi has a micro-SD drive that you can use to boot the computer into any of several Linux operating systems (and Windows 10 IoT Core). All you need is an HDMI monitor (or DVI with an HDMI-to-DVI adapter), a USB keyboard and mouse, and a 5V power supply—and you're off and running.

The Raspberry Pi costs as little as $35.[2] It can be purchased online from electronics vendors such as SparkFun and Adafruit. Most vendors have a host of accessories that have been tested and verified to work with the Raspberry Pi. These include small monitors, miniature keyboards, and cases for protecting the board.

This section explores the origins of the Raspberry Pi 3B, tours the hardware connections, and covers the accessories needed to get started using the Raspberry Pi.

Raspberry Pi Origins

The Raspberry Pi was designed to be a platform to explore topics in computer science. The designers saw the need to provide inexpensive, accessible computers that could be programmed to interact with hardware such as servomotors, display devices, and sensors. They also wanted to break the mold of having to spend hundreds of dollars on a personal computer and thus make computers available to a much wider audience.

The designers observed a decline in the experience of students entering computer science curriculums. Instead of having some experience in programming or hardware, students are entering their academic years having little or no experience working with computer systems, hardware, or programming. Rather, students are well versed

---

[2]History shows new releases of the board can command higher prices until supply catches up with demand. I've seen the latest Raspberry Pi 4B 8GB sell for nearly twice its recommended sale price.

in Internet technologies and applications. One of the contributing factors cited is the higher cost and greater sophistication of the personal computer, which means parents are reluctant to let their children experiment on the family PC.

This poses a challenge to academic institutions, which have to adjust their curriculums to make computer science palatable to students. They have had to abandon lower-level hardware and software topics due to students' lack of interest or ability. Students no longer wish to study the fundamentals of computer science such as assembly language, operating systems, theory of computation, hardware, and concurrent programming. Rather, they want to learn higher-level languages to develop applications and web services. Thus, some academic institutions are no longer offering courses in fundamental computer science.[3] This could lead to a loss of knowledge and skillsets in future generations of computer professionals.

To combat this trend, the designers of the Raspberry Pi felt that, equipped with the right platform, today's youth could return to experimenting with personal computers and electronics as in the days when PCs required a much greater commitment to learning the hardware and system components and programming it in order to meet your needs. For example, the venerable Commodore 64, Amiga, and early Apple and IBM PC computers had very limited software offerings. Having owned a number of these machines, I was exposed to the wonder and discovery of hardware and programming at an early age. Perhaps that is why I find low-cost computing boards so fascinating—they pack a lot of features into a tiny board.

WHY IS IT CALLED RASPBERRY PI?

The name was partly derived from design committee contributions and partly chosen to continue a tradition of naming new computing platforms after fruit (think about it). The Pi portion comes from Python, because the designers intended Python to be the language of choice for programming the computer. However, other programming language choices are available.

The Raspberry Pi is an attempt to provide an inexpensive platform that encourages experimentation. The following sections further explore the Raspberry Pi, discussing topics such as the required accessories and where to buy the boards.

---

[3]Sadly, my alma mater is a fine example of this decline.

Versions That Work with Windows 10 IoT Core

There are currently several versions of Raspberry Pi boards. The latest boards that support Windows 10 IoT Core are the Raspberry Pi 2 Model B and Raspberry Pi 3 Model B. The model nomenclature has to do with the layout and ports available. Essentially, there is a model A variant that does not include Ethernet, whereas the model B variant does.

WINDOWS 10 IOT CORE FOR THE RASPBERRY PI 3B+

Microsoft's latest stable release of Windows 10 IoT Core may not support the latest boards. However, for those of us who want to use the latest boards or the latest features (such as the Raspberry Pi 3B+), Microsoft provides a preview release called the Windows 10 IoT Core Insider Preview. Sometimes, these preview releases are built for specific boards, so choose the one that matches the board or other features that you want.

To download this release, you must have a valid Microsoft account. You can initiate the download by visiting www.microsoft.com/en-us/software-download/windowsiot and logging in for Insider Preview access. Once you log in, you can find the preview images and download them. You can create an account from the page if you do not have one.

This downloads a new .iso file that you mount. Install the .msi and create the SD image in the same manner as the base images, as described in Chapter 2. However, the preview may overwrite existing files, so be sure to back up any files you want to keep.

The Raspberry Pi 2B and 3B are very similar. In fact, they are very hard to tell apart without reading the label on the top of the board. This is because they share the same layout (model B) with the same connectors. The hardware differences are very minor and difficult to spot. Figure 3-1 shows the two boards together, with the Raspberry Pi 2 on the left.

Figure 3-1. *Raspberry Pi 2 and 3 top side*

Can you tell them apart? Hint: Look at the lower-left corner of each board. The Raspberry Pi 3 has a small antenna, whereas the Raspberry Pi 2 has LEDs in the same spot.

While the boards appear nearly identical, the Raspberry Pi 3B (hence Raspberry Pi 3) has a much faster 64-bit quad core processor (Windows 10 IoT Core runs only in 32-bit mode currently) that has shown to be as much as ten times faster than the Raspberry Pi 2. Furthermore, the Raspberry Pi 3 has both Bluetooth and Wi-Fi onboard, whereas the Raspberry Pi 2B has neither. There are a number of smaller changes, but these are by far the most significant features you should use to decide which board to buy.

The underside is easier to distinguish the boards. Figure 3-2 shows the underside of both boards, again, with the Raspberry Pi 2 on the left. Here, you can see the difference at the top-left corner of the boards, where the Raspberry Pi 3 has a small rectangular chip that contains the Wi-Fi and Bluetooth components.

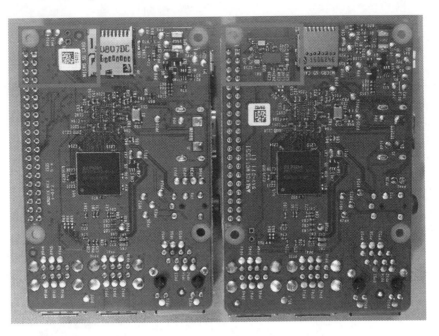

Figure 3-2. *Raspberry Pi 2 and 3 bottom side*

A Tour of the Board

Not much larger than a deck of playing cards, the Raspberry Pi board contains a number of ports for connecting devices. This section presents a tour of the board. If you want to follow along with your board, hold it with the Raspberry Pi logo face up. I work around the board clockwise. Figure 3-3 depicts a drawing of the board with all the major connectors labeled.

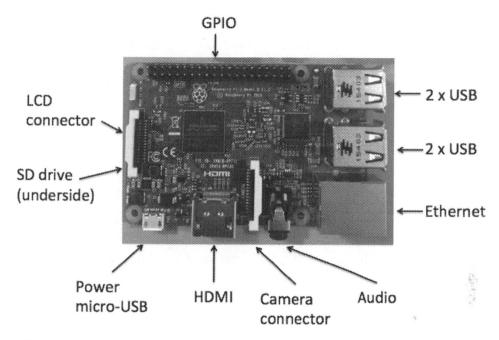

Figure 3-3. *Raspberry Pi 3 Model B*

Let's begin by looking at the bottom edge of the board (looking from above). In the center of the bottom side, you see an HDMI connector. To the left of the HDMI connector is the micro-USB power connector. The power connector is known to be a bit fragile on some boards, so take care plugging and unplugging it. Be sure to avoid putting extra strain on this cable while using your Raspberry Pi. To the right of the HDMI connector is the camera ribbon cable connector and next to that is the audio connector.

On the left side of the board is the LCD ribbon cable. You can use this connector with the Raspberry Pi 7-inch Touch LCD and similar devices. On the underside of the board is the micro-SD card drive. When installed, the SD card protrudes a few millimeters out of the board. If you plan to use a case for your Raspberry Pi, be sure the case provides access to the SD card drive (some do not).

Caution   Because the board is small, it is tempting to use it in precarious places, like in a moving vehicle or on a messy desk. Ensure that your Raspberry Pi is in a secure location. The micro-USB power, HDMI, and SD card slots seem to be the most vulnerable connectors.

On the top edge of the board is the general-purpose input/output (GPIO) header (a double row of 20 pins each), which can be used to attach to sensors and other electronic components and devices. You will work with this connector later in this chapter.

On the right side of the board are two USB connectors with two USB ports each and the Ethernet connector. An external powered USB hub connected to the USB ports on the Raspberry Pi can power some boards, but it is recommended that you use a dedicated power supply connected to the micro-USB connector.

Take a moment to examine the top and bottom faces of the board. As you can see, components are mounted on both sides. This is a departure from most printed circuit boards (PCB) that have components on only one side. The primary reason the Raspberry Pi has components on both sides is that it uses multiple layers for trace runs (the connecting wires on the board). Stacking the trace runs on multiple levels means that you don't have to worry about crossing paths. It also permits the board to be much smaller and enables the use of both surfaces. This is probably the most compelling reason to consider using a case—to protect the components on the bottom of the board and thus avoid shorts (accidental connection of contacts or pins) and can lead to board failure.

Required Accessories

The Raspberry Pi is sold as a bare system board with no case, power supply, or peripherals. Depending on how you plan to use the Raspberry Pi, you need a few commonly available accessories. If you have been accumulating computer and electronic spares like me, a quick rummage through your stores may locate most of what you need.

If you want to use the Raspberry Pi in console mode (no graphical user interface), you need a USB power supply, a keyboard, and an HDMI monitor. The power supply should have a minimal rating of 2.5A or greater. If you want to use the Raspberry Pi with a graphical user interface, you also need a pointing device (such as a mouse).

If you have to purchase these items, stick to the commonly available brands and models without extra features. For example, avoid the latest multifunction keyboard and mouse because they may require drivers that are not available for the various operating system choices for the Raspberry Pi.

You also must have a micro-SD card. I recommend a 16GB or higher version. Recall that the micro-SD is the only onboard storage medium available. You need to put the operating system on the card, and any files you create are stored on the card.

If you want to use sound in your applications, you also need a set of powered speakers that accept a standard 3.5mm audio jack. Finally, if you want to connect your Raspberry Pi to the Internet, you need an Ethernet cable, or if you are using a Raspberry Pi 3, you need a Wi-Fi network.

Recommended Accessories

I highly recommend, at a minimum, adding small 5mm to 10mm rubber or silicone self-adhesive bumpers to the bottom side of the board over the mounting holes to keep the board off your desk. On the bottom of the board are many sharp prongs that can come into contact with conductive materials, which can lead to shorts or, worse, a blown Raspberry Pi. They can also damage your desktop, skin, and clothing. Small self-adhesive bumpers are available at most home-improvement and hardware stores.

If you plan to move the board from room to room or you want to ensure that your Raspberry Pi is well protected against accidental damage, you should consider purchasing a case to house the board. Many cases are available, ranging from simple snap-together models to models made from laser-cut acrylic or even milled aluminum.

Tip   If you plan to experiment with the GPIO pins, or require access to the power test pins or the other ports located on the interior of the board, you may want to consider either using the self-adhesive bumper option or ordering a case that has an open top to make access easier. Some cases are prone to breakage if opened and closed frequently.

Aside from a case, you should also consider purchasing (or pulling from your spares) a powered USB hub. The USB hub power module should be 2A to 2.5A or more. Even though the Raspberry Pi 2 and 3 have four USB ports, a powered hub is required if you plan to use USB devices that draw a lot of power, such as a USB hard drive or a USB toy missile launcher.

Where to Buy

The Raspberry Pi 2 and 3 are plentiful and can be found on many websites serving many continents. Chances are there is an online store available near you. To find out, go to `www.raspberrypi.org/products/raspberry-pi-3-model-b/` and click *Buy Now* and use the drop-down list to look for a country or city near you. Fortunately, those online retailers who stock it offer a host of accessories that are known to work with the Raspberry Pi. The following are some of the more popular online retailers with links to their Raspberry Pi catalog entry:[4]

- *Adafruit*: `www.adafruit.com/category/105`

- *SparkFun*: `www.sparkfun.com/categories/233`

- *The Pi Hut*: `www.thepihut.com`

- *PiShop.us*: `www.pishop.us`

The next section presents a short tutorial on getting started using the Raspberry Pi. If you have already learned how to use the Raspberry Pi, you can skip to the following section to begin learning how to use your board.

Setting Up the Raspberry Pi

The Raspberry Pi is a personal computer with a surprising amount of power and versatility. You may be tempted to consider it a toy or a severely limited platform, but that is far from the truth.[5] With the addition of onboard peripherals like USB, Ethernet, and HDMI video (as well as Bluetooth and Wi-Fi for the Raspberry Pi 3 and later), the Raspberry Pi has everything you need for a lightweight desktop computer. If you consider the addition of the GPIO header, the Raspberry Pi becomes more than a simple desktop computer and fulfills its role as a computing system designed to promote hardware experimentation.

The following sections present a short tutorial on getting started with your new Raspberry Pi, from a bare board to a fully operational platform. A number of excellent works cover this topic in much greater detail. If you find yourself stuck or wanting to

---

[4]You can often find the 2 and 3B listed, but quantities may be limited. Online auction sites sometimes have excellent deals on the older boards.

[5]Especially considering the Raspberry Pi 4B 4GB and 8GB versions.

know more about beginning to use the Raspberry Pi and more about the Raspberry Pi OS (Raspbian) operating system, read *Learn Raspberry Pi with Linux* by Peter Membrey and David Hows (Apress, 2012). If you want to know more about using the Raspberry Pi in hardware projects, an excellent resource is *Practical Raspberry Pi* by Brendan Horan (Apress, 2013).

As mentioned in the "Required Accessories" section, you need a micro-SD card, a USB power supply rated at 2A or better with a male micro-USB connector, a keyboard, a mouse (optional), and an HDMI monitor. However, before you can boot your Raspberry Pi and bask in its brilliance, you need to create a boot image for your micro-SD card.

Choosing a Boot Image (Operating System)

The first thing you need to do is decide which operating system variant you want to use. There are several excellent choices, including the standard Raspberry Pi OS. Each is available as a compressed file called an image or card image. You can find a list of recommended images along with links to download each on the Raspberry Pi foundation download page: `www.raspberrypi.org/downloads`. The following images are available at the site:

- *Raspberry Pi OS* (formally Raspbian): A Debian-based official operating system and contains a graphical user interface (Lightweight X11 Desktop Environment [LXDE]), development tools, and rudimentary multimedia features.

- *Ubuntu MATE*: Features the Ubuntu desktop and a scaled-down version of the Ubuntu operating system. If you are familiar with Ubuntu, you will feel at home with this version.

- *Ubuntu Core*: The developer's edition of the core Ubuntu system. It is the same as MATE with the addition of the developer core utilities.

- *Ubuntu Server*: The developer's edition of the core Ubuntu server system.

- *OSMC* (Open Source Media Center): Build yourself a media center.

- *OpenELEC* (Open Embedded Linux Entertainment Center): Another media center option.

- *PiNet*: A classroom management system. A special edition for educators using the Raspberry Pi in the curriculum.

- *RISC OS*: A non-Linux, Unix-like operating system. If you know what IBM AIX is, or you've used other Unix operating systems, you'll recognize this beastie.

- *Windows 10 IoT Core*: Windows 10 for the IoT. Microsoft's premier IoT operating system.

Tip   If you are just starting with the Raspberry Pi and haven't used a Linux operating system, you should use the Raspberry Pi OS image as it is the most popular choice and more widely documented in examples. Plus, it is the base or default image for Raspberry Pi.

There are a few other image choices, including a special variant of the Raspberry Pi OS image from Adafruit. Adafruit calls their image "occidentals" and includes a number of applications and utilities preinstalled, including Wi-Fi support and several utilities. Some Raspberry Pi examples—especially those from Adafruit—require the occidentals image. You can find out more about the image and download it at `http://learn.adafruit.com/adafruit-raspberry-pi-educational-linux-distro/overview`.

Wow! That's a lot of choices, isn't it? As you can see, the popularity of the Raspberry Pi is very wide and diverse. This makes Windows 10 IoT a huge deal for the platform and Windows users alike. While you may not use these operating systems, it is good to know what choices are available should you need to explore them.

Now let's see how to install the base operating system. As you will see, it is very easy.

Creating the Boot Image

There are two methods for installing the boot image. You can use the automated, graphical user interface platform called New Out Of Box Software (NOOBS[6]), or you can install your image from scratch onto a micro-SD drive. Both require downloading and formatting the micro-SD drive.

---

[6]An unfortunate resemblance to the slightly derogatory slang noob or newbie.

Since you are curious about using the Raspberry Pi native operating system as an experiment or for research, let's stick to the easier method and use NOOBS. Aside from formatting the micro-SD card, everything is automated; nothing requires any complicated commands.

With NOOBS, you download a base installer image that contains Raspberry Pi OS. You can choose to install it or configure NOOBS to download another image and install it. But first, you have to get the NOOBS boot image and copy it to your micro-SD drive.

Begin by downloading the NOOBS installer from `www.raspberrypi.org/downloads/noobs/`. You see two options: an offline and network installer that includes the Raspberry Pi OS image or a network installer that does not contain any operating systems (called NOOBS Lite). The first option (the one with the base image) is what you should use if you are following along with this chapter.

Once you've downloaded the installer (to date about 2.3GB), you need to format a micro-SD card of at least 16GB. You'll use the SD Card Formatter utility for Windows (`www.sdcard.org/downloads`). Simply download the application and install it. Then insert your micro-SD card in your card reader and launch the application. Once you verify that you've selected the correct media, enter a name for the card (I used RASPI) and click the Format button. Figure 3-4 shows the SD Card Formatter application. Notice I named the drive NOOBS to help me remember what is on it.

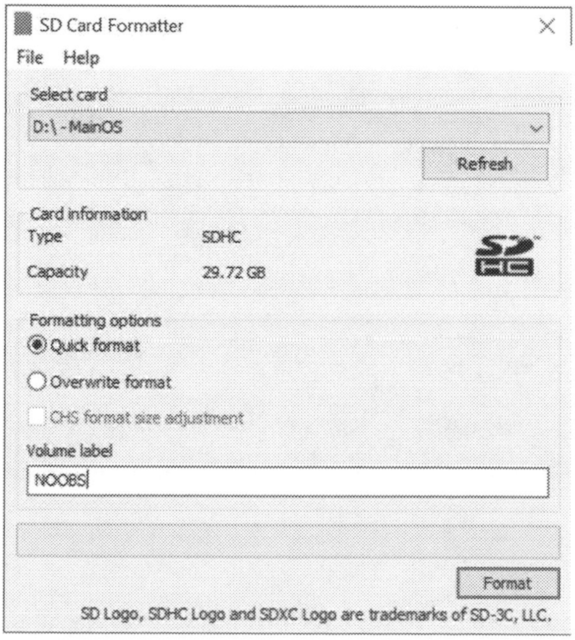

Figure 3-4. *SD Card Formatter*

Once you've formatted the micro-SD card, you now must copy the contents of the NOOBS image to the card. Right-click the file that you downloaded and choose the option to unzip or unarchive the file. This creates a folder containing the NOOBS image. Copy all of those files (not the outside folder) to the SD card and eject it. You are now ready to boot into NOOBS and install your operating system. When this process has finished, safely remove the SD card, and insert it into your Raspberry Pi.

Booting the Board

You are now ready to hook up all of your peripherals. I like to keep things simple and only connect a monitor, keyboard, and (for NOOBS) a mouse. If you want to download an operating system other than Raspberry Pi OS, you also need to connect your Raspberry Pi to your network. If you are planning to use the Wi-Fi option on the Raspberry Pi 3, you'll need to set up your Wi-Fi configuration after you boot up. I'll show you how to do that in a moment.

Once your Raspberry Pi powers on, you see a scrolling display of various messages. This is normal and may scroll for some time before NOOBS starts. When NOOBS is loaded, you see a screen similar to Figure 3-5.

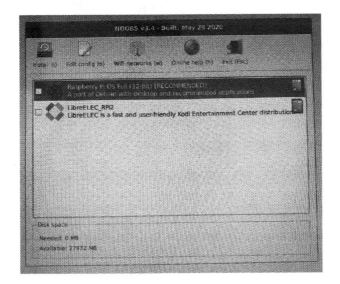

Figure 3-5. *NOOBS startup screen*

Notice the Raspberry Pi OS image in the list of operating systems. To install it, just tick the check box beside the thumbnail and then click the *Install* button. Notice also there are two boxes at the bottom. They set the language and keyboard for use in NOOBS, which does not affect the Raspberry Pi OS setup.

Once you initiate the install, you will be asked to confirm the installation. Once the installation begins, you will see a series of dialogs as Raspberry Pi OS begins its installation to the SD card. This could take a while. The good news is that the dialogs provide a lot of useful information to help you get started. You learn about how to log in to Raspberry Pi OS, tips for configuring and customizing, and suggestions on how to get the most out of your experience.

When installation finishes, click OK on the completed dialog and then wait for the Raspberry Pi to reboot into Raspberry Pi OS. The system boots and automatically logs on as the `pi` user. On first boot, you will also see a Welcome to Raspberry Pi configuration dialog that presents a number of pages where you can customize your installation.

The first dialog asks you to select your country, language, and time zone. Simply make your selections and click *Next*.

On the next dialog page, you will be asked to set the password for the Pi user. Be sure to use something you can remember. Once you enter your new password, enter it again to confirm it and then click *Next*.

On the next dialog page, you have an opportunity to tune the video screen. If you see a black border around your desktop, tick the tick box and then click *Next* or click *Next* without ticking the tick box if your desktop fills the whole screen.

On the next dialog, you will be asked to choose a Wi-Fi network (if using a Raspberry Pi 3B or later). Simply choose the Wi-Fi connection and click *Next*. Then, enter the network password when prompted. You can also skip this step by clicking the *Skip* button and set up your network connection later.

After the network connects, you will be asked if you want to update the software. This is very much like what you would do on Windows 10, so it is always a good idea to do the update. To update the system, click *Next* or, if you want to wait and do it later, click *Skip*. If you elected to update the software, you will see a progress bar for several steps in that process. Once the process is complete, you can click *OK* to the completion dialog and then *Restart* to reboot the computer.

If you decide to make one or more changes later, or if you want to enable certain system features such as the camera, programming interfaces or protocols, and so on, you can use the Raspberry Pi Configuration dialog accessed by the Pi menu (in the upper-left portion of the screen), then click *Preferences* and finally choose *Raspberry Pi Configuration* as shown in Figure 3-6.

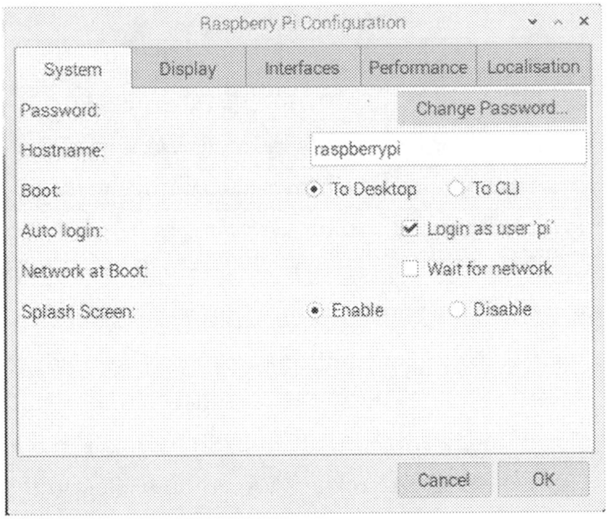

Figure 3-6. *Raspberry Pi OS configuration dialog*

There are five tabs that you can use to change settings for the system. The following list briefly explains each and includes the recommended settings for each. Once you have made your changes, click OK to close the dialog. Depending on which settings you choose, you may be asked to reboot.

- *System*: The board controls for the system. Use this panel to change the root password (highly recommended), hostname (optional), type of boot (use command-line interface [CLI] if you want to set up the Raspberry Pi to boot headless), and automatic login (not recommended), wait for network to boot, and enable or disable the splash screen.

- *Display*: Adjust the display settings such as resolution, overscan, pixel doubling, and screen blanking.

- *Interfaces*: Used to enable system and hardware services such as the camera, SSH (recommended), and hardware interfaces for the GPIO header. Look here if your programs do not work as you expect when using a communication protocol such as I2C or SPI and similar interfaces.

- *Performance*: Used to make changes to how the processor performs. You can choose to overclock (run the CPU faster), but I do not recommend this setting for a Raspberry Pi that hosts IoT solutions.

- *Localization*: Used to set the default language, keyboard, and date and time. If you change nothing else, be sure to set these to your local settings.

To shut down or reboot Raspberry Pi OS, click the Pi, and then choose *Logout*, then *Shutdown*. If you want to shut down from a terminal (command line), use the command `shutdown -h now` to shut down the system.

A Brief Linux Primer

OK, now you have a Raspberry Pi booting Linux (Raspberry Pi OS) into the desktop environment. And although it looks cool, it can be a bit confusing and intimidating. The best way to learn the GUI is to simply spend some time clicking your way through the menus. You'll find the most basic of features, including productivity tools.

However, working with hardware typically requires knowledge of basic commands used in a terminal (also called the command line). This section describes a number of the more basic commands you need to use. This is by no means meant to be a complete or thorough coverage of all of the commands. Rather, it gives you the basics that you need to get started.

Thus, this primer is more like a 15-minute guided tour of an automobile engine. You cannot possibly learn all of the maintenance requirements and internal components in 15 minutes. You would need to have an automotive technician's training or years of experience before you could begin to understand everything. What you get in a 15-minute lightning tour is more of a bird's eye view with enough information to permit you to know where the basic maintenance items are located, not necessarily how they work.

I recommend you read through the following sections to familiarize yourself with the commands that you may need. You can refer back to these sections should you need to recall the command name. Oftentimes, it is simply a matter of learning a different name for the same commands (conceptually) that you're familiar with from Windows. As you will see, many of these commands are familiar in concept, as they also exist on Windows albeit with a different name and parameters.

> **Tip**   If you want to master the Linux command-line commands, tools, and utilities, read the book *Beginning the Linux Command Line* by Sander van Vugt (Apress, 2015).

Let's begin with how to get help about commands.

Getting Help

Linux provides help for all commands by default. While it can be a bit terse, you can always get more information about a command by using the manual command as shown in Listing 3-1. Here, you want more help with the list directory command (ls).

Listing 3-1.   Getting Help for a Man (Manual) Command

```
pi@raspberrypi:~ $ man ls

LS(1) User CommandsLS(1)

NAME
 ls - list directory contents

SYNOPSIS
 ls [OPTION]... [FILE]...

DESCRIPTION
 List  information  about  the FILEs (the current directory by default).
 Sort entries alphabetically if none of -cftuvSUX nor --sort  is  speci-
 fied.

 Mandatory  arguments  to  long  options are mandatory for short options
 too.

 -a, --all
  do not ignore entries starting with .

 -A, --almost-all
  do not list implied . and ..

 ...
```

File and Directory Commands

Like any operating system, some of the most basic commands are those that allow you to manipulate files and directories. These include operations such as copying, moving, and creating files and directories. I list a few of the most common commands in the following sections and provide an example of each. If you want to know more about each, try using the manual command (man) to learn about each. Just use the name of the command you want to know more about as the option. For example, to learn more about ls, enter man ls.

List Directories and Files

The first command you will likely need is the ability to list files and directories. In Linux, we use the ls (list files and directories) command. Without any options, the command lists all of the files and directories in the current location. There are many options available, but the ones I find most helpful are show long listing format (-l), sort the output (-s), and show all files (-a). You can combine these options in a single string, such as -lsa.

The command uses color and highlighting to help distinguish directories from files, executable files, and more. The long listing format also shows you the permissions for the file (the series of rwx values). The first character in the directory list refers to the file type (d means a directory and - is a regular file), the next three characters refer to file owner permissions, the next three are group permissions, and the final three are for other users' permissions. Figure 3-7 shows an example of the ls -lsa command output.

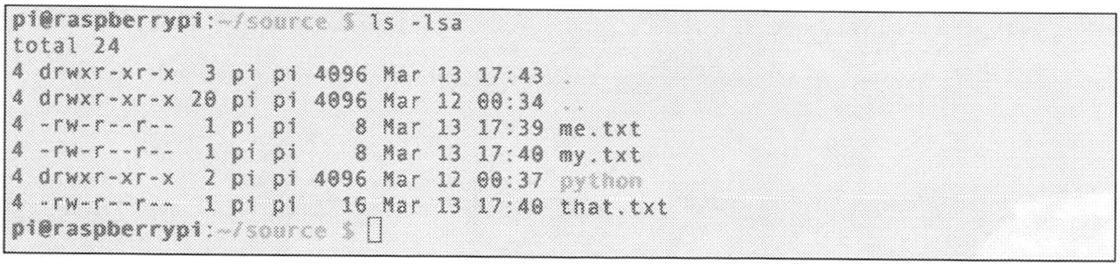

```
pi@raspberrypi:~/source $ ls -lsa
total 24
4 drwxr-xr-x  3 pi pi 4096 Mar 13 17:43 .
4 drwxr-xr-x 20 pi pi 4096 Mar 12 00:34 ..
4 -rw-r--r--  1 pi pi    8 Mar 13 17:39 me.txt
4 -rw-r--r--  1 pi pi    8 Mar 13 17:40 my.txt
4 drwxr-xr-x  2 pi pi 4096 Mar 12 00:37 python
4 -rw-r--r--  1 pi pi   16 Mar 13 17:40 that.txt
pi@raspberrypi:~/source $ []
```

Figure 3-7. *Output of list directories (ls) command*

Change Directory

You can change from one directory to another by using the cd command, which is quite familiar.

```
pi@raspberrypi:~ $ cd source
pi@raspberrypi:~/source $
```

Tip   The Linux path separator is a /, which can take some getting used to.

Copy

You can copy files with the cp command with the usual expected parameters of <from_file> <to_file>, as shown next. You can also use full paths to copy files from one directory to another.

```
pi@raspberrypi:~/source $ ls
me.txt   python
pi@raspberrypi:~/source $ cp me.txt my.txt
pi@raspberrypi:~/source $ ls
me.txt   my.txt   python
```

Tip   Use the * symbol as a wildcard to specify all files (synonymous with *.* in Windows). For example, to copy all of the files from one folder to another, use the cp ./old/* ./new command.

Move

If you want to move files from one folder to another, you can use the mv command with the usual expected parameters of <from> <to>, as shown next. You can also use full paths to move files from one directory to another.

```
pi@raspberrypi:~/source $ ls
me.txt   my.txt   python   this.txt
pi@raspberrypi:~/source $ mv this.txt that.txt
pi@raspberrypi:~/source $ ls
me.txt   my.txt   python   that.txt
```

Create Directories

Creating directories can be accomplished with the `mkdir` command. If you do not specify a path, the command executes in the current directory.

```
pi@raspberrypi:~/source $ ls
me.txt  my.txt  python  that.txt
pi@raspberrypi:~/source $ mkdir test
pi@raspberrypi:~/source $ ls
me.txt  my.txt  python  test  that.txt
```

Delete Directories

If you want to delete a directory, use the `rmdir` command. This command requires that the directory be empty. You will get an error if the directory contains any files or other directories.

```
pi@raspberrypi:~/source $ ls
me.txt  my.txt  python  test  that.txt
pi@raspberrypi:~/source $ rmdir test
pi@raspberrypi:~/source $ ls
me.txt  my.txt  python  that.txt
```

Create (Empty) Files

Sometimes, you may want to create an empty file for use in logging output or just to create a placeholder for editing later. The `touch` command allows you to create an empty file.

```
pi@raspberrypi:~/source $ ls ./test
pi@raspberrypi:~/source $ touch ./test/new_file.txt
pi@raspberrypi:~/source $ ls ./test
new_file.txt
pi@raspberrypi:~/source $ rmdir test
rmdir: failed to remove 'test': Directory not empty
```

Delete Files

If you want to delete a file, use the rm command. There are a number of options for this command, including recursively deleted files in subfolders (-r) and options for more powerful (thorough) cleaning.

```
pi@raspberrypi:~/source $ rm ./test/new_file.txt
pi@raspberrypi:~/source $ ls ./test
```

Caution   You can use the rm command with the force option to remove directories, but you should use such options with extreme caution. Executing sudo rm * -rf in a directory will permanently delete all files!

System Commands

The Linux operating system provides a huge list of system commands to do all manner of operations on the system. Mastering all of the system commands can take quite a while. Fortunately, there are only a few that you may want to learn to use Linux with a minimal of effort.

Show (Print) Working Directory

The system command I use most frequently is the print working directory (pwd) command. This shows you the full path to the current working directory.

```
pi@raspberrypi:~/new_source $ pwd
/home/pi/new_source
```

Command History

The one system command that you may find most interesting and helpful is the history command. This command lists the commands that you have entered over time. So, if you find that you need to issue some command you used a month ago, use the history command to show all of the commands executed until you find the one you need. This is especially helpful if you cannot remember the options and parameters! However, this list is only for the current user. The following is an excerpt of the history for my Raspberry Pi 3:

```
pi@raspberrypi:~/source $ history
  1  sudo apt-get update
  2  sudo apt-get upgrade
  3  sudo shutdown -r now
  4  rpi-update
  5  sudo
  6  sudo rpi-update
  7  sudo apt-get dist-upgrade
  8  sudo shutdown -r now
  9  startx
 10  ls /lib/firmware/brcm
...
```

Tip   Use the *Up* and *Down* keys on the keyboard to call back the last command issued and scroll forward and backward through the history one command at a time.

Archive Files

Occasionally, you may need the ability to archive or unarchive files, which you can do with a system command (utility). The tape archive (tar) command shows the longevity of the Linux (and its cousin/predecessor, Unix) operating system from the days when offline storage included tape drives[7] (no disk drives existed at the time). The following shows how to create an archive and extract it. The first tar command creates the archive and the second extracts it.

```
pi@raspberrypi:~ $ tar -cvf archive.tar ./source/
./source/
./source/test/
./source/my.txt
./source/that.txt
./source/python/
./source/python/blink_me.py
```

---

[7]Anyone remember punch cards?

105

```
./source/me.txt
pi@raspberrypi:~ $ mkdir new_source
pi@raspberrypi:~ $ cd new_source
pi@raspberrypi:~/new_source $ tar -xvf ../archive.tar
./source/
./source/test/
./source/my.txt
./source/that.txt
./source/python/
./source/python/blink_me.py
./source/me.txt
pi@raspberrypi:~/new_source $ ls
source
```

There are a host of options for the tape archive command. The most basic are the create (`-cvf`) and extract (`-xvf`) option strings, as shown in the preceding code. See the manual for the tape archive command if you want to perform more complicated operations.

Administrative Commands

Like the system commands, there is a long list of administrative operations that you may need to perform. I list those operations that you may need to perform for more advanced operations, starting with the run as administrator equivalent command.

Run as Super User

To run a command with elevated privileges, use the sudo command. Some commands and utilities require sudo. For example, to ping another computer, install software, or change permissions, and so forth, you need elevated privileges.

```
pi@raspberrypi:~/new_source $ sudo ping localhost
PING localhost (127.0.0.1) 56(84) bytes of data.
64 bytes from localhost (127.0.0.1): icmp_seq=1 ttl=64 time=0.083 ms
64 bytes from localhost (127.0.0.1): icmp_seq=2 ttl=64 time=0.068 ms
64 bytes from localhost (127.0.0.1): icmp_seq=3 ttl=64 time=0.047 ms
^C
```

```
--- localhost ping statistics ---
3 packets transmitted, 3 received, 0% packet loss, time 1998ms
rtt min/avg/max/mdev = 0.047/0.066/0.083/0.014 ms
```

Change File/Directory Permissions

In Linux, files and directories have permissions, as described in the previous section. You can see the permissions with the list directory command. To change the permissions, use the chmod command as shown in the following code. Here, we use a series of numbers to determine the bits of the permissions. That is, 7 means rwx, 6 means rw, and so forth. For a complete list of these numbers and an alternative form of notation, see the manual for chmod.[8]

```
pi@raspberrypi:~/source $ ls -lsa
total 12
4 drwxr-xr-x  3 pi pi 4096 Mar 13 18:07 .
4 drwxr-xr-x 21 pi pi 4096 Mar 13 18:02 ..
0 -rw-r--r--  1 pi pi 0 Mar 13 18:07 cmd
4 drwxr-xr-x  2 pi pi 4096 Mar 12 00:37 python
pi@raspberrypi:~/source $ chmod 0777 cmd
pi@raspberrypi:~/source $ ls -lsa
total 12
4 drwxr-xr-x  3 pi pi 4096 Mar 13 18:07 .
4 drwxr-xr-x 21 pi pi 4096 Mar 13 18:02 ..
0 -rwxrwxrwx  1 pi pi 0 Mar 13 18:07 cmd
4 drwxr-xr-x  2 pi pi 4096 Mar 12 00:37 python
```

Change Owner

Similarly, you can change ownership of a file with the chown command if someone else created the file (or took ownership). You may not need to do this if you never create user accounts on your Raspberry Pi, but you should be aware of how to do this in order to install some software such as MySQL.

---

[8]For more information, see the numerical permissions section at https://en.wikipedia.org/wiki/Chmod.

```
pi@raspberrypi:~/source $ ls -lsa
total 12
4 drwxr-xr-x  3 pi pi 4096 Mar 13 18:07 .
4 drwxr-xr-x 21 pi pi 4096 Mar 13 18:02 ..
0 -rwxrwxrwx  1 pi pi 0 Mar 13 18:07 cmd
4 drwxr-xr-x  2 pi pi 4096 Mar 12 00:37 python
pi@raspberrypi:~/source $ sudo chown chuck cmd
pi@raspberrypi:~/source $ ls -lsa
total 12
4 drwxr-xr-x  3 pi pi 4096 Mar 13 18:13 .
4 drwxr-xr-x 21 pi pi 4096 Mar 13 18:02 ..
0 -rw-r--r--  1 chuck pi 0 Mar 13 18:13 cmd
4 drwxr-xr-x  2 pi pi 4096 Mar 12 00:37 python
pi@raspberrypi:~/source $ sudo chgrp chuck cmd
pi@raspberrypi:~/source $ ls -lsa
total 12
4 drwxr-xr-x  3 pi pi 4096 Mar 13 18:13 .
4 drwxr-xr-x 21 pi pi 4096 Mar 13 18:02 ..
0 -rw-r--r--  1 chuck chuck 0 Mar 13 18:13 cmd
4 drwxr-xr-x  2 pi pi 4096 Mar 12 00:37 python
```

Tip   You can change the group with the `chgrp` command.

Install/Remove Software

The second most used administrative operation is installing or removing software. To do this on Raspberry Pi OS (and similar Linux distributions), you use the apt-get command, which requires elevated privileges.

Linux maintains a list of header files that contain the latest versions and locations of the source code repositories for all components installed on your system. Occasionally, you need to update these references, and you can do so with the following options. Do this before you install any software. In fact, most documentation for software requires you to run this command. You must be connected to the Internet before running the command, and it could take a few moments to run.

```
sudo apt-get update
```

To install software on Linux, you use the install option (conversely, you can remove software with the remove option). However, you must know the name of the software you want to install, which can be a challenge. Fortunately, most software providers tell you the name to use. Interestingly, this name can be the name of a group of software. For example, the following command initiates the installation of MySQL 5.5 (the latest version is 8.0), which involves a number of packages (shown in bold):

```
pi@raspberrypi:~/source $ sudo apt-get install mysql-server
Reading package lists... Done
Building dependency tree
Reading state information... Done
The following extra packages will be installed:
  libaio1 libdbd-mysql-perl libdbi-perl libhtml-template-perl
  libmysqlclient18
  libterm-readkey-perl mysql-client-5.5 mysql-common mysql-server-5.5
  mysql-server-core-5.5
Suggested packages:
  libclone-perl libmldbm-perl libnet-daemon-perl libsql-statement-perl
  libipc-sharedcache-perl mailx tinyca
The following NEW packages will be installed:
  libaio1 libdbd-mysql-perl libdbi-perl libhtml-template-perl
  libmysqlclient18
  libterm-readkey-perl mysql-client-5.5 mysql-common mysql-server
  mysql-server-5.5 mysql-server-core-5.5
0 upgraded, 11 newly installed, 0 to remove and 0 not upgraded.
Need to get 8,121 kB of archives.
After this operation, 88.8 MB of additional disk space will be used.
Do you want to continue? [Y/n]
```

Shutdown

Finally, you want to shut down your system when you are finished using it or perhaps reboot it for a variety of operations. For either operation, you need to run with elevated privileges (sudo) and use the shutdown command. This command takes several options: use -r for reboot and -h for halt (shutdown). You can also specify a time to perform the operation, but I always use the now option to initiate the command immediately.

To reboot the system, use this command:

```
sudo shutdown -r now
```

To shut down the system, use this command:

```
sudo shutdown -h now
```

Useful Utilities

There are a number of useful utilities that you need at some point during your exploration of Linux. Those that I use most often are described in the following list, which includes editors. There are, of course, many more examples, but these will get you started for more advanced work:

- *Text editor*: `nano` (A simple, easy-to-use text editor. It has a help menu at the bottom of the screen. Some operations may seem odd after using Windows text editors, but it is much easier to use than some other Linux text editors.)

- *File search*: `find` (Locates files by name in a directory or path)

- *File/text search*: `grep` (Locates a text string in a set of files or directory)

- *Archive tools*: `gzip`, `gunzip` (A zip file archive tool (an alternative to tar))

- *Text display tools*: `less`, `more` (Less shows the last portion of a file; more shows the file contents a page (console page) at a time.)

Now that you know more about how to get around in Linux and use the command line, let's look at how you can write a simple program to run on the Raspberry Pi.

Working with Python: Blink an LED

Now that you know a little bit about how to use Raspberry Pi OS, let's take an interesting diversion into the world of programming the Raspberry Pi with Python and working with the GPIO pins. This may seem a bit premature, but I provide this example for those readers who want to experience programming the Raspberry Pi, especially those who want to jump into working with hardware. Thus, I don't explain every detail about the electronic components; however, I present a primer on electronics in Chapter 7. More specifically, I use some components from the Microsoft Internet of Things Pack for the Raspberry Pi from Adafruit (`www.adafruit.com/products/2702`), which is discussed in more detail in Chapter 9.

If you prefer to wait until you've learned more about electronics and the Microsoft Internet of Things Pack for the Raspberry Pi, you can always read through the example and revisit it later once you read through the later chapters. However, you will find a very similar example in Chapters 5 and 6. Reading through this example gives you some insights about what you will accomplish later.

The programming language that you will use is a very easy scripting language called Python.[9] As you will see, the commands are quite intuitive and very expressive. For the purposes of this demonstration, you do not need to be an expert with the language. I provide all of the code and commands you need, explaining each as we go along. Once again, this is a lightning tour rather than a comprehensive study. Let's begin with a description of the project.

PYTHON? ISN'T THAT A SNAKE?

The Python programming language is a high-level language designed to be as close to like reading English as possible while being simple, easy to learn, and very powerful. Pythonistas[10] will tell you the designers have indeed met these goals.

Python does not require a compilation step prior to being used. Rather, Python applications (whose filenames end in `.py`) are interpreted on the fly. This is very powerful, but unless you use a Python development environment, some syntax errors (such as incorrect indentation) are not discovered until the application is executed. Fortunately, Python provides a robust exception-handling mechanism.

---

[9]A plethora of information is available about Python at `www.python.org`.

[10]Python experts often refer to themselves using this term. It is reserved for the most avid and experienced Python programmers.

If you have never used Python or you would like to know more about it, the following are a few good books that introduce the language. A host of resources are also available on the Internet, including the Python documentation pages at `www.python.org/doc/`.

- *Programming the Raspberry Pi* by Simon Monk (McGraw-Hill, 2013)

- *Beginning Python: From Novice to Professional*, 2nd Edition, by Magnus Lie Hetland (Apress, 2008)

- *Python Cookbook* by David Beazley and Brian K. Jones (O'Reilly Media, 2013)

Interestingly, Python was named after the British comedy troupe Monty Python, not the reptile. As you learn Python, you may encounter campy references to Monty Python episodes. I find these references entertaining. Of course, your mileage may vary.

You're going to build a very simple circuit that turns on an LED briefly in a loop that makes the LED appear to blink (you turn it on and then off again repeatedly). This may sound like mad scientist work or something that requires years of electronics training, but it really isn't. You will use only two electronic components—an LED and a resistor—as well as two wires and a breadboard to complete this project.

Tip   If you do not feel comfortable working with electronics, that's OK! Just read the later chapters on electronics (Chapters 6 and 7) and then come back to this chapter.

Hardware Connections

Let's begin by gathering the hardware that you need. The following lists the components that are needed. All of these are available in the Microsoft Internet of Things Pack for the Raspberry Pi from Adafruit. If you do not have that kit, you can find these components separately on the Adafruit website (`www.adafruit.com`), or from SparkFun (`www.sparkfun.com`), or any electronics store that carries electronic components.

- 560 ohm 5% 1/4W resistor (green, blue, brown stripes[11])

- Diffused 10mm red LED (or similar)

---

[11]See `https://en.wikipedia.org/wiki/Electronic_color_code`.

- Breadboard (mini, half, or full sized)

- (2) male-to-female jumper wires

Take a look at the breadboard. You see a center divide (normally a groove or a thick line). This sections the breadboard into two sides. The holes running perpendicular to the center groove are connected together but are not connected to adjacent holes (rows). If you use a half- or full-sized breadboard, you may have power rails, which run horizontally to the center groove, which are connected. Thus, the power rails run parallel to the center groove, and the interior connections run perpendicular to the groove. Now that you know how the breadboard is wired, let's build our circuit.

The only component that is polarized is the LED. Take a look at the LED. You see that one leg (pin) of the LED is longer than the other. This longer side is the positive side.

Begin by placing the breadboard next to your Raspberry Pi and power the Raspberry Pi off, orienting the Raspberry Pi with the label facing you (GPIO pins in the upper-left corner). Next, take one of the jumper wires and connect the female connector to pin 6 on the GPIO. The pins are numbered left to right starting with the lower-left pin. Thus, the left two pins are 1 and 2 with pin 1 below pin 2. Connect the other wire to pin 7 on the GPIO.

Next, plug the resistor into the breadboard with each pin on one side of the center groove. You can choose whichever area you want on the breadboard. Then, connect the LED so that the long leg (sometimes shown as the leg with a bend in drawings) is plugged into the same row as the resistor and the other pin on another row. Finally, connect the wire from pin 6 to the same row as the negative side of the LED and the wire from pin 7 to the row with the resistor. Figure 3-8 shows how all of the components are wired together. Be sure to study this drawing and double-check your connections prior to powering on your Raspberry Pi. Once you're satisfied that everything is connected correctly, you're ready to power on the Raspberry Pi and write the code.

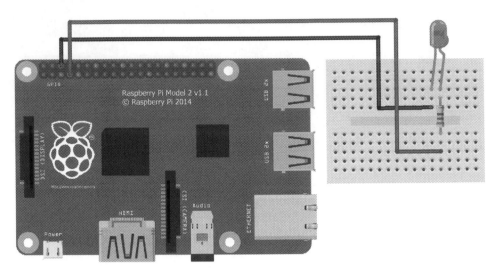

Figure 3-8. *Wiring the LED to a Raspberry Pi*

Writing the Code

The code (a Python script) that you need for this project manipulates one of the GPIO pins on the Raspberry Pi. Recall that you connected the negative side of the LED to pin 6, which is a ground pin. You connected the other side to pin 7. You will write a Python script to turn this pin on (applying power) and off (no power) through a simple command.

Now, create a new directory and open a text editor with the following commands. Use the name blink_me.py for the file.

```
pi@raspberrypi:~ $ mkdir source
pi@raspberrypi:~ $ cd source
pi@raspberrypi:~/source $ mkdir python
pi@raspberrypi:~/source $ cd python
pi@raspberrypi:~/source/python $ nano blink_me.py
```

As you can see, I like to place my source code in folders organized by language, but you can use whatever folder names you'd like. When the editor opens, enter the code as shown in Listing 3-2.

You can skip the comment statements (those that start with #) if you like, but I highly recommend that you get used to documenting your code. You can add your own name if that helps. Notice that I added comments to some of the lines to help understand what the code does. This is another excellent skill to hone.

Listing 3-2. Blink LED Script

```
#
# Windows 10 for the IoT
#
# Raspberry Pi Python GPIO Example
#
# This script blinks an LED placed with the negative lead on pin 6 (GND)
# and the pin 7 connected to a 220 resistor, which is connected to the
# positive lead on the LED.
#
# Created by Dr. Charles Bell
#
import RPi.GPIO as GPIO # Raspberry Pi GPIO library
import sys  # System library
import time # Used for timing (sleep)

ledPin = 7  # Set LED positive pin to pin 7 on GPIO
GPIO.setmode(GPIO.BOARD)# Setup the GPIO numbering mode
GPIO.setup(ledPin, GPIO.OUT)  # Set LED pin as output
GPIO.output(ledPin, GPIO.LOW) # Turn off the LED pin

print("Let blinking commence!")

for i in range(1,20):
  GPIO.output(ledPin, GPIO.HIGH) # Turn on the LED pin
  time.sleep(0.25)
  sys.stdout.write(".")
  sys.stdout.flush()
  GPIO.output(ledPin, GPIO.LOW)  # Turn off the LED pin
  time.sleep(0.25)

GPIO.cleanup() # Shutdown GPIO

print("\nThanks for blinking!")
```

> **Tip** Indentation is important in Python. Indented statements form a code block. For example, to execute multiple statements for an if statement, indent all the lines that you want to execute when the conditions are evaluated as true.

What you see here is a series of comment lines (again, the ones that start with #) followed by some `import` statements that tell Python which modules you want to use. In this case, you use the `GPIO`, `sys` (system), and `time` modules. Following that, you see code to identify the GPIO pin (7) and set up the GPIO code module to initiate pin 7 as an output pin.

On the next line, you print a greeting message and place statements inside a loop that turn on the GPIO pin for the LED (set to high) for a period of time using a delay (in seconds) and then turn off the GPIO pin for the LED (set to low). You loop through 20 times, printing a dot to the screen using the system `stdout` class mechanism. You do this so that you can write the character directly to the screen without buffering (buffering can delay the display output). Finally, you display a message that the process is complete so that you know that it finished.

As you can see, the code is really easy to read. And even if you've never written Python before, you can understand what it is doing. Double-check your code and then save the file (*Ctrl+X*) and reply *Y*. You are now ready to run the code!

Running the Script

Once you've entered the script as written, you are ready to run it. To run the Python script, launch it as follows:

```
pi@raspberrypi:~/source/python $ python ./blink_me.py
You should see the following in the command-line terminal.
pi@raspberrypi:~/source/python $ python ./blink_me.py
Let blinking commence!
.................
Thanks for blinking!
```

Figure 3-9 shows a photo of the program running. Now, did the LED blink? If so, congratulations—you're a Raspberry Pi Python GPIO programmer!

Figure 3-9. *Running the blink_me.py Python script*

If something went wrong, it is likely it's just staring back at you with that one dark LED—almost mockingly. If the LED did not illuminate, shut down the Raspberry Pi (sudo shutdown –h now) and check your connections against Figure 3-8. Be sure the wires are connected to the correct pins and the LED is oriented correctly with the longer pin connected to the resistor and the resistor connected to GPIO pin 7. Also, make sure that you are using the same rows on the breadboard (it is easy to get off by one row or pin). The other pin on the LED should be wired to GPIO pin 6. Once you've corrected any wiring issues, reboot your Raspberry Pi, and try the project again.

Once it is working, try the project a few times by running the python ./blink_ me.py command until the elation passes. If you're an old hand at the Raspberry Pi or electronics, that may be a very short period. If this is all new to you, go ahead and run it again and again, basking in the glory of having built your very first Python script and hardware project!

Summary

There can be little argument that the Raspberry Pi has contributed greatly to the world of embedded hardware and the IoT. With its low cost, GPIO headers, and robust peripheral support, the Raspberry Pi is an excellent choice for building your IoT solutions. Due to its increasing popularity, there are tons of information available for those who want to learn how to work with hardware.

In this chapter, you explored the origins of the Raspberry Pi, including a tour of the hardware and a short primer on how to use its native operating system. You also explored how easy it is to write programs to control hardware on the Raspberry Pi using a Python script.

You learned these things about the Raspberry Pi to help with learning more about the origins of the Raspberry Pi and its native environment so that, once you learn how to write applications in Windows 10 and deploy them to the Raspberry Pi, you can leverage the host of examples written for Linux to implement them in Windows 10. I hope that you have found this chapter aligned toward this goal.

The next chapter returns to Windows 10. You learn how to write the example program in Visual Studio. As you will see, the hardware connections will be similar, but the code and the way you work with the Raspberry Pi will be much more familiar and, in my opinion, easier.

CHAPTER 4

Developing IoT Solutions with Windows 10

Microsoft has produced one of the most advanced integrated development environments (IDE) that easily rivals all competition. Indeed, IDEs on other platforms are often compared to Visual Studio for their depth of features, refinement of tools, and breadth of languages supported.

The feature set in Visual Studio is so vast in fact that it would require a book several times the size of the one you're holding to cover the basics of every feature. Moreover, each language supported would require its own book of similar size. Clearly, mastering all of the features of Visual Studio would require a dedication that few would endure outside of a vocation or research requirement.

Fortunately, most IoT hobbyists and enthusiasts never need to learn every nuance of Visual Studio to develop IoT applications. As you will see, you need only to learn a few of the features, including writing the code, building (compiling), deploying, and debugging.[1] Do not let the sheer size of the features in Visual Studio intimidate you. You're likely to find mastery of the basics is all that you'll ever need. Should you ever need to use the advanced features, you can always learn them when you need them. I find it nice to know that there are advanced tools that can help me develop solutions more easily.

In this chapter, you'll see a demonstration of how to get started using Visual Studio 2019. You will also learn the layout of the GPIO headers for the two low-cost boards and even see how to build, deploy, and test your first Windows 10 IoT Core application. Let's begin with a look at the GPIO headers from both boards.

---

[1]Much like most people will never need or use more than 20% of the features of Microsoft Word.

© Charles Bell 2021
C. Bell, *Windows 10 for the Internet of Things*, https://doi.org/10.1007/978-1-4842-6609-0_4

Working with GPIO Headers

The general-purpose input/output (GPIO) headers permit you to connect hardware, such as electronic circuits, devices, and more, to your board, which you then access via special libraries from your applications. However, you must know what pins are available and what features they support. As you will see, not all pins can be programmed in the same way.

Once you've identified the pins you want to use, you can use those pin numbers or nomenclature to write the code that you need to set up and access the pin and read (or write) to the pin. That is, you can turn pins on or off (applying power or no power), read analog values, write analog values, and more. This allows you to work with both analog and digital sensors. You will discover more about sensors in Chapter 8.

Caution   Whenever you want to connect sensors or circuits to the GPIO header—either directly (not recommended) or via a breakout board (recommended)—you should first shut down your board. This may sound inconvenient when you're working through a project, but it is the best method for ensuring that you do not accidentally short some pins or make the wrong connections.

It is common to use the word "pins" when talking about the header, but you should refrain from using "GPIO pins" when referring to the header itself. As you will see, the pins in the header may be mapped to one of several interfaces (also called a bus), and the GPIO is just one type of interface available. Although you can use the interface pins in your code, the pins that you use to connect to devices (save those that use one of the supported interfaces) are named GPIO pins. However, the physical pin numbering and order may differ among the various boards.

Furthermore, when you use one of the pins in your Visual Studio application, you specify the GPIO number. For example, GPIO 13 on the Raspberry Pi is physically pin number 33. The Visual Studio libraries are designed to use the pin nomenclature rather than the header pin number. Thus, when you refer to GPIO 13, you use 13 as shown in the following code snippet:

```
const int LED_PIN = 13;// GPIO13
auto gpio = GpioController::GetDefault();
...
```

```
pin_ = gpio->OpenPin(LED_PIN);
pin_->Write(pinValue_);
pin_->SetDriveMode(GpioPinDriveMode::Output);
```

I discuss each of the low-cost boards supported by Windows 10 IoT Core in the following sections. As you will see, each board has a different GPIO header layout, including some very important power differences.

Raspberry Pi

The GPIO headers on the Raspberry Pi 2 and 3 have the same layout. The header is located in the upper left and consists of a double row of 20 pins, making a 40-pin header. They are numbered sequentially in pairs starting with the leftmost set of pins.

Figure 4-1 shows the Raspberry Pi next to the header layout. I have oriented the photo to make it easier to see the GPIO pins. Pin 1 is the leftmost pin at the left side of the header (top side shown in the photo). Notice that the GPIO named pins are arranged in a nonsequential pattern.

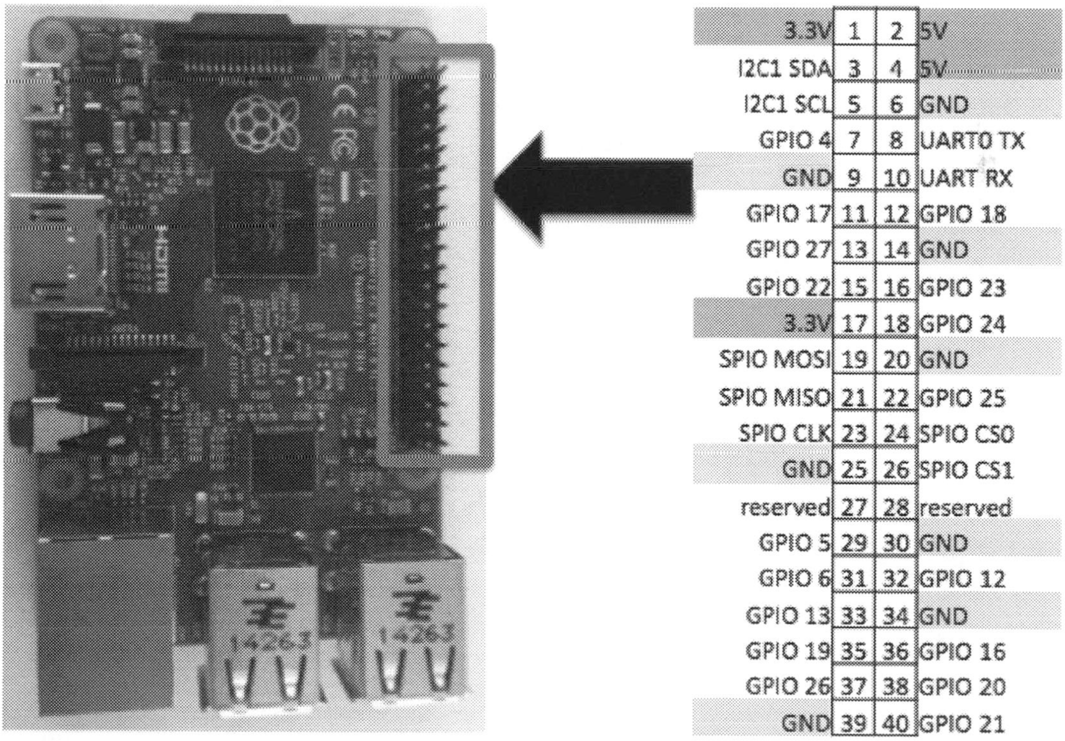

Figure 4-1. *GPIO header (Raspberry Pi 2 and 3)*

The Raspberry Pi GPIO header supports a number of interfaces including an I2C bus, SPIO bus, and serial UART with pins devoted accordingly. You can see these in Figure 4-1. You also see two reserved pins (consider them unusable), eight ground pins, two 3.3V power, and two 5V power pins. This leaves a total of 17 pins that you can use in your applications.

You must take care when using the GPIO pins for reading voltage. On the Raspberry Pi, all pins are limited to 3.3V. Attempting to send more than 3.3V will likely damage your Raspberry Pi. Always test your circuit for maximum voltage before connecting to your Raspberry Pi. You should also limit current to no more than 5mA.

DragonBoard 410C

The GPIO headers on the DragonBoard 410C are located on the top-left side of the board. The GPIO consists of a double row of 20 pins, making a 40-pin header. They are physically numbered sequentially in pairs starting with the left-post pin. Unlike the Raspberry Pi and MinnowBoard Max–compatible boards, the DragonBoard 410C uses female header pins.

Figure 4-2 shows the DragonBoard 410C next to the header layout. Pin 1 is the leftmost pin at the left side of the header (top side shown in the photo). I have oriented the photo to make it easier to see the GPIO pins. The GPIO named pins are arranged in a nonsequential pattern like the Raspberry Pi.

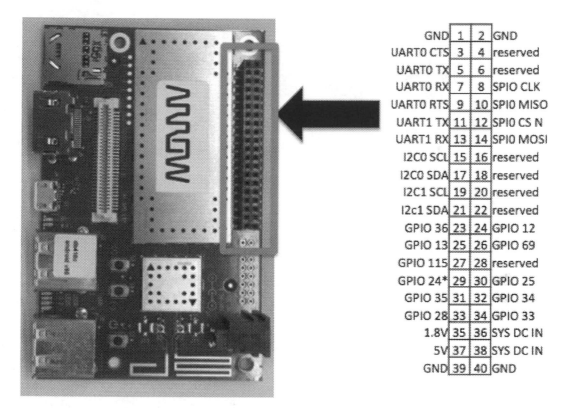

GND	1	2	GND
UART0 CTS	3	4	reserved
UART0 TX	5	6	reserved
UART0 RX	7	8	SPIO CLK
UART0 RTS	9	10	SPIO MISO
UART1 TX	11	12	SPIO CS N
UART1 RX	13	14	SPIO MOSI
I2C0 SCL	15	16	reserved
I2C0 SDA	17	18	reserved
I2C1 SCL	19	20	reserved
I2c1 SDA	21	22	reserved
GPIO 36	23	24	GPIO 12
GPIO 13	25	26	GPIO 69
GPIO 115	27	28	reserved
GPIO 24*	29	30	GPIO 25
GPIO 35	31	32	GPIO 34
GPIO 28	33	34	GPIO 33
1.8V	35	36	SYS DC IN
5V	37	38	SYS DC IN
GND	39	40	GND

Figure 4-2. *GPIO header (DragonBoard 410C)*

The DragonBoard 410C GPIO header supports a number of interfaces including an I2C bus, SPIO bus, and two serial UART with pins devoted accordingly. You can see these in Figure 4-2. You also see four ground pins, one 1.8V power, and one 5V power pin. This leaves a total of 11 pins that you can use in your applications. GPIO 24 can be used for input only. Also, the pins marked SYS DC IN can be used to power the board.

Now that you know more about the GPIO headers on your boards, you can learn how to get started using Visual Studio 2019.

WHAT ABOUT THE MINNOWBOARD?

If you own a MinnowBoard Turbot, you can still use it with Windows 10, but it appears the organization backing the MinnowBoard is no longer present and you may not be able to buy the board much longer. That said, the following shows the GPIO for the MinnowBoard.

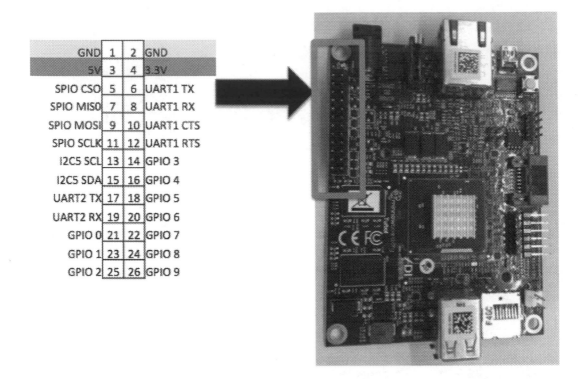

The MinnowBoard Turbot GPIO header supports a number of interfaces including an I2C bus, SPIO bus, and serial UART with pins devoted accordingly. You can see these in Figure 4-2. You also see two ground pins, one 3.3V power, and one 5V power pin. This leaves a total of ten pins that you can use in your applications.

Visual Studio 2019 Primer

The Visual Studio product line is very long lived. In fact, it has been around since the early days of Windows. As you can imagine, the product has undergone a great deal of changes—with new languages, frameworks, platforms, and more added every few years. As I mentioned earlier, Visual Studio 2019 offers a huge array of features for developing applications for Windows using a variety of languages. Visual Studio also supports several platforms.[2]

---

[2]There's even support for Android if you're into that kind of thing.

With all of these features and the many languages, it can be quite intimidating getting started with Visual Studio. In fact, books devoted to Visual Studio (just the features, not languages) can easily exceed hundreds of pages in length. However, you can accomplish quite a lot with a small amount of knowledge.

Tip   If you want to know more about Visual Studio 2019, check out *Getting Started with Visual Studio 2019* by D. Strauss (Apress, 2020). The book contains an introduction to many of the features in Visual Studio.

This section gives you a brief introduction to how to get around in Visual Studio for the purposes of developing applications for Windows 10 IoT Core. Don't feel like you have to learn everything about Visual Studio to enjoy working with Windows 10 IoT Core. This section and the next chapters provide you a guide to getting started writing Windows 10 IoT Core applications by example. But first, let's explore the major features, interface features, and project templates that you will use to write your Windows 10 IoT Core applications.

Note   The following provides a high-level overview of the features that you need to know to write Windows 10 IoT Core applications. As such, it is not a complete tutorial of Visual Studio. If you need more information about Visual Studio, refer to the help system inside the application and the Microsoft Developer Network (MSDN) library of documentation.

Major Features

Visual Studio 2019 is truly the do everything tool for Windows software development. In that respect, Microsoft has established the bar for which all others are measured. Visual Studio is an IDE that places all the tools that you need to develop applications in one interface, making Visual Studio the only tool that you need to develop applications for Windows.

Another aspect to its superiority has to do with the languages supported. For example, there are a variety of programming languages, including C, C#, C++, Visual Basic, Python, and more. Each of these languages can be used to build a host of different applications in one of several frameworks.

> **Note**   Most of the examples for building Windows 10 IoT Core applications are written in either C#, C++, or Visual Basic. Once you master these languages and mechanisms for building projects, learning to do the same in other languages requires only learning the syntax and semantic nuances of the language.

For example, you can build desktop applications (GUI or command line) in several frameworks, including Windows 32 and .NET, create web applications, dynamic libraries for reuse, and more. Of course, you can also build applications for Windows 10 IoT Core or one of several special project templates (prebuilt collections of files and specific settings).[3] Project templates create a special file called a solution (with a filename of `<project_name>.sln`) that contains several types of files.

To create an application, you select a project template and name the solution, write your code, compile the application into an executable, test and debug the application (optional), and finally deploy the application.

I guide you through selecting the right project template later in this section. Writing the source code (developing the functionality) requires modifying one or more of the files in the solution. I discuss the major types of Windows 10 IoT Core projects later in this section.

The compilation step is easy to initiate but can be an iterative process. This is because the compiler performs intensive syntax and semantic checks on the code flagging anything that isn't quite up to the language or framework rules as warnings or errors. Fortunately, you can click each warning or error to zoom to the line of code. But don't worry; compiler errors are normal and are not a sign of inexperience or lack of knowledge—they're just a part of the learning experience. That said, as you learn the language and frameworks, you should encounter fewer issues and errors. Also, do not disregard warnings. While they may not prohibit your code from compiling, it is always a good practice to remove the cause of the warning before completing your solution.

I think it is the interactive debugger where Visual Studio shines the most. In fact, I find the interactive debugger the most useful of all features in Visual Studio. Not only can you run your applications stepping through code one line at a time, you can do so while inspecting variables, memory, the stack (the order of method calls), and more.

---

[3]Project templates are quite extensive and specific to a particular language, framework, and platform. Indeed, you will find project templates are organized in that order.

This helps you test your application and improve quality while finding logic or data errors. You can even remotely debug your Windows 10 IoT Core applications.

Finally, when you are satisfied with the quality of your code, you can use Visual Studio to deploy your application to your Windows 10 IoT device.

The User Interface

At first glance, the Visual Studio 2019 interface appears with several smaller windows arranged inside a larger window. The layout of the windows can change depending on settings you used when installing Visual Studio and can vary slightly from one project template to another. Figure 4-3 shows the layout of the IDE using C++ environment settings. I have placed numbers next to the major components of the IDE. I explain each in more detail.

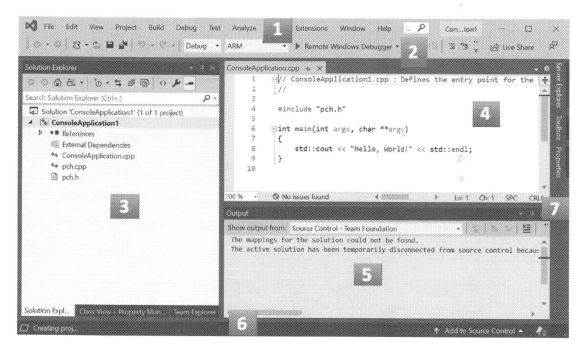

Figure 4-3. *The Visual Studio 2019 interface (C++ settings)*

The IDE components shown include the following:

1. *Menus*: This area contains the system of menus for all the standard features. The features are categorized into several areas, including operations on the project, building, debugging, and more.

2. *Toolbars*: In this area, you see a variety of toolbars that have buttons and other controls for commonly used functions for particular features. You can add and remove (show and hide) using the Tools menu.

3. *Solution Explorer*: This window lists the various files included in the solution. You can double-click any of the files to open the appropriate window to work with that file. You can also use the tree control to drill down into the specifics for each file type. For example, I opened the `ConsoleApplication1.cpp` node and double-clicked the `main()` method, which opened the code editor window and zoomed (placed the cursor) to the method.

4. *Code editor*: This window is where you enter all the code for your source code. You can use this area to edit all manner of files, and the IDE will change its behavior based on the context. That is, it performs operations such as language-specific automatic code completion.

5. *Output window*: This window is used to communicate messages to you from the compiler and other features. For example, look here when you compile your application for warnings and errors.

6. *Information bar*: The bar down at the bottom of the IDE is also used to provide contextual information. Since the focus of the IDE is on the source code editor, data such as cursor location (line position), code line number, and more is shown.

7. *Dock*: To the right (in this layout) are additional windows that are docked or minimized. You can click any of the tabs to open the windows.

Windows

The windows in the IDE can be repositioned and resized. There are a number of states each window can become, including the following. When you expand a window by clicking the window title bar, you see three small icons on each window title bar.

The first one, a small down arrow, when clicked, opens a context menu with options for the disposition of the window. I describe how the other two icons are used in the descriptions of window states:

- *Float*: The window is free to be moved around and floats above (outside) the confines of the IDE. This is also called unpinning.

- *Dock*: The window is restricted to the confines of the IDE but can be repositioned inside the IDE itself. The window remains the size you set. Click the thumbtack (or pin) icon to dock the window. This is also called pinning. You can move the window around in the IDE, and when it hovers near a docking area, an overlay allows you to dock the window in that area, much like the side-by-side feature of Windows.

- *Dock as tabbed document*: The window is reduced to a tab on the tab bar associated with its function. There are multiple tab areas throughout the IDE.

- *Auto hide*: The window opens when you hover over its tab. The window stays open as long as the mouse pointer is within its borders. When you move the mouse away from the window, it reverts to a tab on the tab bar. To auto hide, click the small thumbtack (sometimes called the pin).

- *Hide*: The window closes and is removed from the tab bar. You have to use the menu options to reopen it. Click the X icon to hide the window.

Environment Settings

Recall that during the installation of Visual Studio in Chapter 2, you chose the C++ option for the environment settings. This is another great feature of Visual Studio. Environment settings configure the layout of the IDE for a specific language or framework. There are generic (general) environment settings for language-agnostic layout as well as environment settings for languages such as C# and C++. Figure 4-4 shows the environment settings that are available.

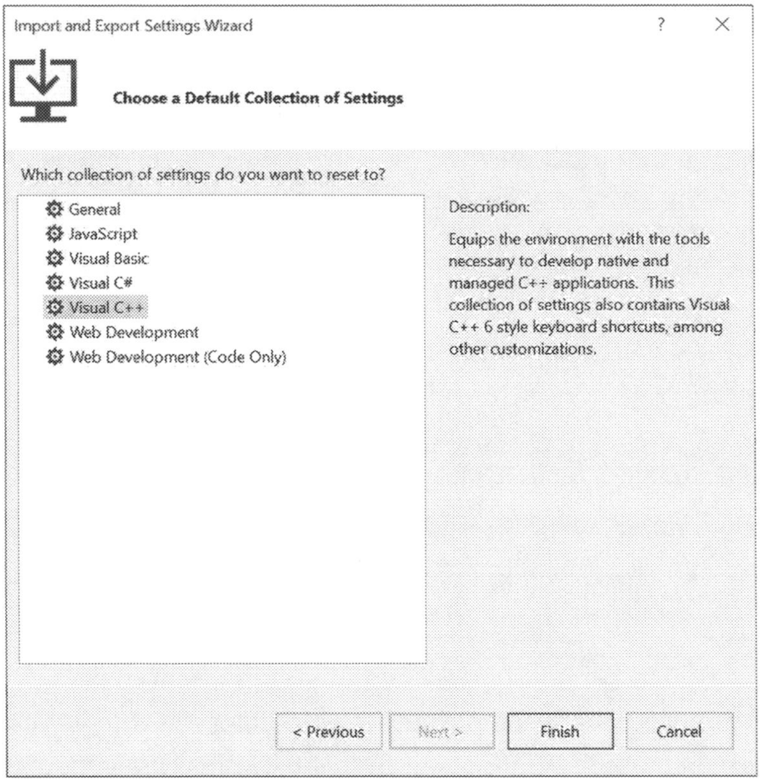

Figure 4-4. *Visual Studio 2019 environment settings*

You can change the environment settings any time you want. Use the *Tools ➤ Import and Export Settings...* menu to open the settings management dialog. You can use this dialog to save the settings (export) or restore settings you've saved (import) by selecting the appropriate radio button.

To change the default environment settings, choose the *Reset all settings* radio button and then click *Next*. The dialog gives you another chance to save your settings. If you do not want to save them, or once you have saved the settings, choose the *No* radio button, and then click *Next*. On the next screen, choose the environment settings (see Figure 4-5) that you want and then click *Finish*.

Common Menu Items

While there are a number of menus in the interface, each with dozens of subitems and submenus, there are a few that you use quite often. The following is a brief overview of the commonly used menu items categorized by the entries on the main menu bar. I add the keyboard shortcut in parentheses where available. Note that `<project>` changes to the name of the currently opened or selected project.

- *File*: File and project operations

  - *New* ➤ *Project (Ctrl+Shift+N)*: Start a new project

  - *Save Selected Items (Ctrl+S)*: Save the file(s) selected

  - *Save All (Ctrl+Shift+S)*: Save all files

- *Project*: Operations on the current project

  - *<project> Properties (Alt+F7)*: Open the properties dialog for the project

- *Build*: Compilation and deployment with two sections—one for the entire solution and one for the current project (a solution can have multiple projects)

  - *Build Solution (F7)*: Build (compile) all projects in the solution

  - *Rebuild Solution (Ctrl+Alt+F7)*: Rebuild (compile) all files in the solution

  - *Clean Solution*: Remove all compiled files and headers and generated files in the solution

  - *Build <project>*: Build (compile) the currently selected project

  - *Rebuild <project>*: Rebuild (compile) all files in the project

  - *Deploy Selection*: Deploy the compiled project to the destination specified in the debug settings

  - *Clean <project>*: Remove all compiled files and headers and generated files in the project

  - *Compile (Ctrl+F7)*: Compile the currently selected file/project

- *Debug*: Interactive debugger

  - *Start Debugging (F5)*: Start the interactive debugger

  - *Start Without Debugging (Ctrl+F5)*: Run the project without the debugger (adds a pause at the end of execution for console applications)

- *Step Into (F11)*: While debugging, enter any method calls one line at a time

- *Step Over (F10)*: While debugging, execute current line (do not step into methods)

- *Toggle Breakpoint (F9)*: Turn a breakpoint on/off

Now that you know a bit more about the interface, let's look at the templates you will use for your Windows 10 IoT Core projects.

Windows 10 IoT Core Project Templates

A project template is a special set of files and settings that are configured for a specific language, framework, or application target. Project templates are arranged by programming language and by project type. The following are some of the more common project types available in the standard Visual Studio installation:

- *Windows*: A very broad category that covers any application type for Windows

- *Web*: Applications built using ASP.NET

- *Office/SharePoint*: Applications for add-ins for the Microsoft productivity tools

- *Cloud*: Builds applications for Windows Azure

- *Windows IoT Core*: Builds applications for deployment to Windows 10 IoT Core devices

---

Note   Some project template categories are only available for certain languages. Similarly, some languages may list fewer project template categories.

See the Visual Studio online documentation (`https://docs.microsoft.com/en-us/visualstudio/windows/?view=vs-2019`) for a complete list of project types available.

To start a new project, click the *File ➤ New ➤ Project...* menu item. An excerpt of the New Project dialog is displayed, as shown in Figure 4-5. There are three drop-down lists across the top of the dialog. The first allows you to narrow the search to a specific programming language. The second allows you to narrow the selection by platform.

The third allows you to you to see the templates installed in a tree view on the left. You can use this view to drill down to the specific language and framework/platform you want. If you click the language itself, you see all the project templates for that language.

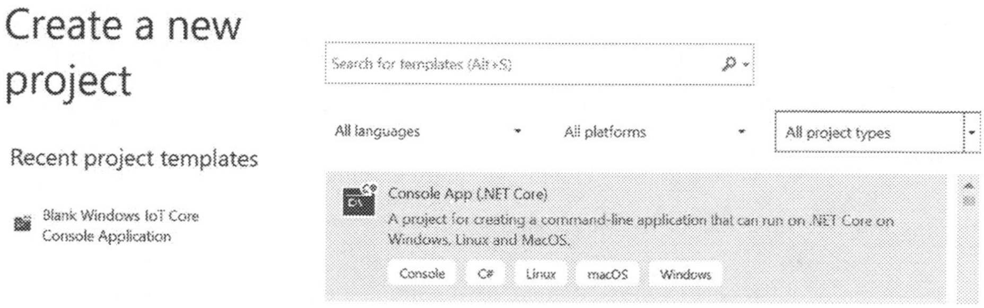

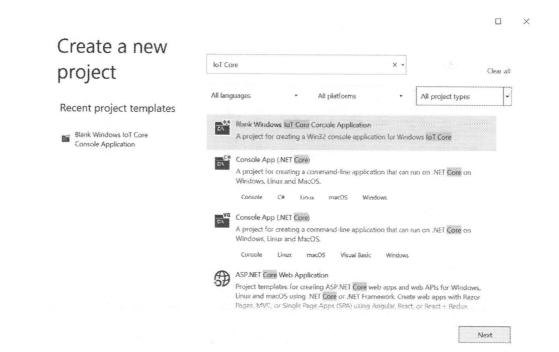

Figure 4-5. *Selecting a Visual Studio project template*

Figure 4-6 shows how you would start a new project for Windows 10 IoT Core. Here, we use the search box to search for IoT Core to find the Windows 10 IoT Core project types. Notice how it now displays all of the IoT project templates. Cool, eh?

Figure 4-6. *Selecting the IoT project templates*

You can also use the drop-down lists to find these templates by choosing the platform as Windows and the project type as UWP (Universal Windows Platform) as shown in Figure 4-7.

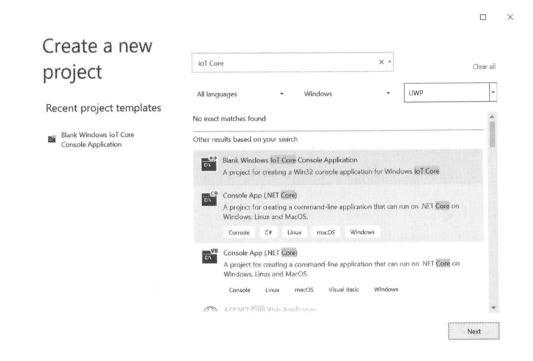

Figure 4-7. *Choosing the platform and project type to find IoT Core project templates*

Once you select the project template you want, click *Next* to configure the project. You can name the project and the location for all of its files. By default, the name of the application and the solution name are the same, but you can change them. Once you are ready to create the project, click *Create*. The IDE opens the project (a project's settings are saved in a file called a solution) and creates the basic source files for you.

There are three basic project templates that you use in this book to create applications for Windows 10 IoT Core. They aren't the only ones you can use, but they are the templates you should use to get started. If you want to build more complicated applications, you may want to consider some of the other project types. The three project templates include the following ordered by complexity. I explain each in the following sections:

- Blank Windows IoT Core Console Application

- Background Application (IoT)

- Blank App (Universal Windows)

Again, there are other project templates that you may want to use, but these three are the ones that you use in this book.

Blank Windows IoT Core Console Application

This project template is the most basic and simplistic of all project templates for Windows 10 IoT Core. It permits you to create a simple text-only console application. If you want to experiment with writing applications or want to create a solution that provides data or produces a report for a headed device, choose this project template. You will use this template in the walk-through of building your first application.

The code for this project template is very simple. The template creates a source file named by default `ConsoleApplication1.cpp`, which you can use to write your application. Use this source file to write your application. You can create additional source files if you want, need, or desire to do so. For example, if you are modeling concepts or creating abstractions for hardware, you may want to add additional source files to contain these models (e.g., classes).

Listing 4-1 shows this file. The project template fills in the `main()` function with a simple `Hello, World!` print statement.

Listing 4-1. Blank Windows IoT Core Console Application

```
// ConsoleApplication1.cpp : Defines the entry point for the console
application.
//

#include "pch.h"

int main(int argc, char **argv)
{
    std::cout << "Hello, World!" << std::endl;
}
```

Background Application (IoT)

This project template is used to create an application that runs in the background (or headless) on your device. It can be used with C++, C#, or Visual Basic. Typically, you'd use this project template to create an application that communicates to other devices, interacts with hardware (and then communicates with other devices), or drives hardware to display or report sensor data to the user.

The project files for this project template are quite different, and there are a few extra files. For the C++ example, the one source file that you work with most is named StartupTask.cpp, which is where you place the code you want to run when the application starts. There is also a corresponding header file named StartupTask.h where you can place function and class primitives for your design.

Listing 4-2 shows the StartupTask.cpp file as created by the new project dialog. You can find this project template quickly by searching for Background Application in the create new project dialog and selecting the language you prefer.

Listing 4-2. Background Application (IoT)

```
#include "pch.h"
#include "StartupTask.h"

using namespace BackgroundApplication2;

using namespace Platform;
using namespace Windows::ApplicationModel::Background;

// The Background Application template is documented at http://
go.microsoft.com/fwlink/?LinkID=533884&clcid=0x409

void StartupTask::Run(IBackgroundTaskInstance^ taskInstance)
{
    //
    // TODO: Insert code to perform background work
    //
    // If you start any asynchronous methods here, prevent the task
    // from closing prematurely by using BackgroundTaskDeferral as
    // described in http://aka.ms/backgroundtaskdeferral
    //
}
```

There is a link in the code for more information about the project template. Also, there are hints in the code comments (designated by //) to help you get started.

Blank App (Universal Windows)

This project template is the most sophisticated of the three project templates that you will use. It can be used with C++, C#, or Visual Basic. This project template allows you to create simple user interfaces using Microsoft's Extensible Application Markup Language (XAML). XAML is similar to XML except you use a special declarative language to create a user interface. It is a very expressive and powerful language that permits you to create sophisticated user interfaces without the need for high overhead graphical libraries.

Tip   To learn more about XAML, see `https://msdn.microsoft.com/en-us/windows/uwp/xaml-platform/xaml-overview?f=255&MSPPError=-2147217396`.

There are several files for this project template. The main files in the C++ version that you will be working with are the XAML files (`MainPage.xaml`, `MainPage.xaml.h`, and `MainPage.xaml.cpp`), which include the XAML code. You also see files named `App.xaml`, `App.xaml.h`, and `App.xaml.cpp`, which are the entry or starting point for the application. There is no specific XAML code here; rather, it contains the code for the *OnLaunched* and *OnSuspending* events. You can use these events to initialize your application or cleanup when it is shut down or suspended. There are a number of other files, but these are explored in later chapters when you see examples of each project template.

Listing 4-3 shows the main page for the XAML source files. You can find this project template quickly by searching for `Blank Application (Universal Windows)` in the create new project dialog and selecting the language you prefer.

Listing 4-3. Blank App (Universal Windows)

```
//
// MainPage.xaml.cpp
// Implementation of the MainPage class.
//
```

```
#include "pch.h"
#include "MainPage.xaml.h"

using namespace App1;

using namespace Platform;
using namespace Windows::Foundation;
using namespace Windows::Foundation::Collections;
using namespace Windows::UI::Xaml;
using namespace Windows::UI::Xaml::Controls;
using namespace Windows::UI::Xaml::Controls::Primitives;
using namespace Windows::UI::Xaml::Data;
using namespace Windows::UI::Xaml::Input;
using namespace Windows::UI::Xaml::Media;
using namespace Windows::UI::Xaml::Navigation;

// The Blank Page item template is documented at https://go.microsoft.com/
fwlink/?LinkId=402352&clcid=0x409

MainPage::MainPage()
{
    InitializeComponent();
}
```

Before we get into writing our first Windows 10 IoT Core application, let's take a look at the examples Microsoft has provided for us.

Example Applications

Microsoft has included a number of sample applications (called Quick-run samples) for you to deploy to your IoT device that do not require you to write any code. Cool, eh? There are four examples to choose from, a simple console application, an Internet radio, a sample game, and an Azure example.

Note   In this example and in future projects, I use the IoT device (or simply device) terms interchangeably.

Let's begin with the easiest of them all—the quintessential "Hello, World!" programming example. Yes, it's been done a billion times and is quite basic in complexity, but in this case it helps us learn more about how to work with our IoT device.

Begin by powering on your board and waiting until it is running, and you see the welcome screen. Then, use the Windows 10 IoT Core dashboard to connect to your device. Recall, to connect to the Device Portal, simply right-click the board in the *My Devices* tab of the dashboard and choose *Device Portal* or double-click it and choose the *Open Windows Device Portal in browser* link. You may need to log in to continue.

In the Device Portal, click the small triangle on the left side next to *Apps* and then click *Quick-run examples*. Then, you can click the *Hello World* example to see its details and a sample screenshot (may not be displayed immediately). Figure 4-8 shows the Device Portal with the example applications shown.

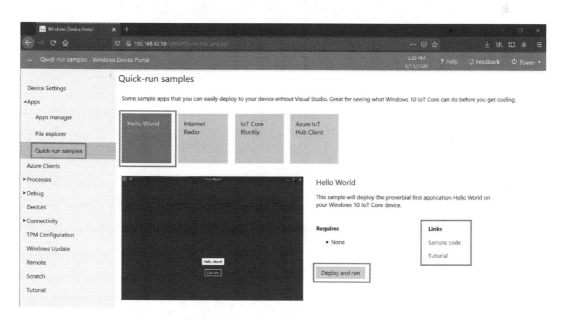

Figure 4-8. *Quick-run examples (Device Portal)*

Notice on the right side of the screen is a button, `Deploy and run`, that you can use to install the application on your device. To the right of that are two links you can use to see the source code and read a tutorial on the project. Go ahead and click those if you're curious, but let's talk about installing and running the example.

For now, let's click the `Deploy and run` button and wait while the application is deployed. You will see feedback in the Device Portal on the installation. This may take a few moments to complete. Once complete, the button will be replaced with the text, `Sample installed and launched successfully`. Your device should display the application as shown in Figure 4-9. The application will fill the screen, but the figure shows an excerpt for clarity.

Figure 4-9. *Hello World application*

As you can see, it is in the center of the screen, and we see a label and a button. Go ahead and click it (if using a mouse connected to your device) or press *Enter* on the keyboard for the device. When you click the button, the label will change as shown in Figure 4-10.

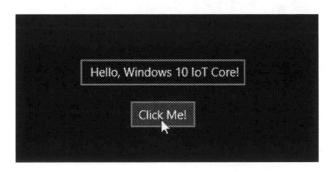

Figure 4-10. *Hello World application (button clicked)*

OK, so that wasn't all that exciting, but it does illustrate how a console application could look like running on the device. The code for this example is written in C# and is found in the link on the Device Portal (`https://github.com/ms-iot/samples/tree/develop/HelloWorld/CS`). Listing 4-4 shows the code for the main page. You can explore all of the code if you are curious how the C# version of the project is written.

Listing 4-4. Hello World Code

```
<Page
    x:Class="HelloWorld.MainPage"
    xmlns="http://schemas.microsoft.com/winfx/2006/xaml/presentation"
    xmlns:x="http://schemas.microsoft.com/winfx/2006/xaml"
    xmlns:local="using:HelloWorld"
    xmlns:d="http://schemas.microsoft.com/expression/blend/2008"

    mc:Ignorable="d">

    <Grid Background="{ThemeResource ApplicationPageBackgroundThemeBrush}">
        <StackPanel HorizontalAlignment="Center" VerticalAlignment="Center">
            <TextBox x:Name="HelloMessage" Text="Hello, World!" Margin="10"
            IsReadOnly="True"/>
            <Button x:Name="ClickMe" Content="Click Me!"  Margin="10"
            HorizontalAlignment="Center" Click="ClickMe_Click"/>
        </StackPanel>
    </Grid>
</Page>
```

This brings up a very important skill you will need to master—how to manage applications running on your device. If you click *Apps manager* in the Device Portal, you will see a screen that displays all of the applications running on your device as shown in Figure 4-11.

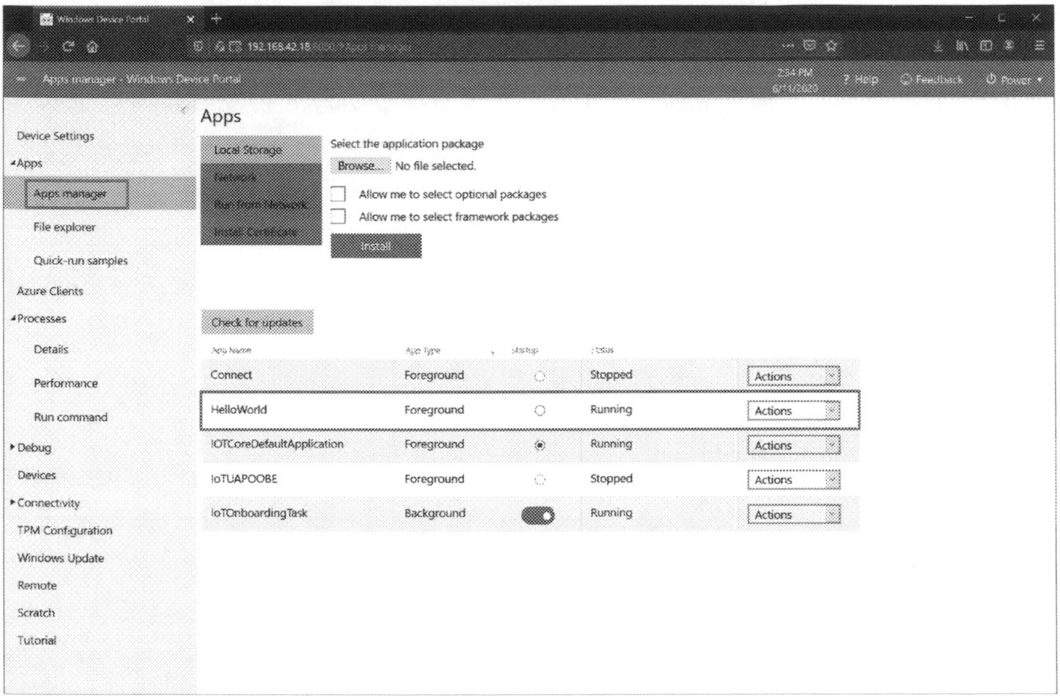

Figure 4-11. *Apps manager*

If you deployed the *HelloWorld* application, you should see it in the list with the status of *Running* as shown earlier. To stop the application, click the *Actions* drop-down and choose *Stop*. This will stop the application, and the list will change to display the new status as shown in Figure 4-12.

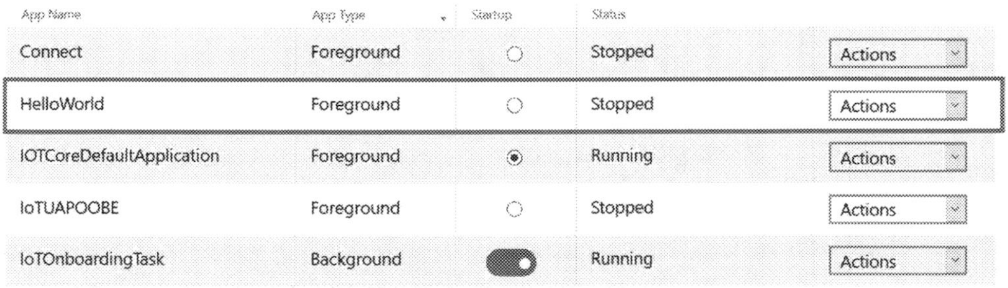

Figure 4-12. *HelloWorld application stopped*

You can also use the drop-down to start (if stopped), stop (if running), uninstall, and display the details of the application. For example, Figure 4-13 shows the details of the HelloWorld application. Click *Close* to close the dialog.

Figure 4-13. *HelloWorld details (Device Portal)*

If you enjoyed working with the Hello World example, you can uninstall it as described earlier. If you're still curious, be sure to check out the other three (the Azure example requires an Azure account, which we will discuss in Chapter 16).

Now that you've had a brief overview of Visual Studio, let's see these features in action by writing your very first Windows 10 IoT Core application.

Example Project: Temperature Conversion

Now let's get our hands into some code and see how to write a basic application for your Windows 10 IoT Core device. I know you've been itching to get started, so grab a stockpile of your favorite diet crushing snacks and kid-approved beverages of choice, recline your chair,[4] and let's write some code! OK, so it won't be quite so exciting, nor will it be a lengthy project, but it is good to get in the spirit all the same.

The project is an adaptation of a classic computer science homework assignment: converting temperature from Celsius to Fahrenheit. But you're going to mix it up a bit and make the application a bit more interactive. That is, you ask the user which scale they want to use as the base temperature, prompt for the temperature, and finally convert and print the result.

In case you've forgotten or perhaps never gave it another thought after the midterm and final exam, the formulas that you use include converting Celsius to Fahrenheit and Fahrenheit to Celsius as follows. You use these in the code that you write for the application:

```
Celsius = (5/9) X (fahrenheit - 32)
Fahrenheit = ((9/5) X Celsius) + 32.0
```

---

[4]A common pose that is very bad for your posture but some insist it exudes the proper attitude of a serious coder. Recline at your own risk.

To keep things as simple as possible, you'll write a Windows IoT Core console application that you can deploy to your Windows 10 device and run from a remote login (SSH). This keeps the code simple and allows you to focus on the mechanics of building and deploying the application.

Now that you know what you want to build, let's get started!

Create the Project

You need to create a new project in Visual Studio to contain your code. If you're following along, go ahead and launch Visual Studio and create a new project. You can use the File ➤ *New Project...* menu option, or you can click the *Create a New Project* in the welcome dialog or press *Ctrl+Shift+N*.

When the new project dialog appears, you need to select the language, platform, and project type. For this project, you want to search for the template using the phrase Windows IoT Core Console Application. Then choose the Blank Windows IoT Core Console Application entry from the list, as shown in Figure 4-14.

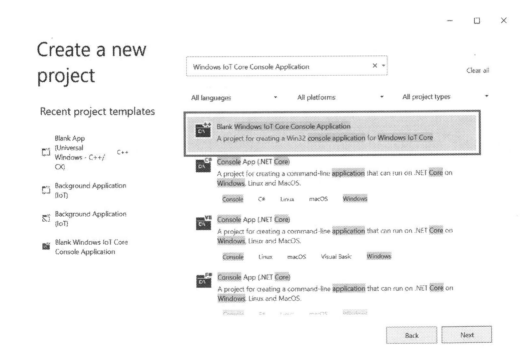

Figure 4-14. *Create a new project dialog*

Once you have selected the correct template, click Next to display the Configure your new project dialog as shown in Figure 4-15.

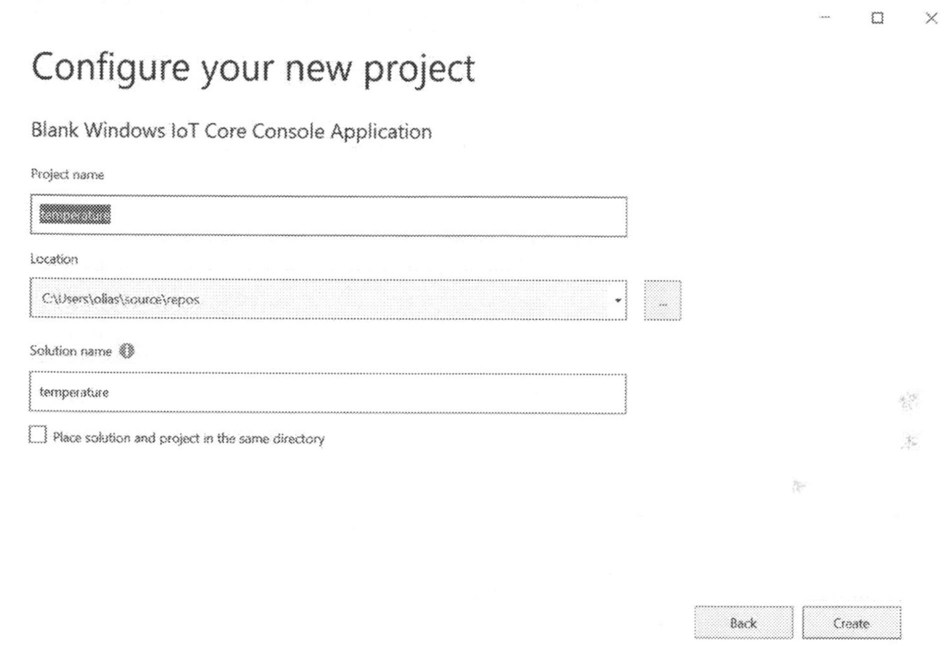

Figure 4-15. *Configure your new project dialog*

On the left side of the dialog, you can choose the name of the application (use *TemperatureConverter*) and optionally a directory of where to store the solution (all the source code and related files). If you type TemperatureConverter in the *Project name* box, the dialog will fill in the same name for the *Solution name*. Once you are satisfied with the name and location, click *Create* to create the project. You may also be asked for the target and minimum platform versions. You should choose the latest version of each. Once you click *OK*, the project creation process may take a moment to run. Once complete, you see the blank files for the solution, as shown in Figure 4-16.

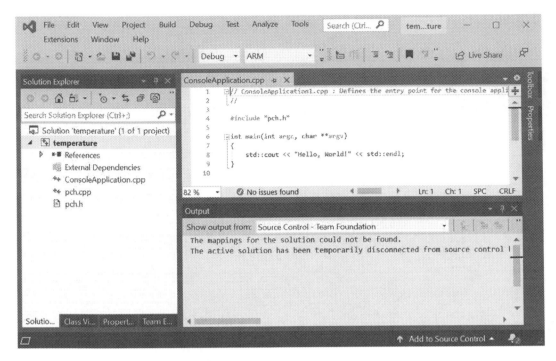

Figure 4-16. *Blank console application project*

Now that we have a starting point, let's write some code!

Write the Code

Now it is time to write the code. In this project, you place all the source code in the `int main(int argc, char **argv)` method inside the `ConsoleApplication.cpp` source file. To open this file (if it is not already open in the code editor), select the project in the *Solution Explorer* on the left and navigate to the `ConsoleApplication.cpp` file. Then, double-click that file. Visual Studio opens and refocuses the code editor window to the method, as shown in Figure 4-16.

Did you notice something interesting in the source code? Yep, Microsoft has populated the `main()` method with the ubiquitous "Hello, World" code. In this case, it is the statement that sends the character string to the standard out (`std::cout`) or console. Thus, if you were to run this code without modification, the application will simply print that string and exit. You want to do something similar since you are only working with the console, but with a bit more sophistication than that simple output statement.

Don't worry too much about whether you've written C++ code before. I provide all the statements that you need in this chapter. I present a brief tutorial of the C++ language in the next chapter. For now, just enter the code as shown in Listing 4-5. I'll walk you through what the code does in a moment.

Listing 4-5. TemperatureConverter Example Code

```
//
// Windows 10 for the IoT Second Edition
//
// Example C++ console application to demonstrate how to build
// Windows 10 IoT Core applications.
//
// Created by Dr. Charles Bell
//

#include "pch.h"

using namespace std;

int main(int argc, char **argv)
{
    double fahrenheit = 0.0;
    double celsius = 0.0;
    double temperature = -.0;
    char scale{ 'c' };

    cout << "Welcome to the temperature conversion application.\n";
    cout << "Please choose a starting scale (F) or (C): ";
    cin >> scale;
    if ((scale == 'c') || (scale == 'C')) {
        cout << "Converting value from Celsius to Fahrenheit.\n";
        cout << "Please enter a temperature: ";
        cin >> celsius;
        fahrenheit = ((9.0 / 5.0) * celsius) + 32.0;
        cout << celsius << " degrees Celsius = " << fahrenheit <<
            " degrees Fahrenheit.\n";
    }
```

```cpp
    else if ((scale == 'f') || (scale == 'F')) {
        cout << "Converting value from Fahrenheit to Celsius.\n";
        cout << "Please enter a temperature: ";
        cin >> fahrenheit;
        celsius = (5.0 / 9.0) * (fahrenheit - 32.0);
        cout << fahrenheit << " degrees Fahrenheit = " << celsius <<
            " degrees Celsius.\n";
    }
    else {
        cout << "\nERROR: I'm sorry, I don't understand '" << scale << "'.";
        return -1;
    }
}
```

The code begins with a number of comment lines. These are designated with the //
symbol at the start of the line. In fact, you can place the // anywhere on the line, making
everything to the right a comment. The compiler ignores these statements so you can
write whatever you like. The very first thing that you may notice is the following line of
code. This is part of the preamble of the code. The first line is an include directive that tells
the compiler to include the pch.h (precompiled header file), which Visual Studio supplies.

```cpp
#include "pch.h"
```

Next is a shorthand notation I like to use. It allows you to avoid typing the namespace
over and over again. In this case, you're using the namespace std:: a lot. The following
line of code permits you to omit that—provided that what you are referencing (cout,
cin) can be found unambiguously in the namespace. That is, if there were multiple
namespaces used, cout and cin would have to be unique to one of them; otherwise,
you'd still have to provide the namespace std::.

```cpp
using namespace std;
```

Those two lines are the only code outside of the main() method. The main() method
is the entry point for a C++ application. When the application is run, code inside this
method is executed first. The code you want for this method is shown in Listing 4-5.

You begin by asking the user what scale she wants to use. You will prompt the user
to enter a C for Celsius or an F for Fahrenheit. Take a look at the code again. If the user
specifies either a C or c, you calculate the conversion using Celsius as the base. If the user

specifies either an F or f, you calculate the conversion using Fahrenheit as the base. If something else was entered, you exit with an error. You include additional prompts to ask the user for the base temperature. Following that, you use the formulas to perform the calculations.

You may be wondering what all of those cout and cin and << and >> thingies do. Essentially, the cout << string phrase tells the compiler you want to take a string and print it to the screen (standard out or stdout). The cin >> variable phrase tells the compiler to read information entered by the user and store it in the variable named. Go through the code again and read it until you understand how this works. Again, I explain C++ coding in more detail in the next chapter.

Once you've entered all the code, go ahead and save the project using either the save on the toolbar or the *File* ➤ *Save TemperatureConverter* menu item. Next, let's build and test the code!

Build and Test Your Code

To build the code, you first must choose the build type and the architecture. Figure 4-17 shows where these options are on the toolbar. Go ahead and choose Debug as the build type and x86 or x64 as the architecture to match your PC. You choose x86 or x64 here because you will first build and test the application on your PC before deploying it to the Raspberry Pi. This shows the true power of UWP—the ability to run the same code on different architectures.

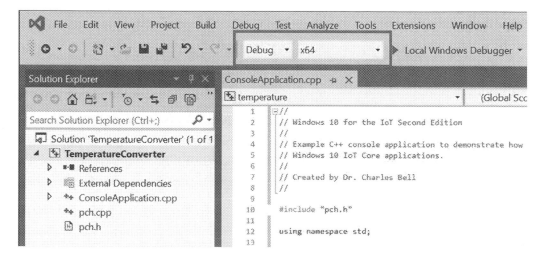

Figure 4-17. *Choose a build platform*

Tip   It doesn't matter which board you use because writing code and deploying it is hardware neutral. The only difference in choosing platforms for building your code concerns the architecture selection. For Raspberry Pi and DragonBoard 410C, choose ARM; for the MinnowBoard Max–compatible boards, choose x86 in the Solutions Platform drop-down box.

Once *Debug* and *x86* or *x64* are selected, press *Ctrl+Shift+B* or choose *Build ➤ Build Solution* from the menu. This compiles your application and opens the output window to show you the results of the build. As you can see in Figure 4-18, the solution built without errors. If you encounter errors, check the code against Listing 4-5 and correct any lines that do not compile.

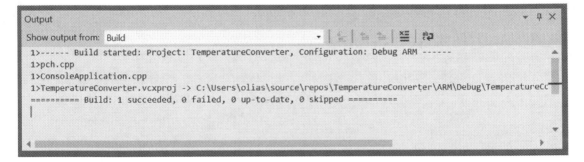

Figure 4-18. *Build and test your code*

Test the Application

To test the application, you can simply run it with the debugger (but without any breakpoints) by pressing *Ctrl+F5* or choosing *Debug ➤ Start without Debugging.* This opens a console and runs the application on your PC and waits to close the console. Note that if you changed your code since the last build, you may be asked to rebuild the solution/project. If you choose the *F5* (with debugging) option, the console will close when the application terminates. Figure 4-19 shows what the application should look like.

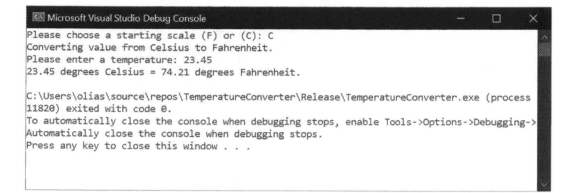

Figure 4-19. *Testing the TemperatureConverter application*

You see what you expected: the application prompts you for some information and, depending on your selection, converts the temperature entered. Go ahead and run the application several times until the joy has passed or you are convinced it is working correctly. When you're done, congratulate yourself on having written your first Windows 10 application!

There is just one more skill we need to master—the interactive debugger.

Using the Interactive Debugger

The interactive debugger allows you to step through your code one line at a time, view the state of variables, see the stack, and much more. You must set the build type to debug in order to use debugging.

The key concepts involved are the use of breakpoints or places in the code where you want the debugger to stop at when executing. You can set breakpoints on most lines of code. To do so, click the leftmost portion of the code editor next to the line you want a breakpoint. You can also position the cursor on the line and press *F9* to set (turn on) the breakpoint. A small red circle appears, which indicates that a breakpoint has been set. To clear a breakpoint, just click the red circle again or position the cursor on the line and press *F9*. Figure 4-20 shows the code editor window with a breakpoint set on the first print statement.

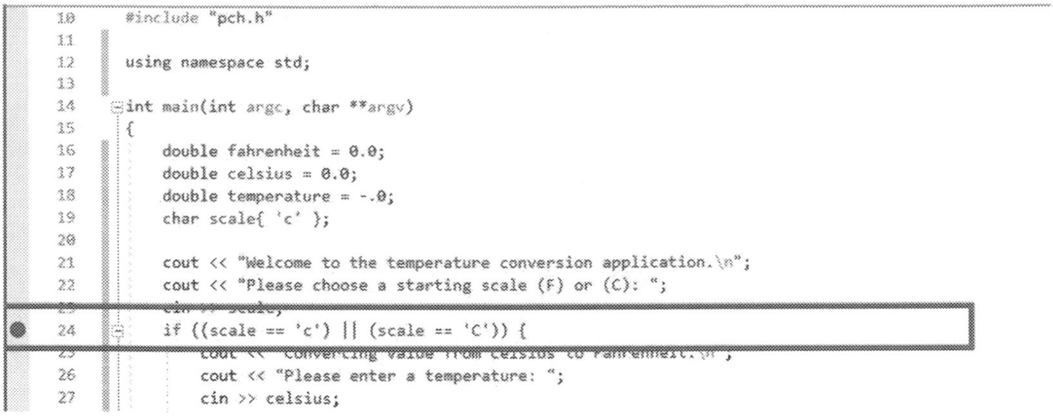

```
10      #include "pch.h"
11
12      using namespace std;
13
14   ⊟int main(int argc, char **argv)
15    {
16          double fahrenheit = 0.0;
17          double celsius = 0.0;
18          double temperature = -.0;
19          char scale{ 'c' };
20
21          cout << "Welcome to the temperature conversion application.\n";
22          cout << "Please choose a starting scale (F) or (C): ";
         cin >> scale;
● 24   ⊟      if ((scale == 'c') || (scale == 'C')) {
              cout << "Converting value from celsius to Fahrenheit.\n";
26              cout << "Please enter a temperature: ";
27              cin >> celsius;
```

Figure 4-20. *Interactive debugger: setting breakpoints in the code editor*

To start the debugging session, press *F5* or choose the *Debug* ➤ *Start Debugger* menu. This causes the application to start a new console window so that you can interact with the application. The debugger stops on the first line of code (where the breakpoint is set). You can then step through the code one line at a time by pressing *F10* or by choosing the *Debug* ➤ *Step Into* menu.

When the debugger is running, several new windows open, including diagnostic tools that show memory and CPU usage, a tabbed window that allows you to watch variables (and more), and another tabbed window that allows you to see the call stack, breakpoints, and more. Figure 4-21 shows an excerpt of running the debugger with our sample project.

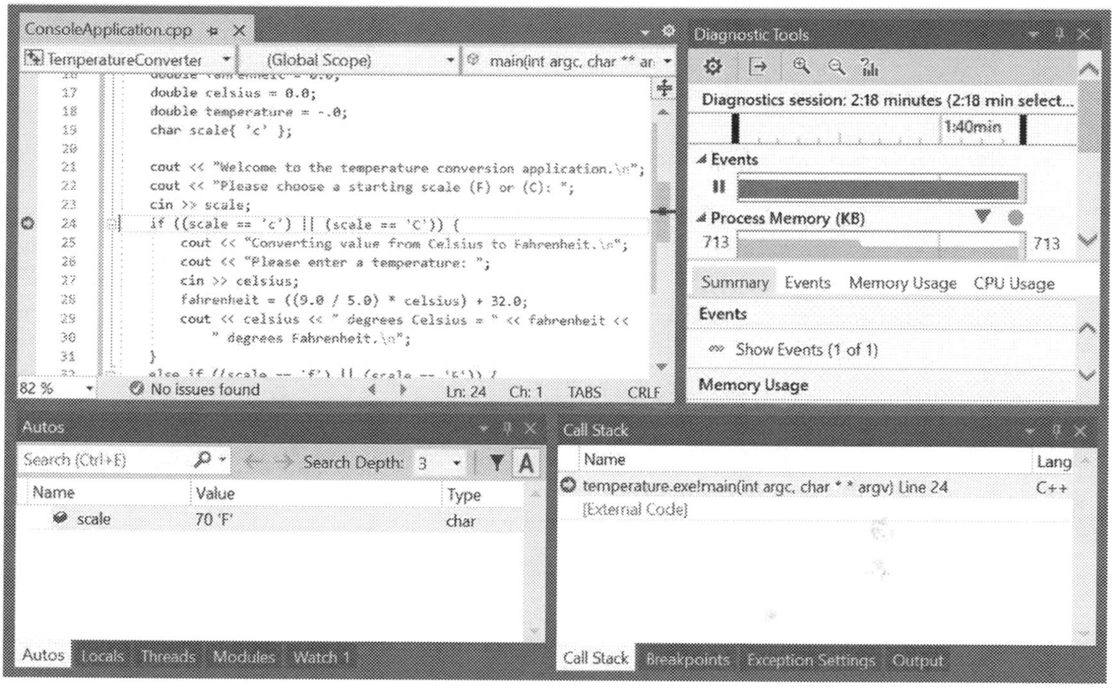

Figure 4-21. *Interactive debugger: inspection*

Go ahead and step through the application. Take some time to tour the various windows. Pay attention to the watch window and note how the variables change values. Clearly, there is far more that you can do with debugging Windows applications than what I have described here. Indeed, I consider the interactive debugger the most important feature of Visual Studio: one that requires some time and experience using to be able to fully utilize it. However, what I've shown here is the bare essentials for getting started.

Tip   To learn more about the interactive debugger, see `https://docs.` `microsoft.com/en-us/visualstudio/debugger/debugger-feature-` `tour?view=vs-2019`.

It is also possible to remotely debug an application running on your Windows 10 IoT device. I will demonstrate how to do this in a later chapter. Remote debugging allows you to run the application on the Windows 10 IoT device and monitor its progress with the interactive debugger on your PC. Due to the nature of console applications, if you were to try remote debugging this project, you would not get far. This is because the application prompts the user for input, and there is no way to connect to the application

153

to interact with it. Thus, remote debugging is best used for cases where you want to see how the code works on the device—checking variables, logic, and so forth, much like the interactive debugger.

You are now ready to set up your Windows 10 IoT device and deploy the application.

Set Up Your Windows 10 IoT Device

You're ready to power on your device and get it ready for deployment. All you need to do is set up the device as described in Chapter 2 and power it on. Once powered on, take note of the IP address. You need this to deploy your application in the next step.

Deployment Options

This is the step that most people have issues mostly due to the fact that you must configure the project settings correctly or the deployment will fail. Once you know what to change (and it isn't so obvious), deployment is easy.

While Visual Studio 2019 is designed to allow you to build and deploy UWP applications from the IDE, deploying other IoT Core projects like a console application requires a different process. In fact, there are several ways to deploy Windows 10 IoT Core applications. The following summarizes the options depending on what project type you use. We will see a demonstration of most of these options with more examples in later chapters. As you will see, the UWP application deployment is by far the easiest.

- *UWP applications*: Use Visual Studio to build and deploy the application by specifying the IP address of the device. This is the most streamline option available. However, the process for C++ applications differs slightly from C# applications. We will see examples of both.

- *Non-UWP applications*: You can use the remote debugger settings to build and deploy your application to a directory of your choosing. This requires making some selections in the project debug properties and requires you to start the remote debugger on the device.

- *Manual copy*: You can use the Device Portal to copy your executable and its dependency files one at a time to the device. This can be a bit tedious, but it does work for the most basic of applications like a console application.

- *Manual package install*: You can use the Device Portal application manager to install your application. However, this requires a deployment package project be added to your solution, and you build the package, sign it, then deploy it and its dependencies to the device. This is the most complicated of the available options. Due to the complexity of this option and that it is not one of the recommended deployment methods, we omit this option in our discussion. However, you can research more about it in the Windows IoT Core documentation.

Deploying UWP Applications

In this section, we will see the basics of deploying UWP applications. Rather than discuss a new project and its code, we will use a blank C# application since it is easy to create. There is an ulterior motive here as well. In order to deploy console and other IoT Core (non-UWP) applications, you need to have the remote debugger enabled on the device. Unfortunately, the remote debugger is not installed by default. Fortunately, you can install it easily by deploying a blank UWP application. Let's see how to do that.

The process for deploying UWP applications is easy. In fact, if you work with UWP applications, you won't need to worry too much about deploying (installing) applications and focus on controlling them with the application manager in the Device Portal.

Begin by starting a new, blank C# application using the Background Application (IoT). Recall, you can search the new project templates for key phrases like "background application" as shown in Figure 4-22.

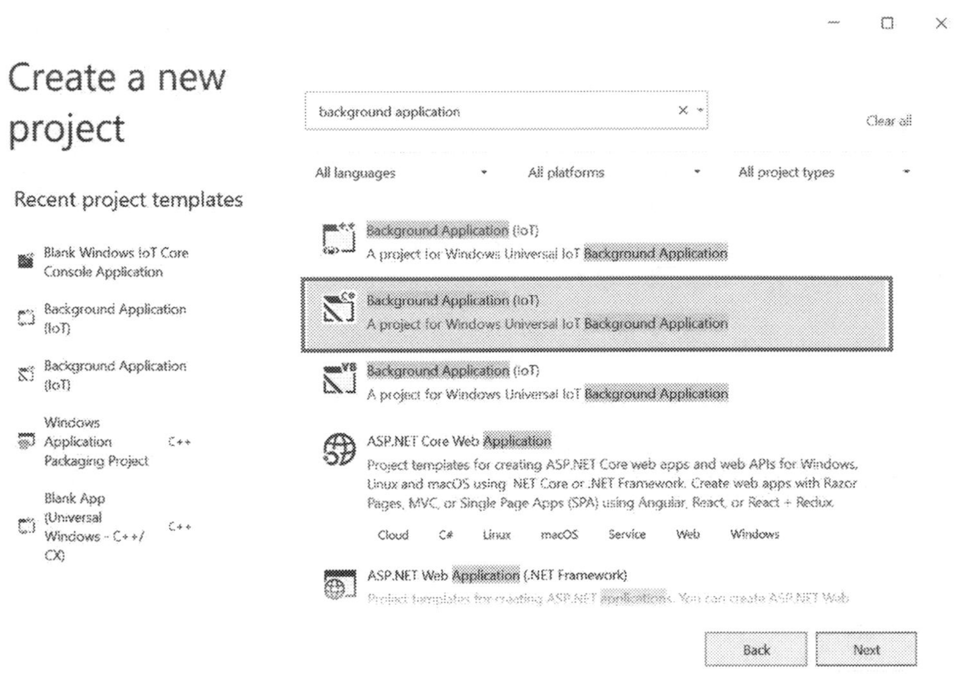

Figure 4-22. *Create a new, blank C# application*

Name the solution BlankCSharp. Recall, you may need to choose the Windows IoT Core version as shown in Figure 4-23. You should choose the latest (the default).

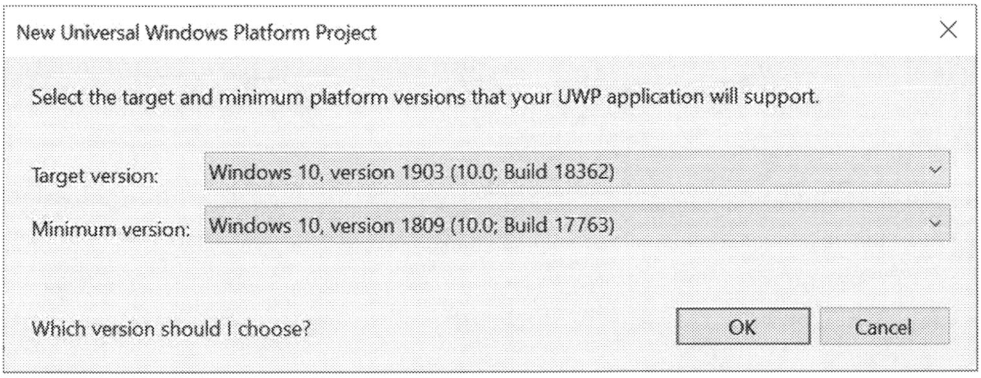

Figure 4-23. *Choose Windows IoT version*

Next, we're going to build the application. But you must choose the debug option for the platform (ARM) in order to include the debugger. Then, click the *Device* drop-down and choose *Remote Machine* as shown in Figure 4-24.

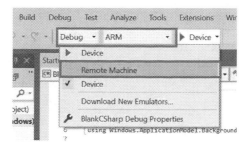

Figure 4-24. *Choose debug build for ARM platform*

This opens a dialog for choosing the device for the deployment as shown in Figure 4-25. If the device is shown in the available devices, choose it. Otherwise, simply copy the IPv4 address from the dashboard and paste it into the address box. Then click *Select* to select it.

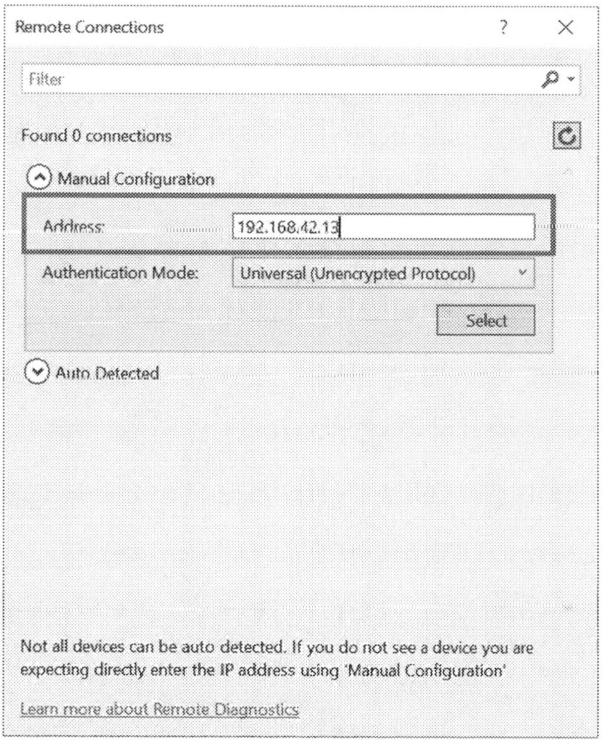

Figure 4-25. *Remote Connections dialog*

At this point, you can click the *Build ➤ Deploy Solution* as shown in Figure 4-26, and Visual Studio will deploy your application for you.

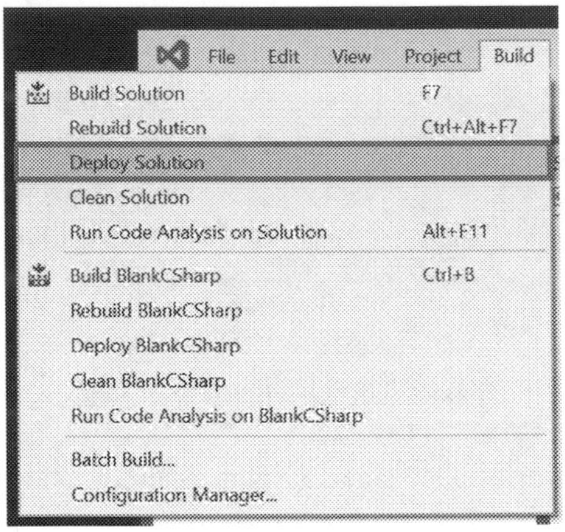
File
Edit
View
Project
Build
Build Solution F7
Rebuild Solution Ctrl+Alt+F7
Deploy Solution
Clean Solution
Run Code Analysis on Solution Alt+F11
Build BlankCSharp Ctrl+B
Rebuild BlankCSharp
Deploy BlankCSharp
Clean BlankCSharp
Run Code Analysis on BlankCSharp
Batch Build...
Configuration Manager...

Figure 4-26. *Deploy Solution*

You can watch the progress in the information box by clicking the *Show output from* drop-down and choose *Deployment* as shown in Figure 4-27.

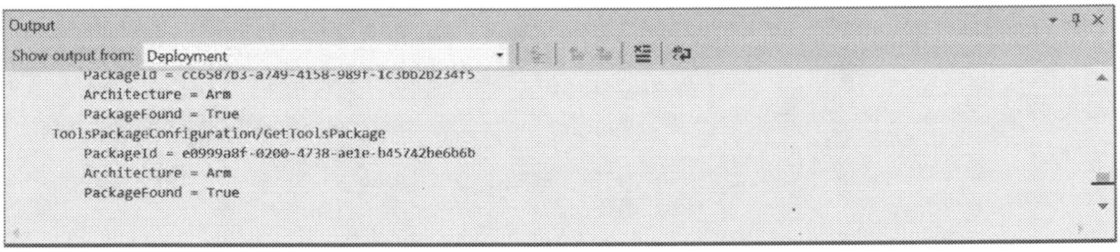
Output
Show output from: Deployment
Architecture = Arm
PackageFound = True
ToolsPackageConfiguration/GetToolsPackage
PackageId = e0999a8f-0200-4738-ae1e-b45742be6b6b
Architecture = Arm
PackageFound = True

Figure 4-27. *Deployment progress*

Caution If you previously started the remote debugger on the device, you would need to stop it prior to deploying your application. The deployment will fail if the remote debugger is running.

Once the deployment is complete, you can visit the Device Portal and look at the applications as shown in Figure 4-28.

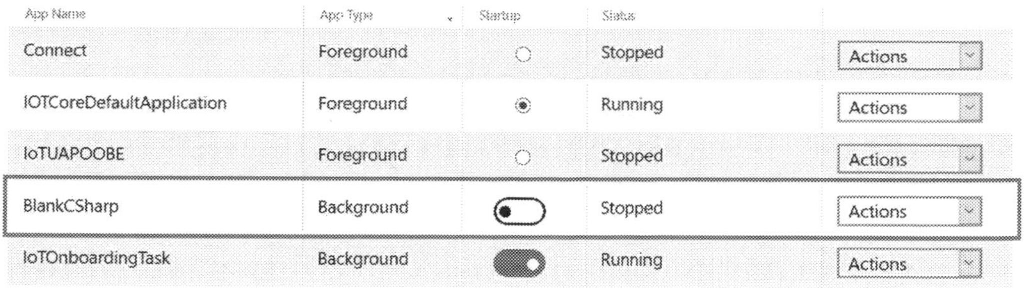

App Name	App Type		Startup	Status	
Connect	Foreground		○	Stopped	Actions ▾
IOTCoreDefaultApplication	Foreground		◉	Running	Actions ▾
IoTUAPOOBE	Foreground		○	Stopped	Actions ▾
BlankCSharp	Background		◉━	Stopped	Actions ▾
IoTOnboardingTask	Background		━◉	Running	Actions ▾

Figure 4-28. *Applications (Device Portal)*

You do not need to start the application because it doesn't do anything. Once you see the application in this list, you're done. You've just deployed your first UWP application (and the remote debugger). We will see how to deploy more interesting applications in later chapters. Now, let's talk about deploying our console application.

Non-UWP Applications

While this procedure covers any non-UWP application, we will use the TemperatureConverter console application as your example.

At first, it may seem like you cannot deploy console applications because the deploy option in the Build menu isn't shown (it isn't enabled). That is because it requires making some modifications to the debug settings for the project, changing the configuration manager settings for the build, and starting the remote debugger on the device.

Tip   The remote debugger is not installed by default. If you haven't already done so, make sure you do this first before trying to build and deploy your console application.

Begin by starting the remote debugger on the device. Open the Device Portal and click *Debug* ➤ *Debug Settings* on the left side of the portal. To start the remote debugger, click the *Start* button on the bottom right as shown in Figure 4-29.

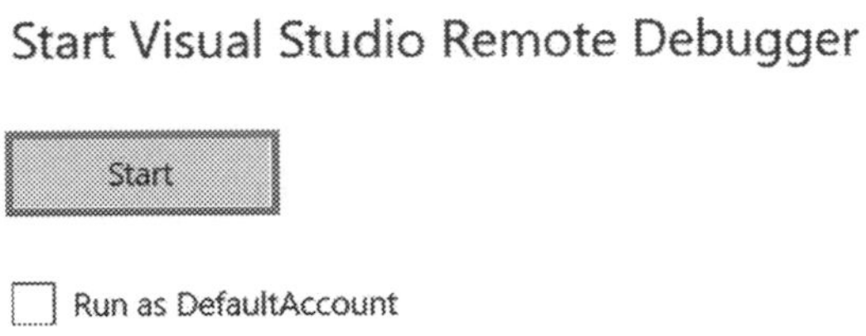

Start Visual Studio Remote Debugger

Start

Run as DefaultAccount

Figure 4-29. *Starting the remote debugger (Device Portal)*

Once the remote debugger is started, notice the address that the remote debugger reports as shown in Figure 4-30. Note this address because you will need it in the next step.

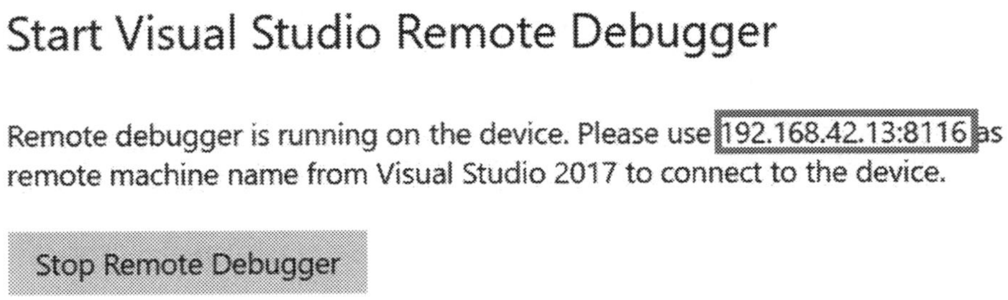

Start Visual Studio Remote Debugger

Remote debugger is running on the device. Please use 192.168.42.13:8116 as remote machine name from Visual Studio 2017 to connect to the device.

Stop Remote Debugger

Figure 4-30. *Address for remote debugger (Device Portal)*

Now you can return to your console application. Open the debug settings by opening the project options and clicking the *Debugging* entry in the list. Figure 4-31 shows the completed dialog.

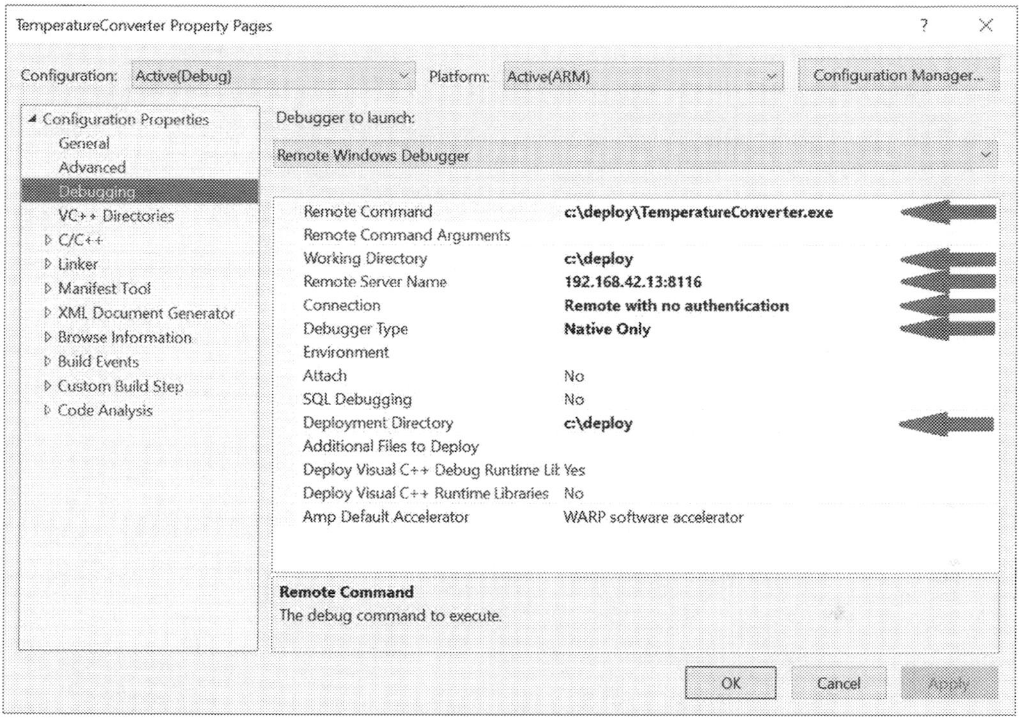

Figure 4-31. *Project properties (debug builds)*

The current configuration is at the top of the dialog. Make sure that everything corresponds to the debug build and ARM platform. There are six settings you must change. I describe each next:

- *Remote Command*: Set this to the path and executable of your application. Use the following working directory set.

- *Working Directory*: Enter a working directory to deploy your application. The directory is created for you. Remember to add the trailing slash or else you may get a warning when you deploy the application.

- *Remote Server Name*: Set this to the IP address or hostname of your device along with port 8116 with :8116. For example, my Raspberry Pi was on 192.168.42.13, so I entered 192.168.42.13:8116. This is the most commonly misunderstood setting.

- *Connection*: Set this to Remote with no authentication.

- *Debugger Type*: Choose Native Only. You only want to deploy the native *ARM* application.

- *Deployment Directory*: Set this to be the same as the working directory. You can choose a different directory, but if you do, make sure that the remote command reflects the difference in path.

Once these settings are entered, click *Apply* and then close the dialog. You are almost ready to deploy.

Next, pull down the *Debug* drop-down box and choose *Configuration Manager....* This will open a dialog that shows all of the build types. You need to tick the *Deploy* tick box for the platform as shown in Figure 4-32. Then close the dialog.

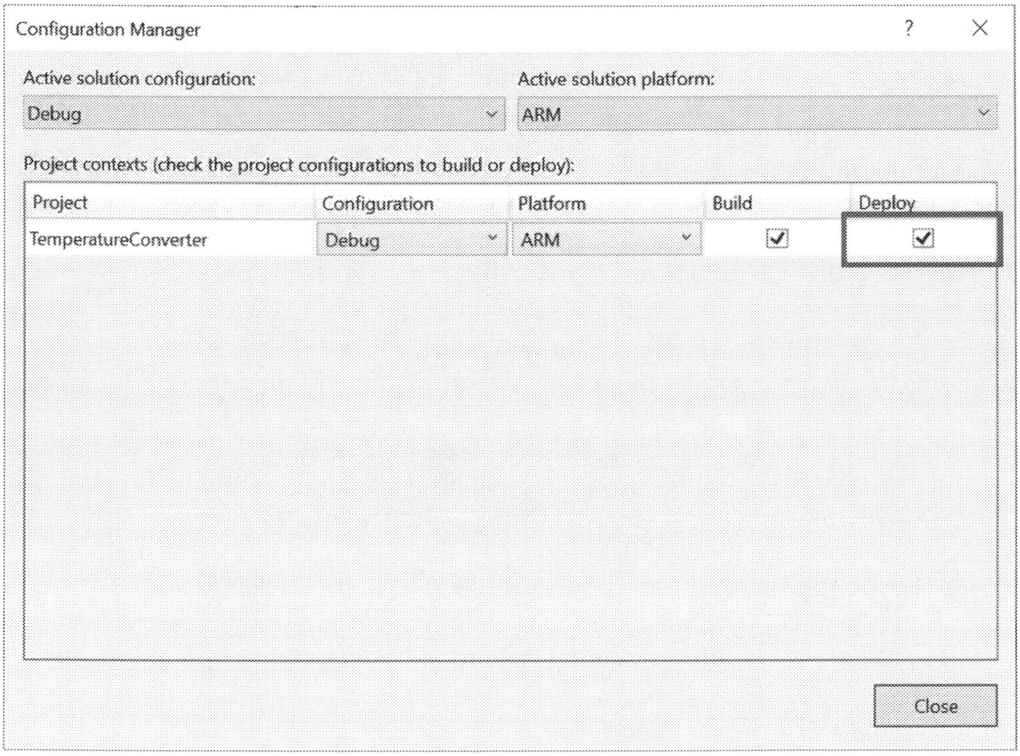

Figure 4-32. *Setting the Configuration Manager deploy option*

Now we can deploy our application! Simply click *Build* ➤ *Deploy TemperatureConverter*. You can follow the progress in the information window.

Note   Applications deployed in this manner will not show up in the Device Portal.

Once the deployment is complete, you can open an SSH terminal connection and see your application is installed as shown in Listing 4-6. See the preceding PuTTY terminal example for how to open an SSH connection.

Listing 4-6. Verifying Console Application Deployment (SSH Terminal)

```
administrator@MINWINPC C:\Data\Users\administrator>cd \deploy

administrator@MINWINPC C:\deploy>dir
 Volume in drive C is MainOS
 Volume Serial Number is BE3A-0FBA

 Directory of C:\deploy

06/27/2020  05:01 PM    <DIR>          .
06/27/2020  05:01 PM    <DIR>          ..
06/27/2020  03:55 PM            71,680 TemperatureConverter.exe
03/18/2019  03:50 PM         1,384,912 ucrtbased.dll
               2 File(s)      1,456,592 bytes
               2 Dir(s)     284,262,400 bytes free

administrator@MINWINPC C:\deploy>TemperatureConverter.exe
Welcome to the temperature conversion application.
Please choose a starting scale (F) or (C): C
Converting value from Celsius to Fahrenheit.
Please enter a temperature: 21.5
21.5 degrees Celsius = 70.7 degrees Fahrenheit.

administrator@MINWINPC C:\deploy>
```

It worked. Cool! Go ahead and experiment with the application or take a look around the device directories. You'll see many that are familiar to you.

Now, let's talk about the manual copy deployment option.

Manual Copy

This option is a manual copy of the executable files to a location on the device. If you know all of the files needed, you can use this option to deploy your application, but as you will see, it requires a bit of manual manipulation.

For this demonstration, we will use the same debug version of the TemperatureConverter application (but it works with the release version too). Recall, we select *Debug* and *ARM* in the debug drop-down boxes and then build the solution using the *Build* ➤ *Build Solution*. This will create a folder named Debug (or Release for release builds) in the project directory. Inside that directory is another named ARM and inside that is the executable named TemperatureConverter.exe. It is this file that we will upload to the device. Remember this path as you will need it in the next step.

There is a limitation in how you can create folders and have them visible in the Device Portal. For example, if you create a new folder on the device, you may not see it in the Device Portal. Thus, we will need to upload the file to a specific set of folders. If you want it in another location, you will have to move it to the alternative location. We will see an example of this in this demonstration. Let's upload the file first.

Open Dashboard and then right-click the device in My Devices and open it in the Device Portal.

In the Device Portal, click *Apps* ➤ *File explorer*. You will see a list of folders that are on the device as shown in Figure 4-33.

Figure 4-33. *List of folders on the device (Device Portal)*

Notice on the right are some columns that permit you to save (download), delete, and rename files. Those options will be handy in cleaning up files and folders on your device.

Now, let's copy the file to the documents folder. Click the Documents folder. Then, click the *Choose File* button and navigate to the ARM binary you built earlier. Once selected, click the *Upload* button to upload the file as shown in Figure 4-34.

Directory path

User Folders \ Documents \

Directory contents

Type	Name		Date Created	File Size	Save	Delete	Rename
🗋	desktop.ini		10/27/2018, 2:06:23 AM	190.0 byt...	🖫	🗑	🖉

🗋 Upload a file to this directory

Browse... TemperatureConverter.exe

Upload

Figure 4-34. *Uploading a file (Device Portal)*

Now, go back to the dashboard and right-click the device in My Devices and choose *Copy IPv4 Address*. Open a new SSH connection.

Let's create a new folder for the application and move it there from the Documents folder. However, once you do this, you will no longer see the application in the Device Portal. Thus, some may want to leave the application in the Documents folder. The following lists the steps we will take to move the application. Listing 4-7 demonstrates these steps in a terminal.

1. Create a folder in the root of the C: folder named c:\deploy\ temperature.

2. Move the TemperatureConverter.exe file to the new folder.

3. Change to the new folder and test the application.

It should be noted here that the folder path to the preceding Documents folder is not in the administrator user's home directory. It is in a different location, specifically, under C:\Data\Users\DefaultAccount\. If you list the files in that directory, you will see the folders as displayed in the Device Portal.

Listing 4-7. Move Application to a New Folder

```
login as: Administrator
Administrator@192.168.42.13's password:
Microsoft Windows [Version 10.0.17763.107]
Copyright (c) Microsoft Corporation. All rights reserved.

administrator@MINWINPC C:\Data\Users\administrator>mkdir c:\deploy\
temperature

administrator@MINWINPC C:\Data\Users\administrator>move c:\Data\Users\
DefaultAcc
ount\Documents\TemperatureConverter.exe c:\deploy\temperature\.
        1 file(s) moved.

administrator@MINWINPC C:\Data\Users\administrator>cd c:\deploy\temperature

administrator@MINWINPC c:\deploy\temperature>TemperatureConverter.exe
Welcome to the temperature conversion application.
Please choose a starting scale (F) or (C): F
Converting value from Fahrenheit to Celsius.
Please enter a temperature: 78.22
78.22 degrees Fahrenheit = 25.6778 degrees Celsius.

administrator@MINWINPC c:\deploy\temperature>
```

While this is a bit clunky, it does work for the simplest of cases. Fortunately, most of the examples in this book use the UWP and non-UWP deployment methods.

Summary

Developing applications for Windows 10 IoT Core requires the use of Visual Studio and the choice of several programming languages. Visual Studio is loaded with advanced features that meet the needs of even the most serious developers for the most complex solutions.

Fortunately for you, you do not need to master every nuance of the IDE. Indeed, you need only learn the most basics of getting around in the IDE, including starting new projects, coding, compiling, testing, and debugging, and deploying your applications to your Windows 10 IoT Core devices.

In this chapter, you explored the Visual Studio 2019 interface, which included learning how the windows are laid out in the IDE and the sample project templates used in this book to write applications for your Windows 10 IoT Core devices. You learned the basic features that you will use for most of the book's projects. You also had a detailed overview of a basic Windows 10 IoT Core application using C++.

In the next chapter, you'll take a closer look at developing Windows 10 IoT Core applications by using C++. I present a short tutorial on the major language constructs and walk you through an application that blinks an LED.

## CHAPTER 5

Windows 10 IoT Development with C++

Now that you have a basic understanding of how to use Visual Studio 2019, you can learn more about some of the languages you may encounter when developing your IoT solutions. One of those languages is C++—a very robust and powerful language that you can use to write very powerful applications. Mastering C++ is not a trivial task and indeed could take someone several years to be fully knowledgeable of all of its features.[1] However, you do not need to achieve a Zen-like harmony with C++ to be able to write applications for Windows 10 IoT Core. You saw this in action in the last chapter. In fact, if you are just getting started programming or know little about C++, all you need to get going is knowledge of the fundamentals of the language and how to use it in Visual Studio.

This chapter presents a crash course on the basics of C++ programming in Visual Studio, including an explanation about some of the most commonly used language features. As such, this chapter provides you with the skills you need to understand the growing number of IoT project examples available on the Internet. The chapter concludes with a walk-through of another C++ example project that shows you how to interact with hardware. Specifically, you will implement the LED project you saw in Chapter 3. Only this time, you'll be writing it as a Windows 10 IoT Core application. So, let's get started!

Tip   If you are not interested in using C++ in your IoT solutions, or you already know the basics of C++ programming, feel free to skim through this chapter. I recommend working through the example project at the end of the chapter, especially if you've not written IoT applications.

---

[1]But who has the time?

© Charles Bell 2021

C. Bell, *Windows 10 for the Internet of Things*, https://doi.org/10.1007/978-1-4842-6609-0_5

Getting Started

Microsoft's implementation of C++ is named Visual C++ and is often referred to as VC++ or MSVC. Visual C++ conforms to the latest standards.[2] Early versions of Visual C++ were sold as a separate product. Visual C++ today is integrated in the Visual Studio product. As you saw in Chapter 2, you must explicitly select Visual C++ during installation of Visual Studio because it is not installed by default. However, as you saw, it is very easy to add optional features in Visual Studio.

While the formal product name is Visual C++, and it is designed to work with the Microsoft Windows operating system, learning the basics of the Visual C++ language is no different than learning C++ that is offered for other platforms. Thus, in this section and throughout the rest of the book, I use C++ as shorthand for Visual C++. Most of the material in the next section applies to C++ in general and is not specific to Visual C++. Indeed, except for the Microsoft frameworks and the components of Visual Studio, the basics of the C++ language apply universally. Let's learn more about the origins and merits of C++.

The C++ programming language has been around for over 39 years and is one of the most widely used programming languages among all platforms. There are many reasons for this. C++ was designed to permit programmers to express hardware, machine, and programming concepts as close to the hardware level as possible. The C++ language therefore is designed to be very expressive with concepts close to actual hardware. Indeed, well-designed and well-written C++ applications are typically faster and more efficient than those written in other languages.[3] Part of this is due to the fact that C++ can be used to program a wide array of solutions—from computer games to highly sophisticated scientific analysis applications to device drivers and even entire frameworks. The longevity of C++ is a testament to the resiliency of the language.

However, there has been some misunderstanding in the history of C++ for which Visual C++ was not immune. Early releases of the C++ language specification occurred during a time when the C language was popular. While it is true C++ was based on C and to this day you can use a number of C features (but not nearly as many now that C++ has evolved from those early days), C++ has become a completely different language.

---

[2]For more about compatibility with C++ standards, see `https://docs.microsoft.com/en-us/previous-versions/hh567368(v=vs.140)?redirectedfrom=MSDN`.

[3]However, I have seen the opposite happen for poorly written code. Language features can never overcome poor programming.

The evolution included the addition of static type safety (every object, name, value, etc. must have a type known at compile time), full object-oriented programming features (classes, inheritance, etc.), and a host of improvements designed to make programming easier and faster (e.g., the standard template library). What started out as "C with classes" (circa 1979), which included the beginnings of object-oriented programming, the language was renamed C++ (circa 1983) to distinguish it from C.

Finally, C++ programs are compiled into executable files. More specifically, the code you write is translated into binary code for the platform chosen. This is another reason C++ is potentially faster than other languages because there is no interpreter involved. For example, the Python language is an interpreted language that, while there is a building stage, the code is turned into an intermediate form that is platform agnostic, which is then executed by the interpreter. The power of compiling the code directly into binary code with the gains in speed and efficiency is why many people choose C++ over other languages.

Should you require more in-depth knowledge of C++ (Visual C++), there are a number of excellent books on the topic. The following is a list of a few of my favorites:

- *Beginning C++* by Ivor Horton (Apress, 2014)

- *The C++ Programming Language* by Bjarne Stroustrup (Addison-Wesley Professional, 2013)[4]

- *C++ Recipes: A Problem-Solution Approach* by Bruce Sutherland (Apress, 2015)

Another excellent resource is Microsoft's documentation on MSDN. The following are some excellent resources for learning Visual C++:

- Visual C++ in Visual Studio 2015 (`https://docs.microsoft.com/en-us/previous-versions/60k1461a(v=vs.140)?redirectedfrom=MSDN`)

- C++ Language Reference (`https://docs.microsoft.com/en-us/cpp/cpp/cpp-language-reference?redirectedfrom=MSDN&view=vs-2019`)

---

[4]Creator of the C++ language.

C++ Crash Course

Now let's learn some of the basic concepts of C++ programming. You begin with the building blocks of the language, such as comments, variables, and basic control structures, and then move into the more complex concepts of data structures and libraries.

While the material may seem to come at you in a rush (hence, the crash part), this crash course on C++ covers only the most fundamental knowledge of the language and how to use it in Visual Studio. It is intended to get you started writing C++ Windows 10 IoT Core applications. If you find you want to write more complex applications than the examples in this book, I encourage you to acquire one or more of the resources listed earlier to learn more about the intriguing power of C++ programming.

The following sections present many of the basic features of C++ programming that you need to know in order to understand example projects for Windows 10 IoT Core and vital to successfully implementing the C++ projects in this book.

The Basics

There are a number of basic concepts about the C++ programming language that you need to know in order to get started. In this section, I describe some of the fundamental concepts used in C++, including how the code is organized, how libraries are used, namespaces, and how to document your code. Before you begin, let's take a look at a slightly different version of the temperature converter application you saw in Chapter 4. Listing 5-1 shows the code rewritten slightly to use functions.

Listing 5-1. TemperatureConverter Code Example Rewrite

```
//
// Windows 10 for the IoT Second Edition
//
// Example C++ console application rewrite.
//
// Created by Dr. Charles Bell
//
#include "pch.h"

using namespace std;
```

```cpp
double convert_temp(char scale, double base_temp) {
    if ((scale == 'c') || (scale == 'C')) {
        return ((9.0 / 5.0) * base_temp) + 32.0;
    }
    else if ((scale == 'f') || (scale == 'F')) {
        return (5.0 / 9.0) * (base_temp - 32.0);
    }
    return 0.0;
}

int main(int argc, char** argv) {
    double temp_read = 0.0;
    char scale{ 'c' };
    cout << "Welcome to the temperature conversion application.\n";
    cout << "Please choose a starting scale (F) or (C): ";
    cin >> scale;
    cout << "Please enter a temperature: ";
    cin >> temp_read;
    if ((scale == 'c') || (scale == 'C')) {
        cout << "Converting value from Celsius to Fahrenheit.\n";
        cout << temp_read << " degrees Celsius = " <<
            convert_temp(scale, temp_read) << " degrees Fahrenheit.\n";
    }
    else if ((scale == 'f') || (scale == 'F')) {
        cout << "Converting value from Fahrenheit to Celsius.\n";
        cout << temp_read << " degrees Fahrenheit = " <<
            convert_temp(scale, temp_read) << " degrees Celsius.\n";
    }
    else {
        cout << "\nERROR: I'm sorry, I don't understand '" << scale << "'.";
        return -1;
    }
    return 0;
}
```

Wow, that's quite a change! While the functionality is exactly the same, the code looks very different. The following describe the C++ concepts I have implemented in this example.

Functions

Notice that I've added a new function named `convert_temp()` that converts the temperature based on the scale chosen. This effectively moves that logic out of the `main()` function, thereby simplifying the code. This technique is a key technique you use when writing C++ applications. More specifically, in C++, applications are always built using functions.

Recall that the `main()` function is the starting or initial execution for the C++ console project. Traditional C++ applications (such as the console application) must have a main function.

Notice the main function again. Here, you see the function name is preceded by a type (integer). This tells the C++ compiler that this method returns an integer value. On Windows GUI applications (not command-line interface applications), it is common practice to not return a value from `main()` but to be pedantic; you should do so as I have in the preceding example.

```
int main(int argc, char **argv)
```

Next, you see the name, main, followed by a list of parameters enclosed in parentheses. For the `main()` function, the parameters are fixed and are used to store any command-line arguments provided by the user. In this case, you have the number of arguments stored in `argc` and a list (array) of the arguments stored in `argv` (which is actually a double pointer—more on pointers later).

A function in C++ is used as an organizational mechanism to group functionality and make your programs easier to maintain (functions with hundreds of lines of code are very difficult to maintain), improve comprehensibility, and localize specialized operations in a single location, thereby reducing duplication.

Functions are used in code to express the concepts of the functionality that they provide. Notice how I used the `convert_temp()` function. Here, I declared it as a function that returned a `double` and takes a character and a `double` as input. As you can see, the body of the function (defined inside the curly braces) uses the character as the scale in the same way as you do in main and uses the double parameter as the target (or base) temperature to convert.

> **Tip** Function parameters and values passed must match on type and order when called.

Notice also that I placed it in the line of code that prints the value to the screen. This is a very common practice in C++ (and other programming languages). That is, you use the function to perform some operation, and rather than store the result in a variable, you use it directly in the statements (code).

Curly Braces

Notice that both methods are implemented with a pair of curly braces that define the body of the function. Curly braces in C++ are used to define a block of code or simply to express grouping of code. Curly braces are used to define the body of functions, structures, classes, and more. Notice that you use them everywhere, even in the conditional statements (see the if statements).

> **Tip** Some C++ programmers prefer to place the starting curly brace on the same line as the line of code to which it belongs like I did in the example. However, others prefer the open curly brace placed on the next line. Neither preference matters to the compiler; rather, this is an example of code style. You should choose the style you like best.

Including Libraries

If you recall from earlier examples, there were some lines of code at the top of the source code that indicated something was to be included. These are called preprocessor directives and often look like the following. They are called preprocessor directives because they signal the compilation process to perform some tasks before the code is compiled.

```
#include "pch.h"
```

The directive does what it sounds like: it tells the compiler to include that file along with your source code. When the compiler encounters this directive, it "includes" that source file with your source code and compiles it.

The #include directive is one of the fundamental mechanisms that support modularity in C++. That is, you can create a library of source code that provides some functionality that resides in one or more separate source code files. Even if you do not create a new library, you can use modularity to split your source code into separate parts that form some high-level abstraction. More specifically, you would place like functionality together, making the code easier to maintain or allowing more than yourself to work on it at the same time. However, you would not use modularity to separate random sections of code—that would gain you nothing except confusion as to where the bits of code reside.

The file that the preceding example includes is called a header file. A header file is named .h and contains only the declaration of the code. You can think of it as a blueprint or pattern for the code. A header file often contains only the primitives of the code that you will use, hence making it possible for the compiler to resolve any references to the features in the header file. A separate companion file called the source file is named .cpp (you can also use .cc, but most prefer .cpp) and contains the actual code for the features. In the preceding example, there are two files: one named pch.h and the other pch.cpp.

Using Namespaces

Notice the line of code that begins with using in the example. This is another preprocessor directive. The using directive tells the compiler that you are using the namespace std. A namespace is a special organizational feature that allows you to group identifiers (names of variables, constants, etc.) under a group that is localized to the namespace. Using the namespace tells the compiler to look in the namespace for any identifier you've used in your code that is not found.

```
using namespace std;
```

For example, the following cout statement is included in the std namespace. Had I not added the using directive, the compiler would not know what cout was. Notice in the second line that I could have included the namespace std followed by two colons. This tells the compiler to look in the std namespace for cout or more correctly use the cout in the std namespace.

```
cout << "Hello, World!\n";
std::cout << "Right back at you!";
```

Namespaces can also help you avoid duplication of identifiers. Had I wanted to, I could have created a new module (library, source file) that contains a new namespace and reused the cout identifier. However, care must be taken when reusing identifiers among namespaces because the compiler may not know which namespace I want to use. Creating namespaces is a bit advanced for a crash course, but I encourage you to consider using them for applications that grow beyond a single source code and header file.

Finally, namespaces associated with libraries of classes often form hierarchies that you can chain together. For example, if you wanted to use the namespace inside the Windows Foundations library named Collections, you would refer to it as follows. This is a very common occurrence in Windows C++ applications. In fact, you use several namespaces in our example project.

```
using namespace Windows::Foundation::Collections;
```

Comments

One of the most fundamental concepts in any programming language is the ability to annotate your source code with text that not only allows you to make notes among the lines of code but also forms a way to document your source code.[5]

To add comments to your source code, use two slashes, // (no spaces between the slashes). Place them at the start of the line to create a comment for that line, repeating the slashes for each subsequent line. This creates what is known as a block comment, as shown. Notice that I used a comment without any text to create whitespace. This helps with readability and is a common practice for block comments.

```
//
// Windows 10 for the IoT Second Edition
//
// Example C++ console application rewrite.
//
// Created by Dr. Charles Bell
//
```

---

[5]If you ever hear someone claim, "My code is self-documenting," be cautious when using their code. There is no such thing. Sure, plenty of good programmers can write code that is easy to understand (read), but all fall short of that lofty claim.

You can also use the double slash to add a comment at the end of a line of code. That is, the compiler ignores whatever is written after the double slash to the end of the line. You see an example of this next. Notice that I used the comment symbol (double slash) to comment out a section of code. This can be really handy when testing and debugging, but generally discouraged for final code. That is, don't leave any commented out code in your deliverable (completed) source code. If it's commented out, it's not needed!

```
if (size < max_size) {
  size++;
} //else {
//   return -1;
//}
```

Writing good comments and indeed documenting your code well is a bit of an art form, one that I encourage you to practice regularly. Since it is an art rather than a science, keep in mind that your comments should be written to teach others what your code does or is intended to do. As such, you should use comments to describe any preconditions (or constraints) of using the code, limitations of use, errors handled, and a description of how the parameters are used and what data is altered or returned from the code (should it be a function or class member).

Variables and Types

No program would be very interesting if you did not use variables to store values for calculations. Variables are declared with a type and, once defined with a specific type, cannot be changed. Since C++ is strongly typed, the compiler ensures that anywhere you use the variable, it obeys its type, for example, that the operation on the variable is valid for the type. Thus, every variable must have a type assigned.

There are a number of simple types that the C++ language supports (often called *built-in types*). They are the basic building blocks for more complex types. Each type consumes a small segment of memory which defines not only how much space you have to store a value but also the range of values possible.[6]

For example, an integer consumes 4 bytes, and you can store values in the range –2,147,483,648 to 2,147,483,647. In this case, the integer variable is signed (the highest bit

---

[6]For a complete list, see `https://docs.microsoft.com/en-us/cpp/cpp/data-type-ranges?redirectedfrom=MSDN&view=vs-2019`.

is used to indicate positive or negative values). An unsigned integer can store values in the range 0 to 4,294,967,295.

You can declare a variable by specifying its type first and then an identifier. The following shows a number of variables using a variety of types:

```
int num_fish = 0;              // number of fish caught
double max_length {0.0};       // length of the longest fish in feet
char fisherman[25];            // name of the fisherman
char rod_used[40];             // name or type of rod used
```

Notice also that I have demonstrated how to assign a value to the variable in the declaration. I demonstrate two widely used techniques: using a simple assignment and using the initialization mechanism available since C++11 (meaning it is the C++ standard adopted in 2011) and newer.

The assignment operator is the equal sign. All assignments must obey the type rules. That is, I cannot assign a floating-point number (e.g., 17.55) to an integer value.

There is an alternative mechanism to use for initialization. Rather than use the int x = 4; syntax, we can use the C++ initialization mechanism using curly braces (called an initializer list) that contain the value you want to assign. The following shows an example:

```
int x {14};
```

Note that you can include the assignment operator with the curly braces (the compile will not complain), but that is considered sloppy and discouraged. For example, the following code will compile, but it is considered a bad form:

```
int y = {15};
```

Table 5-1 shows a list of the commonly used built-in types that you use in your applications.

Table 5-1. *Commonly Used Types in C++*

Symbol	Size in Bytes	Range
bool	1	False or true
char	1	−128 to 127 by default
signed char	1	−128 to 127
unsigned char	1	0 to 255
short	2	−32,768 to 32,767
unsigned short	2	0 to 65,535
int	4	−2,147,483,648 to 2,147,483,647
unsigned int	4	0 to 4,294,967,295
long	4	−2,147,483,648 to 2,147,483,647
unsigned long	4	0 to 4,294,967,295
float	4	3.4E +/− 38 (7 digits)
long long	8	−9,223,372,036,854,775,808 to 9,223,372,036,854,775,807
unsigned long long	8	0 to 18,446,744,073,709,551,615
double	8	1.7E +/− 308 (15 digits)

It is always a good practice to initialize your variables when you declare them. It can save you from some nasty surprises if you use the variable before it is given a value (although the compiler will complain about this). For example, it is possible for code to work correctly in debug mode (since variables may be initialized by the debugger) but fail in release mode.

There is also a convenient automatic type keyword (auto) that you can use to permit the compiler to choose the correct type. This is helpful for things like loops as follows but also helps with maintainability as well as permitting advanced reuse through templates:

```
auto j {3};    // integer is used here
for (auto i=0; i < 10; ++i) {
  cout << i + j << "\n";
}
```

Here, you see you create the variable j with the auto keyword, which given the initialization is an integer results in j being an integer. Similarly, you use the auto keyword in the for loop counting variable. Since it was assigned an integer, the variable i will be an integer.

Arithmetic

You can perform a number of mathematical operations in C++, including the usual primitives, but also logical operations and operations used to compare values. Rather than discuss these in detail, I provide a quick reference in Table 5-2 that shows the operation and an example of how to use the operation.

Table 5-2. *Arithmetic, Logical, and Comparison Operators in C++*

Type	Operator	Description	Example
Arithmetic	+	Addition	`int_var + 1`
	-	Subtraction	`int_var - 1`
	*	Multiplication	`int_var * 2`
	/	Division	`int_var / 3`
	%	Modulus	`int_var % 4`
	-	Unary subtraction	`-int_var`
	+	Unary addition	`+int_var`
Logical	&	Bitwise and	`var1&var2`
	|	Bitwise or	`var1|var2`
	^	Bitwise exclusive	`var1^var2`
	~	Bitwise compliment	`~var1`
	&&	Logical and	`var1&&var2`
	||	Logical or	`var1||var2`

(*continued*)

Table 5-2. (*continued*)

Type	Operator	Description	Example
Comparison	==	Equal	expr1==expr2
	!=	Not equal	expr1!=expr2
	<	Less than	expr1<expr2
	>	Greater than	expr1>expr2
	<=	Less than or equal	expr1<=expr2
	>=	Greater than or equal	expr1>=expr2

Bitwise operations produce a result on the values performed on each bit. Logical operators (and, or) produce a value that is either true or false and are often used with expressions or conditions.

Finally, C++ has a concept called *constants*, where a value is set at compile time. There are two types of constants. One, signified by using the const keyword, creates a value (think variable) that will never be changed. The other, signified by using the constexpr keyword, creates a function whose body or functionality is evaluated at compile time. The following are examples of constants in C++:

```
const int fish_catch_limit {7};   // Creates a constant variable whose
                                        value cannot change
constexpr double square(double z) { return z*z; }  // A constant expression
```

Now that you understand variables and types, the operations permitted on them, and expressions, let's look at how you can use them in flow control statements.

Flow Control Statements

Flow control statements change the execution of the program. They can be conditionals that cause one section of code to execute vs. another (also called gates). These conditionals use expressions that, when evaluated, restrict execution to only those cases where the expression is true. There are special constructs that allow you to repeat a block of code (loops) as well as functions to switch context to perform some special operations. You've already seen how functions work, so let's look at conditional and loop statements.

Conditionals

Conditional statements allow you to direct execution of your programs to sections (blocks) of code based on the evaluation of one or more expressions. There are two types of conditional statements in C++—the if statement and the switch statement.

You have seen the if statement in action in our example code. Notice in the example that you can have one or more (optional) else phrases that you execute once the expression for the conditions evaluates to false. You can chain if/else statements to encompass multiple conditions where the code executed depends on the evaluation of several conditions. The following shows the general structure of the if statement:

```
if (expr1) {
  // execute only if expr1 is true
} else if ((expr2) || (expr3)) {
  // execute only if expr1 is false *and* either expr2 or expr3 is true
} else {
  // execute if both sets of if conditions evaluate to false
}
```

Although you can chain the statement as much as you want, use some care here because the more else/if sections you have, the harder it becomes to understand, maintain, and avoid logic errors in your expressions.

If you have a situation where you want to execute code based on one of several values for a variable or expression that returns a value (such as a function or calculation), you can use the switch statement. The following shows the structure of the switch statement:

```
switch (eval) {
  case <value1> :
    // do this if eval == value1
    break;
  case <value2> :
    // do this if eval == value2
    break;
  default :
    // do this if eval != any case value
    break;  // Not needed, but good form
}
```

The case values must match the type of the thing you are evaluating. That is, case values must be the same type as eval. Notice the break statement. This is used to halt evaluation of the code once the case value is found. Otherwise, each successive case value will be compared. Finally, there is a default section for code that you want to execute, should eval fail to match any of the values.

Tip   Code style varies greatly in how to space/separate these statements. For example, some indent the case statements, some do not.

Loops

Loops are used to control the repetitive execution of a block of code. There are three forms of loops that have slightly different behavior. All loops use conditional statements to determine whether to repeat execution or not. That is, they repeat as long as the condition is true. The three types of loops are while, do, and for. I explain each with an example.

The while loop has its condition at the "top" or start of the block of code. Thus, while loops only execute the body if and only if the condition evaluates to true on the first pass. The following illustrates the syntax for a while loop. This form of loop is best used when you need to execute code only if some expression(s) evaluates to true, for example, iterating through a collection of things whose number of elements is unknown (loop until you run out of things in the collection).

```
while (expression) {
   // do something here
 }
```

The do loop places the condition at the "bottom" of the statement, which permits the body of the loop to execute at least once. The following illustrates the do loop. This form of loop is handy for cases where you want to execute code that, depending on the results of that execution, may require repetition, for example, repeatedly asking the user for input that matches one or more known values, repeating the question if the answer doesn't match.

```
do {
  // do something here - always done once
} while (expression);
```

The for loop is sometimes called counting loops because of their unique form. The for loop allows you to define a counting variable, a condition to evaluate, and an operation on the counting variable. More specifically, for loops allow you to define stepping code for a precise number of operations. The following illustrates the structure of the for loop. This form of loop is best used for a number of iterations for a known number (either at runtime or as a constant) and commonly used to step through memory, count things, and so forth.

```
for (<init> ; <expression> ; <increment>) {
// do something
}
```

The <init> section or counting variable declaration is executed once and only once. The <expression> is evaluated on every pass. The <increment> code is executed every pass except the last. The following is an example for loop:

```
for (int i; i < 10; i++) {
    // do something here
}
```

Now let's look at some commonly used data structures.

Basic Data Structures

What you have learned so far about C++ allows you to create applications that do simple to moderately complex operations. However, when you start needing to operate on data (either from the user or from sensors and similar sources), you need a way to organize and store data and operations on the data in memory. The following introduces three data structures in order of complexity: arrays, structures, and classes.

Arrays allocate a contiguous area of memory for multiple storage of a specific type. That is, you can store several integers, characters, and so forth, set aside in memory. Arrays also provide an integer index that you can use to quickly access a specific element. The following illustrates how to create an array of integers and iterate through them with a for loop. Array indexes start at 0.

```
int num_array[10] {0,1,2,3,4,5,6,7,8,9};  // an array of 10 integers
for (int i = 0; i < 10; ++i) {
  cout << "num_array[" << i << "] = " << num_array[i] << "\n";
}
```

Notice the ++i in the for loop. This is a shorthand for i = i + 1 and is very common in C++. You can also define multiple dimensional arrays (arrays of arrays). Arrays can be used with any type or data structure.

If you have a number of data items that you want to group together, you can use a special data structure called, amazingly, struct. A struct is formed as follows:

```
struct <name> {
  // one or more declarations go here
};
```

You can add whatever declarations you want inside the struct body (defined by the curly braces). The following shows a crude example. Notice that you can use the structure in an array.

```
struct address {
  char first_name[30];
  char last_name[30];
  int street_num;
  char street_name[40];
  char city[40];
  char state[2];
  char zip_code[12];
};

address address_book[100];
```

Arrays and structures can increase the power of your programs by allowing you to work with more complex data types. However, there is one data structure that is even more powerful: the class.

A class is more than a simple data structure. You use classes to create abstract data types and to model concepts that include data and operations on data. Like structures, you can name the class and use that name to allocate (instantiate) a variable of that type. Indeed, structs and classes are closely related.

You use classes to break your programs down into modules. More specifically, you place the definition of a class in a header file and the implementation in a source file. The following shows the header file (myclass.h) for a simple and yet trivial class to store an integer and provide operations on the integer:

```cpp
class MyClass {
  public:
    MyClass();
    int get_num();
    void inc();
    void dec();
  private:
    int num;
};
```

Notice several things here. First, the class has a name (MyClass), a public section where anything in this area is visible (and usable) outside of the class. In this case, there are three functions. The function with the same name as the class is called a constructor, which is called whenever you instantiate a variable of the class (type). The private section is only usable from functions defined in the class (private or public).

The source code file (myclass.cpp) is where you implement the methods for the class as follows:

```cpp
#include "myclass.h"

MyClass::MyClass() {
  num = 0;
}

int MyClass::get_num() {
  return num;
}

void MyClass::inc() {
  ++num;
}

void MyClass::dec() {
  --num;
}
```

Notice that you define the methods in this file prefixed with the name of the class and two colons (MyClass::). While missing in this example, you can also provide a destructor (noted as ~MyClass) that is executed when the class instantiation is deallocated. Finally, notice at the top is the `#include` preprocessor directive to include the header file so that the compiler knows how to compile this code (using the class header or declaration). You can then use the class in the program, as follows:

```
#include <iostream>
#include "myclass.h"

using namespace std;

int main(int argc, char **argv) {
  MyClass c = MyClass();
  c.inc();
  c.inc();
  cout << "contents of myclass: " << c.get_num() << "\n";
}
```

Notice how you use the class. This is actually allocating memory for the class—both data and operations. Thus, you can use classes to operate on things or provide functionality when you need it, saving you time and making your programs more sophisticated. Classes are used to form libraries of functionality that can be reused. Indeed, you have entire suites of libraries built using classes.

As you may have surmised, classes are the building blocks for object-oriented programming, and as you learn more about using classes, you can build complex libraries of your own.

Pointers

Pointers are one of the most difficult things for new programmers to understand. However, the following attempts to explain the basics of using pointers. There is a lot more that you can do with pointers, but this is the fundamental concept of simple pointers.

A pointer (also called a *pointer variable*) stores the memory address of a variable or data. Thus, a pointer "points to" a section of memory. You declare by type and the * symbol. All pointers must be typed, and any operation on what the pointer points to must obey the condition of that type. When you access the thing the pointer "points to," you call that dereferencing and use the * symbol to tell the compiler you want the value of the thing the pointer is "pointing to."[7] The following shows how you can declare a pointer and then dereference it. Note that you expect int_ptr to be assigned a value; otherwise, the code may not compile or exhibit side effects.

```
int *int_ptr; // pointer to an integer
int i = *int_ptr; // Store what int_ptr is pointing to
```

To store an address in a pointer variable, you use the & symbol (also called the *address of operator*). The following shows an example:

```
int *int_ptr = &i; // Store address of i in int_ptr
```

You can perform arithmetic and comparison on pointers. You can add or subtract an integer to change the address of the pointer (the actual value of the pointer variable, not the thing the pointer points to) by multiples of the size of the type. For example, adding 1 to an integer pointer advances (increases) the memory value by 4 bytes.

You can also compare pointers to determine equality and subtract one pointer from another to find distance (in bytes) between the pointers. This could be handy for calculating distance for contiguous memory segments. When performing arithmetic on pointers, you should use parentheses to avoid nasty mistakes with operator precedence. For example, the following code is not equivalent. The second line increments the thing that the pointer points to, but the third line increments the value of the pointer variable (memory address). Be careful when performing math on pointers because you could unexpectedly end up dereferencing portions of memory.

```
*int_ptr = 10;        // set the thing that the pointer points to = 10
i = *int_ptr + 1;     // add one to the thing that the pointer points to
i = *(int_ptr + 1);   // add 4 bytes to the pointer variable (size of
                         integer) - ERROR? points to nowhere!
```

Finally, always use nullptr to initialize a pointer variable when the address is not known, as follows:

```
int *int_ptr {nullptr};
```

---

[7] I think you get the point.

Now that you know how to declare pointers, dereference them to retrieve the value of the thing the pointer points to and to find the address to store in a pointer variable; let's see pointers in action. The following demonstrates how pointers are used. What follows is an overly simplified example where the memory locations use fictitious values, but it illustrates the concept of pointers as memory addresses. You will follow several code statements as they execute using Visual Studio and the debug watch window . Figure 5-1 shows the initial state for our demonstration.

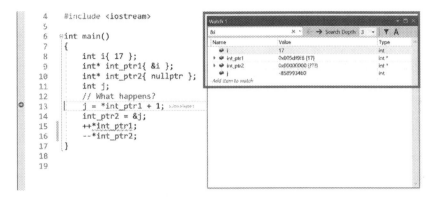

Figure 5-1. *Pointer illustration: initial state*

The values shown are in hexadecimal and shown using only two bytes for integers and pointers to integers. You also show data saved with the low byte first.

Take a few moments to study the figure until you are confident you see how each value is stored. Here, you see you have two integer variables and two pointers to integer variables. The figure shows how each is allocated in memory. Notice that you see the variable i is stored in memory with the value 17, the first integer pointer stored with a value of the address of i the second integer pointer stored with the value of nullptr, and the integer variable j stored without an initial value.

Now let's see what happens when the first line of code is executed. Figure 5-2 shows the results in our memory space.

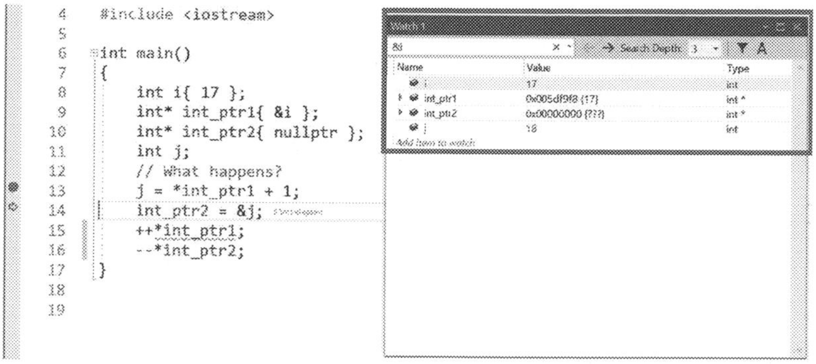

Figure 5-2. *Pointer illustration: step 1*

Here, you see that the variable j was assigned the value dereferenced from int_ptr1 plus one. To execute this statement, the compiler dereferenced int_ptr1 by using its value (memory address) to get the value from that memory location (17 or 0x11) and then adding 1 (18 or 0x12) and storing it in memory.

Now let's see what happens when you execute the next statement. Figure 5-3 shows the results in our fictional memory space.

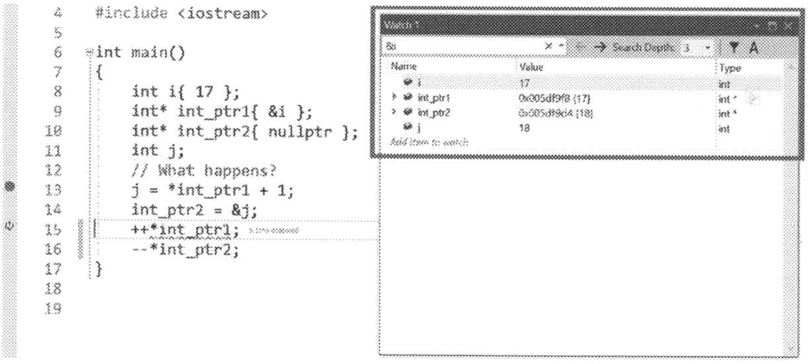

Figure 5-3. *Pointer illustration: step 2*

Here, you see the address of the variable j is stored in the pointer variable int_ptr2.

If you are following along and the changes make sense, you are seeing and comprehending how pointers work. If you are not entirely certain, take a few moments and work through the figures and code again until you are convinced it is working correctly. If you are in this frame of mind, do not be disappointed (or frustrated) as learning how pointers work takes some time to get your mind around. These figures are designed to help you understand how they work. Just make sure you see how the values are changing before proceeding.

Tip   There are entire books written about pointers! If you want a more in-depth look at pointers or want to dive into the details of how to use pointers, see the book *Understanding and Using C Pointers* by Richard Reese (O'Reilly, 2013). The book was written for C programmers, and although some of the data is outdated, it is an excellent study on pointers.

Now let's see what happens when you execute the next statement. Figure 5-4 shows the results in our fictional memory space.

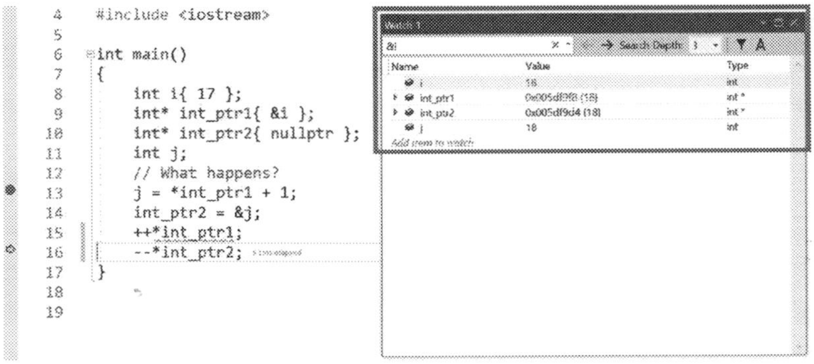

Figure 5-4.   *Pointer illustration: step 3*

Here, you see the statement executed increments the value that int_ptr1 points to by 1 (the ++ operator). In the drawing, you see the compiler dereferences int_ptr1 and added 1 to it. Thus, the value 17 (0x11) becomes 18 (0x12).

Tip   There is one more pointer-related concept that you will encounter—the hat or caret symbol (^). Visual C++ as a special pointer handler that automatically destroys the allocated memory when it is no longer in use uses this symbol. For more information about the caret symbol and objects, see `https://docs.microsoft.com/en-us/cpp/extensions/handle-to-object-operator-hat-cpp-component-extensions?redirectedfrom=MSDN&view=vs-2019`.

Now let's see what happens when you execute the last statement. Figure 5-5 shows the results in our fictional memory space.

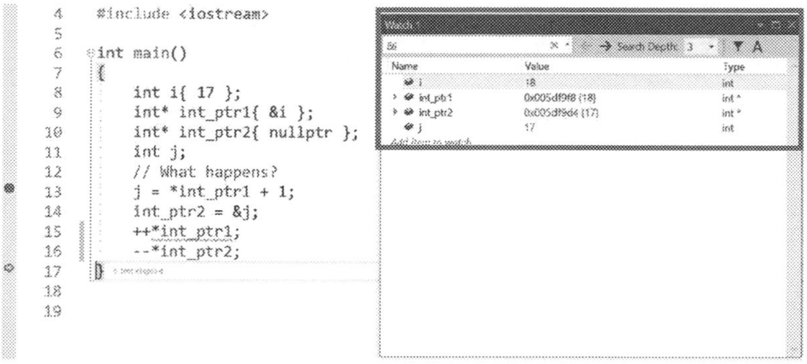

Figure 5-5. *Pointer illustration: step 4*

Here, you see a very similar operation takes place as the last statement, only this time you decrement the value. You see the statement executed decrements the value that int_ptr2 points to by 1 (the -- operator). In the figure, you see the compiler dereferences int_ptr2 (which contains the memory address of j) and subtracted 1 from it. Thus, the value 18 (0x12) becomes 17 (0x11).

Wow! That was a wild ride, wasn't it? I hope that this short crash course in C++ has explained enough about the sample programs shown so far that you now know how they work. This crash course also forms the basis for understanding the other C++ examples in this book.

OK, now it's time to see some of these fundamental elements of C++ in action. Let's look at the blink an LED application you saw in Chapter 3, only this time you're going to write it for Windows 10 IoT Core!

Blink an LED, C++ Style

OK, let's write some C++ code! This project is the same concept as the project from Chapter 3 where you used Python to blink an LED on your Raspberry Pi. Rather than simply duplicate that project, you'll mix it up a bit and make this example a headed application (recall that a headed application has a user interface). The user interface presents the user with a greeting, a symbol that changes color in time with the LED, and a button to start and stop the blink timer.

Rather than build the entire application at once by presenting you a bunch of code, you will walk through this example in two phases. The first phase builds the basic user interface. The second phase adds the code for the GPIO. By using this approach, you can test the user interface on your PC, which is really convenient.

Recall that the PC does not support the GPIO libraries (there is no GPIO!), so if you built the entire application, you would have to test it on the device, which can be problematic if there are serious logic errors in your code. This way, you can ensure that the user interface is working correctly and therefore eliminate any possible issues in that code before you deploy it.

Before you get into the code for the user interface, let's see what components you will use and then set up the hardware.

Required Components

The following lists the components that you need. All of these are available in the Microsoft Internet of Things Pack for the Raspberry Pi from Adafruit. If you do not have that kit, you can find these components separately on the Adafruit website (`www.adafruit.com`), from SparkFun (`www.sparkfun.com`), or any electronics store that carries electronic components.

- 560 ohm 5% 1/4W resistor (green, blue, brown stripes[8])

- Diffused 10mm red LED (or similar)

- Breadboard (mini, half, or full sized)

- (2) male-to-female jumper wires

You may notice that this is the same set of components you used in Chapter 3.

Set Up the Hardware

Begin by placing the breadboard next to your Raspberry Pi. Power off the Raspberry Pi, orienting it with the label facing you (GPIO pins in the upper left). Next, take one of the jumper wires and connect the female connector to pin 6 on the GPIO. The pins are numbered left to right starting with the lower-left pin. Thus, the left two pins are 1 and 2 with pin 1 below pin 2. Connect the other wire to pin 7 on the GPIO.

---

[8]`https://en.wikipedia.org/wiki/Electronic_color_code`

Tip   The only component that is polarized is the LED. The longer side is the positive side.

Next, plug the resistor into the breadboard with each pin on one side of the center groove. You can choose whichever area you want on the breadboard. Next, connect the LED so that the long leg is plugged into the same row as the resistor and the other pin on another row. Finally, connect the wire from pin 6 to the same row as the negative side of the LED and the wire from pin 7 to the row with the resistor. Figure 5-6 shows how all of the components are wired together. Be sure to study this drawing and double-check your connections prior to powering on your Raspberry Pi. Once you're satisfied that everything is connected correctly, you're ready to power on the Raspberry Pi and write the code.

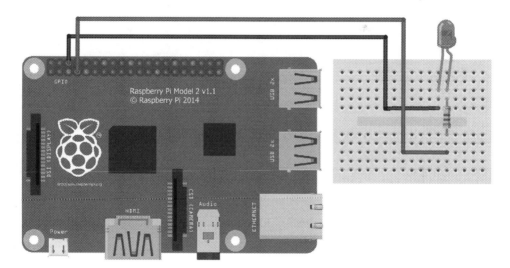

Figure 5-6.   *Wiring the LED to a Raspberry Pi*

Since you are building a headed application, you'll also need a keyboard, mouse, and monitor connected to the Raspberry Pi.

OK, now that you have your hardware set up, it's time to start writing the code.

Write the Code: User Interface

Begin by opening a new project template. Use the C++ *Blank App (Universal Windows)* template. This template creates a new solution with all of the source files and resources you need for a UWP application. Figure 5-7 shows the project template that you need. Use BlinkCPPStyle for the project name. Remember, you may need to select the IoT Core version when you create the project.

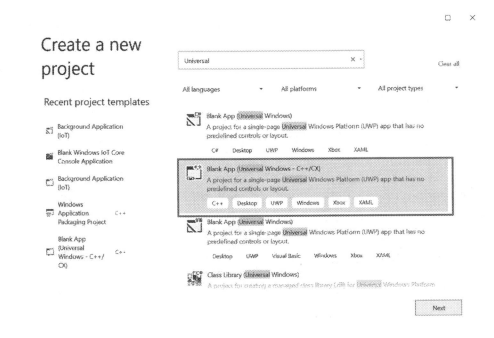

Figure 5-7. *New project dialog: blank application*

A number of files have been created. First, add the XAML code in the MainPage.xaml file. Simply click the file in the project list to display the GUI and code. You may need to resize the code portion to see it all. Listing 5-2 shows the bare XAML code placed in the file by default. I've added a note that shows where to add new code.

Listing 5-2. Bare XAML Code (MainPage.xaml)

```
<Page
    x:Class="BlinkCPPStyle.MainPage"
    xmlns="http://schemas.microsoft.com/winfx/2006/xaml/presentation"
    xmlns:x="http://schemas.microsoft.com/winfx/2006/xaml"
```

```
    xmlns:local="using:BlinkCPPStyle"
    xmlns:d="http://schemas.microsoft.com/expression/blend/2008"

    mc:Ignorable="d"
    Background="{ThemeResource ApplicationPageBackgroundThemeBrush}">

    <Grid>

    </Grid>
</Page>
```

Recall that the XAML file is used to define a user interface in a platform-independent way using an XML-like language. In this project, I demonstrate some of the more basic controls: a text box, a button, and an ellipse (circle) placed inside a special control called a stacked panel. The stacked panel allows you to arrange the controls in a vertical "stack," making it easier to position them. As you can see in the listing, you want to place your XAML user interface items in the <Grid></Grid> section.

In this example, you want a text box at the top and a circle (ellipse) to represent the LED that you use to turn on (change to green) and off (change to gray) to correspond with the hardware on/off code that you will add later. You also need a button to toggle the blink operation on and off. Finally, you'll add another text box to allow you to communicate with the user about the state of the GPIO code (that you'll add later).

Now let's add the code. Since the stacked panel is a container, all of the controls are placed inside it. Listing 5-3 shows the code that you want to add (shown in bold).

Listing 5-3. Adding XAML Code for the User Interface: MainPage.xaml

```
<Page
    x:Class="BlinkCPPStyle.MainPage"
    xmlns="http://schemas.microsoft.com/winfx/2006/xaml/presentation"
    xmlns:x="http://schemas.microsoft.com/winfx/2006/xaml"
    xmlns:local="using:BlinkCPPStyle"
    xmlns:d="http://schemas.microsoft.com/expression/blend/2008"

    mc:Ignorable="d"
    Background="{ThemeResource ApplicationPageBackgroundThemeBrush}">

    <Grid Background="{ThemeResource ApplicationPageBackgroundThemeBrush}">
        <StackPanel Width="400" Height="400">
```

```xml
            <TextBlock x:Name="title" Height="60" TextWrapping="NoWrap"
                    Text="Hello, Blinky C++ Style!" FontSize="28"
                    Foreground="Blue"
                    Margin="10" HorizontalAlignment="Center"/>
            <Ellipse x:Name="led_indicator" Fill="LightGray"
            Stroke="Gray"  Width="75"
                    Height="75" Margin="10" HorizontalAlignment="Center"/>
            <Button x:Name="start_stop_button" Content="Start" Width="75"
            ClickMode="Press"
                    Click="start_stop_button_Click" Height="50"
                    FontSize="24"
                    Margin="10" HorizontalAlignment="Center"/>
            <TextBlock x:Name="status" Height="60" TextWrapping="NoWrap"
                    Text="Status" FontSize="28" Foreground="Blue"
                    Margin="10" HorizontalAlignment="Center"/>
        </StackPanel>
    </Grid>
</Page>
```

Notice the button control. Here, you have an event that you want to associate with the button named `start_stop_button_Click`, which you assigned via the `Click` attribute. That is, when the user clicks it, a method named `start_stop_button_Click()` is called.

XAML provides a great way to define a simple, easy user interface with the XML-like syntax. However, it also provides a mechanism to associate code with the controls. The code is placed in another file called a source-behind file, including a header and source file named `MainPage.xaml.h` and `MainPage.xaml.cpp`. Recall that you place declarations in the header file and the body of the code in the source file.

If you were typing this code in by hand, you notice a nifty feature of Visual Studio—context-sensitive help called IntelliSense that automatically completes the code you're typing and provides drop-down lists of choices. For example, when you type in the button control and then type `Click=`, a drop-down box appears, allowing you to create the event handler (a part of the code that connects to the XML). In fact, it creates the code in the `MainPage.xaml.cpp` (and `MainPage.xaml.h`) file for you. If you copy and pasted the code, you will not get this option and would have to type in the code manually. However, I will show you the code so that you can complete it yourself.

Let's see the code for the button control starting with the header file (`MainPage.xaml.h`). Listing 5-4 shows the code you need to add in bold. Notice that you place everything in the private section because it is only used by the `BlinkCPPStyle` (application) class.

Listing 5-4. Adding the Declarations: MainPage.xaml.h

```
//
// MainPage.xaml.h
// Declaration of the MainPage class.
//

#pragma once

#include "MainPage.g.h"

namespace BlinkCPPStyle
{
    /// <summary>
    /// An empty page that can be used on its own or navigated to within
        a Frame.
    /// </summary>
    public ref class MainPage sealed
    {
    public:
        MainPage();
    private:
        // Add references for color brushes to paint the led_indicator
            control
        Windows::UI::Xaml::Media::SolidColorBrush^ greenFill =
            ref new Windows::UI::Xaml::Media::SolidColorBrush(Windows::UI::
            Colors::Green);
        Windows::UI::Xaml::Media::SolidColorBrush^ grayFill =
            ref new Windows::UI::Xaml::Media::SolidColorBrush(Windows::UI::
            Colors::LightGray);
```

```
    // Add the start and stop button click event header
    void start_stop_button_Click(Platform::Object^ sender,
        Windows::UI::Xaml::RoutedEventArgs^ e);

    // Variables for blinking
    bool blinking{ false };
  };
}
```

OK, there are a few extra bits here that may not be very obvious why they're here. Recall that you want to paint the LED control green and gray for on and off. To do that, you need a reference (the keyword ref is used to create an object of the referenced wrapper or type) to the green and gray brush resources. Thus, I create a new object (using the caret for cleanup) from the Windows user interface colors namespace. This is a common way to express brushes for painting controls (but not the only way).

You also add the header for the button click event—start_stop_button_Click()— as well as a boolean member variable that you use to trigger the LED timer.

Let's see the source code file (MainPage.xaml.cpp) in Listing 5-5 where you use these variables and fill in the code for the event. Again, only the code in bold is new. The project template provided the rest of the code.

Listing 5-5. Adding Code for the Event: MainPage.xaml.cpp

```
//
// MainPage.xaml.cpp
// Implementation of the MainPage class.
//

#include "pch.h"
#include "MainPage.xaml.h"

using namespace BlinkCPPStyle;

using namespace Platform;
using namespace Windows::Foundation;
using namespace Windows::Foundation::Collections;
using namespace Windows::UI::Xaml;
using namespace Windows::UI::Xaml::Controls;
using namespace Windows::UI::Xaml::Controls::Primitives;
```

```
using namespace Windows::UI::Xaml::Data;
using namespace Windows::UI::Xaml::Input;
using namespace Windows::UI::Xaml::Media;
using namespace Windows::UI::Xaml::Navigation;

// The Blank Page item template is documented at https://go.microsoft.com/
fwlink/?LinkId=402352&clcid=0x409

MainPage::MainPage()
{
    InitializeComponent();
}

void BlinkCPPStyle::MainPage::start_stop_button_Click(Platform::Object^
sender,
    Windows::UI::Xaml::RoutedEventArgs^ e)
{
    blinking = !blinking;
    if (blinking) {
        led_indicator->Fill = greenFill;
        start_stop_button->Content = "Stop";
    }
    else {
        led_indicator->Fill = grayFill;
        start_stop_button->Content = "Start";
    }
}
```

Notice that you added code that inverts the blinking variable (toggles between false and true), and depending on the value, you turn the LED indicator control green (meaning the LED is on) or gray (meaning the LED is off). You also change the label of the button to correspond with the operation. That is, if the button is labeled Start, the LED indicator is off, and when clicked, the label changes to Stop and the LED indicator is turned on.

That's it! You've finished the user interface. Go ahead and build the solution, correcting any errors that may appear. Once compiled, you're ready to test it.

Test and Execute: User Interface Only

That was easy, wasn't it? Better still, since this is a universal app, you can run this code on your PC. To do so, choose debug and x86 (or x64 for 64-bit machines) from the platform box and press *Ctrl+F5*. Figure 5-8 shows an excerpt of the output (just the control itself). I used an alternative contrast for easier viewing. The default background may be black when you run it on your PC.

Figure 5-8 shows what happens when you click the button. Cool, eh? Figure 5-9 shows an excerpt of the output when the timer is turned on.

Figure 5-8.   *The user interface: timer off*

Figure 5-9.   *The user interface: timer on*

You may be wondering where the blink part is. Well, you haven't implemented it yet. You will do that in the next phase.

Add the GPIO Code

Now, let's add the code to work with the GPIO header. For this phase, you cannot run the code on your PC because the GPIO header doesn't exist, but you can add code to check the GPIO header status—hence, the extra text box in the interface.

Note   The following is a bit more complicated and requires changes to the header and source files. Thus, I will walk through the code changes one part at a time. Henceforth, for brevity, I present excerpts of the files that you will be editing.

Now, let's add the code you need in the header file (`MainPage.xaml.h`). Listing 5-6 shows the code you need to add in context. The new code is marked in bold. Here, you add a new method to initialize the GPIO, `InitGPIO()`; a new event named `OnTick()`, which you use with the timer object; a reference to the timer object; a variable to store the pin value; and finally a constant set to GPIO 4 (hardware pin #7) and a pointer handler to a pin variable.

Listing 5-6. Adding Code for the GPIO: MainPage.xaml.h

```
namespace BlinkCPPStyle
{
    public ref class MainPage sealed
    {
    public:
        MainPage();
    private:
        // Add references for color brushes to paint the led_indicator
           control
        Windows::UI::Xaml::Media::SolidColorBrush^ greenFill =
            ref new Windows::UI::Xaml::Media::SolidColorBrush(Windows::UI::
            Colors::Green);
        Windows::UI::Xaml::Media::SolidColorBrush^ grayFill =
```

```
        ref new Windows::UI::Xaml::Media::SolidColorBrush(Windows::UI::
        Colors::LightGray);

    // Add the start and stop button click event header
    void start_stop_button_Click(Platform::Object^ sender,
        Windows::UI::Xaml::RoutedEventArgs^ e);

    // Add a constructor for the InitGPIO class
    void InitGPIO();

    // Add an event for the timer object
    void OnTick(Platform::Object^ sender, Platform::Object^ args);

    // Add references for the timer and a variable to store the GPIO
      pin value
    Windows::UI::Xaml::DispatcherTimer^ timer;
    Windows::Devices::Gpio::GpioPinValue pinValue = Windows::Devices::G
    pio::GpioPinValue::High;

    // Variables for blinking
    bool blinking{ false };
    const int LED_PIN = 4;   // physical pin#7
    Windows::Devices::Gpio::GpioPin^ pin;
    };
}
```

Now let's see the code for the new event (in `MainPage.xaml.cpp`). You begin by adding some namespaces, as shown in Listing 5-7. The new ones are in bold. Here, you added namespaces for enumeration, the GPIO, and concurrency (for the timer).

Listing 5-7. Adding Namespaces for the GPIO: MainPage.xaml.cpp

```
using namespace BlinkCPPStyle;

using namespace Platform;
using namespace Windows::Foundation;
using namespace Windows::Foundation::Collections;
using namespace Windows::UI::Xaml;
using namespace Windows::UI::Xaml::Controls;
using namespace Windows::UI::Xaml::Controls::Primitives;
```

```
using namespace Windows::UI::Xaml::Data;
using namespace Windows::UI::Xaml::Input;
using namespace Windows::UI::Xaml::Media;
using namespace Windows::UI::Xaml::Navigation;
using namespace Windows::Devices::Enumeration;  // Add this
using namespace Windows::Devices::Gpio;          // Add this
using namespace concurrency;                     // Add this
```

Next, you need to add some code to the constructor for the MainPage class as follows. I show the lines you add in bold in Listing 5-8. Notice that you have a new syntax, namely, the use of ref new, which creates a new instance of the DispatcherTimer class.

Listing 5-8. Adding DispatcherTimer Code for the GPIO: MainPage.xaml.cpp

```
MainPage::MainPage()
{
    InitializeComponent();
    InitGPIO();
    if (pin != nullptr) {
        timer = ref new DispatcherTimer();
        TimeSpan interval;
        interval.Duration = 500 * 1000 * 20;
        timer->Interval = interval;
        timer->Tick += ref new EventHandler<Object^>(this,
        &MainPage::OnTick);
    }
}
```

You have added a bit of code here to set up the GPIO header calling the new method (you'll add that shortly) and code that checks to see if the pin variable is allocated (has a value); you instantiate the DispatcherTimer class, set an interval, and add the event handler named OnTick.

Next, let's add the new InitGPIO() method. You can place this after the constructor or at the end of the file as shown in Listing 5-9. You use a new syntax, -> (arrow), that dereferences a pointer to access a method or attribute.

Listing 5-9. Adding InitGPIO Code for the GPIO: MainPage.xaml.cpp

```cpp
void MainPage::InitGPIO()
{
    auto gpio = GpioController::GetDefault();
    if (gpio == nullptr) {
        pin = nullptr;
        status->Text = "No GPIO Controller!";
        return;
    }
    pin = gpio->OpenPin(LED_PIN);
    pin->Write(pinValue_);
    pin->SetDriveMode(GpioPinDriveMode::Output);
    status->Text = "You're good to go!";
}
```

OK, there's some stuff going on here. Like the constructor, you check the status of the GPIO, but this time if the GPIO is null (nullptr), you set the text of the status text box with an error. This shows how easy it is to add code to affect the XAML controls. You also see code to open the GPIO pin, set a value (in this case, high or positive voltage), set the mode, and then update the status text box with a success message.

Next, you need to complete the OnTick() method code as shown in Listing 5-10.

Listing 5-10. Adding OnTick Code for the GPIO: MainPage.xaml.cpp

```cpp
void MainPage::OnTick(Object^ sender, Object^ args)
{
    if (pinValue == Windows::Devices::Gpio::GpioPinValue::High) {
        pinValue = Windows::Devices::Gpio::GpioPinValue::Low;
        pin->Write(pinValue_);
        led_indicator->Fill = grayFill;
    }
    else {
        pinValue = Windows::Devices::Gpio::GpioPinValue::High;
        pin->Write(pinValue_);
        led_indicator->Fill = greenFill;
    }
}
```

Here is where the real operation of the code happens. In this code, if the pin is set to high (on), you set it to low (off) and paint the LED control gray. Otherwise, you set the pin to high (on) and paint the LED control green.

Note   You could change this color to match the color of your LED if you wanted. Just remember to change the brush accordingly in the header file.

Finally, you need to add code to the start_stop_button_Click() method to start and stop the timer. Listing 5-11 shows the code you need to add.

Listing 5-11. Adding Calls to the OnTick Code for the GPIO: MainPage.xaml.cpp

```cpp
void BlinkCPPStyle::MainPage::start_stop_button_Click(Platform::Object^
sender,
    Windows::UI::Xaml::RoutedEventArgs^ e)
{
    blinking = !blinking;
    if (blinking) {
        timer->Start();
        led_indicator->Fill = greenFill;
        start_stop_button->Content = "Stop";
    }
    else {
        timer->Stop();
        led_indicator->Fill = grayFill;
        start_stop_button->Content = "Start";
    }
}
```

That's it! Now, let's build the solution and check for errors. You should see something like the list shown in Listing 5-12 in the output window.

Listing 5-12. Build Output Messages (Typical)

```
1>------ Build started: Project: BlinkCPPStyle, Configuration: Debug ARM
------
1>pch.cpp
1>App.xaml.cpp
1>MainPage.xaml.cpp
1>XamlTypeInfo.Impl.g.cpp
1>XamlTypeInfo.g.cpp
1>BlinkCPPStyle.vcxproj -> C:\Users\olias\source\repos\BlinkCPPStyle\ARM\
Debug\BlinkCPPStyle\BlinkCPPStyle.exe
========== Build: 1 succeeded, 0 failed, 0 up-to-date, 0 skipped ==========
```

Before we deploy our application, let's review one critical aspect of deploying C++ UWP applications; they require using the remote debugger to deploy applications. Yes, unlike the C# application we saw previously, deploying C++ application requires a bit more work. But don't worry; once you understand this, you can avoid all manner of frustration trying to use the deploy option on the build menu.

OK, now you're ready to deploy the application to your device. Go ahead—set up everything and power on your device.

Deploy and Execute: Completed Application

Once your code compiles, you're ready to deploy the application to your Raspberry Pi (or another device). Recall from Chapter 4, you have to set up the debug settings to specify the IP address of your Raspberry Pi. Fortunately, unlike the console application, you only have to change two items as indicated in Figure 5-10. Remember to choose ARM for the platform. Click *Apply* and then *OK* to close the dialog.

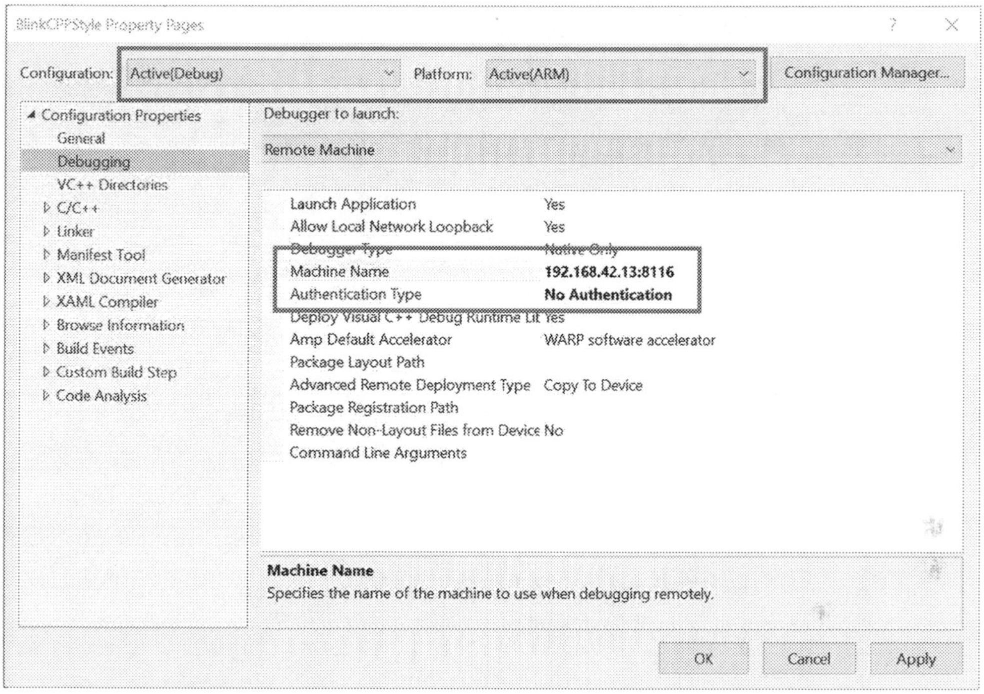

Figure 5-10. *Setting debug settings for deployment*

Once you have these set, you can power on your Raspberry Pi, and once it is booted, go to the Device Portal, and turn on the remote debugger, as shown in Figure 5-11.

Figure 5-11. *Turning on the remote debugger*

Be sure to note the IP address and port reported by the remote debugger as shown in Figure 5-12.

Start Visual Studio Remote Debugger

Remote debugger is running on the device. Please use 192.168.42.13:8116 as remote machine name from Visual Studio 2017 to connect to the device.

Stop Remote Debugger

Figure 5-12. *IP address and port from the remote debugger*

To deploy this application, we have to do some trickery as mentioned because this project template (C++ UWP) does not contain the wizardry we need for deployment. But that's OK because we can use the remote debugger.

There are two options for using the remote debugger. You can use the *Start Debugging* or the *Start Without Debugging* options from the *Debug* menu. Both will deploy your application to your device, but the first option will enter an interactive, remote debugging session. If you just want to deploy the application and run it, use the *Start Without Debugging* option.

Go ahead and do that now. Use the *Debug* ➤ *Start Without Debugging* option. When complete, you'll get messages from the output window (choose the *Deployment* option) similar to those shown in Listing 5-13.

Listing 5-13. Deploy Output Messages (Typical)

```
Starting remote deployment...
Deploy: START
DeployAsync: START
Reading package recipe file "C:\Users\olias\source\repos\BlinkCPPStyle\ARM\
Debug\BlinkCPPStyle\BlinkCPPStyle.build.appxrecipe"...
    Target.MachineClientId = af5015c2-68a1-43dc-8681-1fd18f9eec04
...
Registering the application to run from layout...
    Target.MachineClientId = af5015c2-68a1-43dc-8681-1fd18f9eec04
...
Deployment complete (0:00:09.945). Full package name:
"BlinkCPPStyle_1.1.0.0_arm__mvq1akfegk7g6"
    Target.DeviceFamily = Windows.IoT
    DeploymentSucceeded = True
```

```
TimeToDeploy = 9945
HasSharedCode = False
Target.Id = 1024
ProjectGuid = {27ac5400-179b-46b2-98b0-539bad41fb1e}
Project.TargetPlatformVersion = 10.0.18362.0
Project.TargetPlatformMinVersion = 10.0.17763.0
```
DeployAsync: END (Success, 0:00:10.331)
Deploy: END (Success, 0:00:10.332)

After a few moments, you will see the application start on your device. If you have all of the wiring complete, you can click the button and observe the LED blinking until you stop it. Cool, eh?

If the LED is not blinking, double-check your wiring and ensure that you chose pin 4 in the code and pin 7 on the pin header (recall that the pin is named GPIO 4, but it is pin #7 on the header for the Raspberry Pi).

Only...how do you stop it? Go to the Device Portal, click *Apps*, then *Apps manager*, and then click the *Actions* drop-down box for the application and click *Stop*. Figure 5-13 shows the settings.

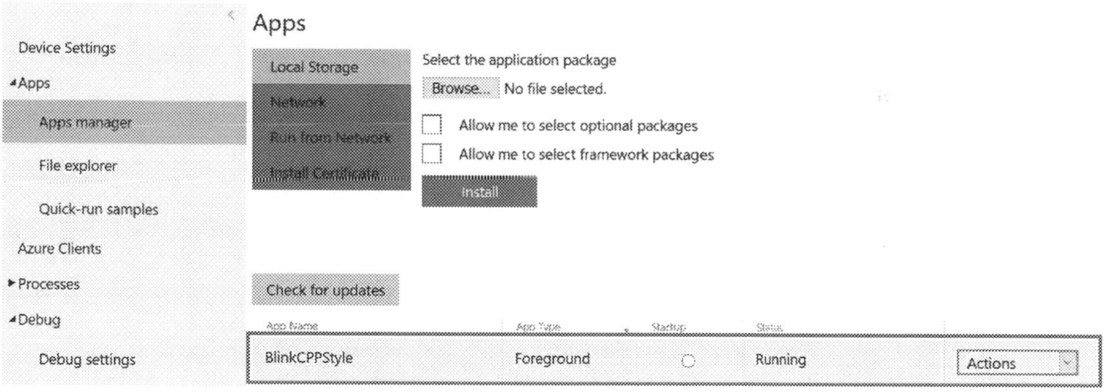

Figure 5-13. *Stopping an application on the device (Device Portal)*

Note   If the app deployed successfully but doesn't show in the drop-down list, try disconnecting and reconnecting. If that doesn't work, try rebooting your device. And if that doesn't work, make sure the remote debugger is running.

If you want to start the application again, you do not need to redeploy it. The application is installed on the device. To start it, go back to the Device Portal, click *Apps*, then *Apps manager*, and then click the *Actions* drop-down box for the application and click *Start*.

Should you want to uninstall the application, go back to the Device Portal, click *Apps*, then *Apps manager*, and then click the *Actions* drop-down box for the application and click *Uninstall*.

If you want to debug your application, you can by setting a breakpoint in the code and then use the *Debug* ➤ *Start Debugging* menu, which will deploy the application, start it, and connect your Visual Studio debugger. It's a nice way to debug application. Go ahead and try it out yourself!

C++ Application Deployment Troubleshooting

If you're like me, things sometimes go wonky and just don't work, or they present you with an interesting but nearly indecipherable error message. I present a couple of these incidents you may encounter along with actions you can take to prevent or correct them.

Application Already/Not Running

If you get an error dialog saying the application is already or not running, you can ignore (close) it. This can happen if you've deployed the application numerous times or have previous versions installed.

Deployment Errors ... Continue?

You may see an error dialog informing you of one or more errors during deployment. Most times, you can click *Yes* to continue with the deployment. This can happen if you're overwriting (redeploying) an application, you've changed the application name but not the package name, and so on. Take a look at the deployment errors in the output window once it finishes to determine if the error is something that you need to fix. Most times, it will be a warning or inconsequential. The acid test is whether the application starts on your device.

Missing Framework

If you get an error during deployment that the VCLibs.ARM... is missing, your deployment will fail, and no amount of fiddling will correct this. Worse, the error message isn't very helpful as shown in the following:

```
DEP0800: The required framework "C:\Program Files (x86)\Microsoft SDKs\
Windows Kits\10\ExtensionSDKs\Microsoft.VCLibs\14.0\.\AppX\Debug\ARM\
Microsoft.VCLibs.ARM.Debug.14.00.appx" failed to install. [0x80073CF9] AppX
Deployment operation failed with error 0
```

What this is trying to tell you is that your application package is missing the Visual C++ library and that redistributable library is not already on the device. Fortunately, we can manually install this on our device by copying the file to our device and using the Windows PowerShell to install it.

First, locate the missing file, which should have the name shown in the error message. On my PC, the file was installed in C:\Program Files (x86)\Microsoft SDKs\Windows Kits\10\ExtensionSDKs\Microsoft.VCLibs\14.0\Appx\Debug\ARM. To upload it to your device, use the Device Portal and click *Apps*, then *File manager*. Double-click the *Documents* entry in the list and then the *Choose File* button to locate the file. Once selected, click the *Upload* button.

Next, open the PowerShell via the dashboard on *My Devices* by right-clicking the device and choosing *Open in PowerShell*. You will be asked to log in.

Once logged in, use the Add-AppxPackage to install the missing framework as shown in the following:

```
[192.168.42.13]: PS C:\Data\Users\administrator\Documents> Add-AppxPackage
C:\Data\Users\DefaultAccount\Documents\Microsoft.VCLibs.arm.Debug.14.00.appx
```

That's it! Now, you can retry your deployment.

Failure to Unregister Application

If you have installed the application many times and attempted to uninstall it, or it crashed and became corrupt, or you accidentally (maybe intentionally) removed the application files, you can see an error message during deployment like the following:

```
DEP0900: Failed to unregister application, login to the device and remove
the package by name found in C:\Data\Users\Administrator\AppData\Local\
Packages>
```

What this means is either the application is still registered on the device or some of its files are present, but it is not registered. To fix this, open an SSH terminal to the device and remove the application directory and all of its files like shown as follows. In this example, I had a derelict package on my device left over from a previous session,[9] and it caused my deployment of a newer application with the same package name to fail. In this case, the package was named BlinkCPPStyle_mvq1akfegk7.

```
[192.168.42.13]: PS C:\Data\Users\Administrator\AppData\Local\Packages>
rmdir /s BlinkCPPStyle_mvq1akfegk7g6
BlinkCPPStyle_mvq1akfegk7g6, Are you sure (Y/N)? Y.
```

Once you've removed the folder, you can deploy your application.

Tip   As a challenge, you can modify this project to add a *Close* or an *Exit* button to stop the application.

Summary

If you are learning how to work with Windows 10 IoT Core and don't know how to program with C++, learning C++ can be a daunting challenge. While there are many examples on the Internet you can use, very few are documented in such a way as to provide enough information for someone new to C++ to understand or much less get started or even compile and deploy the sample!

This chapter has provided a crash course in Visual C++ that covers the basics of the things you encounter when examining most of the smaller example projects. You discovered the basic syntax and constructs of a Visual C++ application, including a walk-through of building a real C++ application that blinks an LED. Through the course of that example, you learned a little about XAML, including how to wire events to controls, and even a little about how to use the dispatcher timer.

In the next chapter, you discover another programming language called C#. You implement the same example project you saw in this chapter, so if you want to see how to do it in C#—read on!

---

[9]From the first edition of this book.

CHAPTER 6

Windows 10 IoT Development with C#

Now that you have a basic understanding of how to use Visual Studio 2019, you can learn more about some of the languages you may encounter when developing your IoT solutions. One of those languages is C# (pronounced "see sharp"[1])—a very robust and powerful object-oriented language that you can use to write managed Windows .NET and UWP applications. Mastering C# is not a trivial task, but it is not quite as challenging as other programming languages.

WHAT IS .NET?

In short, the .NET Framework is a huge library designed to provide a layer above the operating system for building Windows applications. The .NET Framework supports a number of languages, including C#, as well as a number of platforms. As you will see when you deploy your IoT application to your device, Visual Studio includes a subset of the framework for use on Windows 10 IoT Core. Some of the classes within the namespaces that we will use derive from the .NET Framework.

If you are used to using C++ or Java, you may find C# to be familiar, but more verbose than C++ and, to some extent, Java. That is, it may seem like you're typing a lot more or adding more lines of code. The libraries we will use have more verbose naming conventions (names or identifiers are longer), and we will use a few more lines of code in the process. However, as you will see, C# source code reads easier than some other

---

[1]Not "see-hash" or, worse, "see-hashtag"—both of which may show you're uninformed, so don't do that.

© Charles Bell 2021
C. Bell, *Windows 10 for the Internet of Things*, https://doi.org/10.1007/978-1-4842-6609-0_6

languages, making it easier to understand and modify in the future. It also helps when debugging your code. You don't have to guess what a library class and method may do because the name is more descriptive (in general).

Some find learning C# easier than other programming languages because if you are familiar with C++ or especially Java, some of what you will learn is similar. Essentially, C# is an improvement on languages like C++ and Java. In fact, you may only need a little knowledge of the fundamentals of the language and how to use it in Visual Studio to become proficient in creating Windows 10 IoT Core applications. Moreover, C# is featured in more examples and documentation from Microsoft that it is fairly safe to say it is the premier language for IoT applications.

This chapter presents a crash course on the basics of C# programming in Visual Studio including an explanation about some of the most commonly used language features. As such, this chapter provides you with the skills you need to understand the growing number of IoT project examples available on the Internet. The chapter concludes with a walk-through of a C# example project that shows you how to interact with hardware. Specifically, you will implement the LED project you saw in Chapter 3. Only this time, you'll be writing it as a C# Windows 10 IoT Core application. So, let's get started!

Tip   If you are not interested in using C# in your IoT solutions, or you already know the basics of C# programming, feel free to skim through this chapter. I recommend working through the example project at the end of the chapter, especially if you've not written IoT applications.

Getting Started

The C# language has been around since the first introduction of the .NET Framework (pronounced "dot net"). In fact, C# was specifically designed to be the object-oriented programming language of choice for writing .NET applications. C# was released in 2000 with the release of the .NET Framework. The latest version of C# is version 8.0 and is sometimes called Visual C# (but Microsoft seems to prefer C#).

You may be thinking that C# and .NET may restrict the types of applications you can write, but that also is not true. You can use C# to write a host of applications from

Windows 10 IoT Core to desktop to web applications and beyond. As you will see, you can also write C# console applications like you did in Chapter 3 with C++.

C# was heavily influenced by Java and C++. Indeed, if you have programmed in Java or C++, C# will seem familiar to you. What sets C# apart is it is a purely object-oriented language. That is, you must write all of your programs as an object using a class. In fact, the easy-to-understand class syntax[2] makes developing your object-oriented programming easier. This gives C# a powerful advantage over C++ and other languages that have more complex syntax where objects are largely optional.

While some may be tempted to think C# is another flavor of C++, there are some serious differences. The most important is the use of an execution manager to keep the C# applications in a protected area (called a managed application). This protected area ensures that the C# code has access only to portions that the execution manager permits—those defined in the code. One of the great side effects of this arrangement is automatic garbage collection—freeing of allocated memory that is no longer used (or in scope). While you can choose to make a C# application an unmanaged application, the practice is discouraged (and largely unnecessary for the vast majority of use cases).

Tip   For more information about the differences between C++ and C#, see
`www.differencebetween.info/difference-between-cplusplus-and-csharp`.

Another very important difference between C# and C++ is how they are made into an *executable* (compiled). C++ is compiled and linked to form an executable program with object code native to the platform, but C# is compiled in two steps: the first is a platform-independent intermediate object code called the *common intermediate language* (CIL) and the second when that is converted to native object code by the *just-in-time* compiler (JIT). The JIT compiler operates in the background and, as the name implies, prepares the code for execution when needed. In fact, you cannot tell the JIT compiler is running.

When a CIL is built, the compiler includes all of the references needed to execute the application. This may include framework files, dynamic libraries, and metadata needed by the system. This is called an *assembly*. You need not think of this other than an executable, but technically it is a small repository (hence, assembly).

---

[2]As opposed to other languages like C++ with differing notation based on how an object is instantiated.

This mechanism allows C# applications to execute within a cordoned off area of memory (called a *managed application*) that the .NET *common language runtime* (CLR) can monitor and protect other applications from harm. A side effect allows C# applications to have garbage collection (by the CLR) freeing memory automatically based on scope and use. Figure 6-1 shows a pictorial example of the way C# applications are compiled and executed in phases.

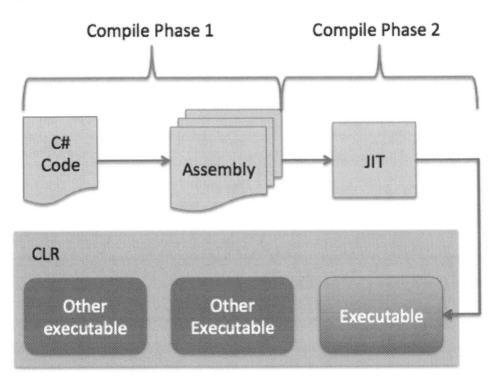

Figure 6-1. *How C# applications are compiled and executed*

Here, you see the two phases of compilation. Phase 1 occurs when you compile the application in Visual Studio. Phase 2 occurs when you execute the application, placing the executable in the CLR for execution. Notice that I depict other applications running in the same CLR, each protected from the others by the CLR's managed features.

Note   While C# is technically compiled in two phases, the rest of the chapter focuses on compilation as executed from Visual Studio.

Should you require more in-depth knowledge of C#, there are a number of excellent books on the topic. Here is a list of a few of my favorites. While some are a little dated (they cover Visual Studio 2015 or 2017), they provide excellent resources for the language:

- *The C# Player's Guide* by R. B. Whitaker (Starbound Software, 2015)

- *Microsoft Visual C# Step by Step* (Developer Reference) 9th Edition by John Sharp (Microsoft Press, 2018)

- *Programming C# 8.0: Build Cloud, Web, and Desktop Applications* by Ian Griffiths (O'Reilly, 2020)

For those already familiar with C# (and those that will be by the time they finish this chapter) and anyone who wants to hone their C# skills, you should check out *Getting Started with Advanced C#* by Sarcar, Vaskaran (Apress, 2020).

Another excellent resource is Microsoft's documentation on MSDN. The following are some excellent resources for learning C#:

- Getting Started with C# (`https://docs.microsoft.com/en-us/dotnet/csharp/getting-started/`)

- C# Programming Guide (`https://docs.microsoft.com/en-us/dotnet/csharp/programming-guide/`)

Now that you know some of the origins and unique features of C# and the .NET CLR, let's learn about the syntax and basic language features for creating applications.

C# Crash Course

Now let's learn some of the basic concepts of C# programming. Let's begin with the building blocks of the language, such as classes, methods, variables, and basic control structures, and then move into the more complex concepts of data structures and libraries.

While the material may seem to come at you in a rush (hence the crash part), this crash course on C# covers only the most fundamental knowledge of the language and how to use it in Visual Studio. It is intended to get you started writing C# Windows 10 IoT Core applications. If you find you want to write more complex applications than the examples in this book, I encourage you to acquire one or more of the resources listed earlier to learn more about the intriguing power of C# programming.

C# Fundamentals

There are a number of basic concepts about the C# programming language that you need to know in order to get started. In this section, I describe some of the fundamental concepts used in C#, including how the code is organized, how libraries are used, namespaces, and how to document your code.

C# is a case-sensitive language, so you must take care when typing the names of methods or classes in libraries. Fortunately, Visual Studio's IntelliSense feature recognizes mistyped case letters, which allows you to choose the correct spelling from a drop-down list as you write your code. Once you get used to this feature, it is very hard to live without it.

Namespaces

The first thing you may notice is that C# is an object-oriented language and that every program you write is written as a class. Applications are implemented with a namespace that has the same name. A namespace is a special organizational feature that allows you to group identifiers (names of variables, constants, etc.) under a group that is localized to the namespace. Using the namespace tells the compiler to look in the namespace for any identifier you've used in your code that is not found.

You can also create namespaces yourself, as you see in the upcoming example source code. Namespaces may contain any number of classes and may extend to other source files. That is, you can define a namespace so that it spans several source files.

Tip   Source files in C# have a file extension of `.cs`.

Classes

The next thing you may notice is a class definition. A class is more than a simple data structure. You use classes to model concepts that include data and operations on the data. A class can contain private and public definitions (called *members*) and any number of operations (called *methods*) that operate on the data and give the class meaning.

You can use classes to break your programs down into more manageable chunks. That is, you can place a class you've implemented in its own `.cs` file and refer to it in any of the code provided there aren't namespace issues, and even then you simply use the namespace you want.

Let's look at a simple class named Vector implemented in C#. This class manages a list of double variables hiding the data from the caller while providing rudimentary operations for using the class. Listing 6-1 shows how such a class could be constructed.

Listing 6-1. Vector Class in C#

```
class Vector
{
    private double[] elem;
    private int sz;

    public Vector(int s)
    {
        elem = new double[s];
        sz = s;
    }

    ~Vector() { /* destructor body */ }

    public int size() { return sz; }

    public double this[int i]
    {
        get { return elem[i]; }
        set { elem[i] = value; }
    }
}
```

This is a basic class that is declared with the keyword class followed by a name. The name can be any identifier you want to use, but the convention is to use an initial capital letter for the name. All classes are defined with a set of curly braces that define the body or structure of the class.

By convention, you list the member variables (also called *attributes*) indented from the outer curly braces. In this example, you see two member variables that are declared as private. You make them private to hide information from the caller. That is, only member methods inside the class itself can access and modify private member variables. Note that derivatives (classes built from other classes—sometimes called a *child class*) can also access protected member variables.

Next, you see a special method that has the same name as the class. This is called the *constructor*. The constructor is a method that is called when the class is instantiated (used). You call the code that defines the attributes and methods a class, and when executed, you call the resulting object an instance of the class.

In this case, the constructor takes a single integer parameter that is used to define the size of the private member variable that stores the array of double values. Notice how this code is used to dynamically define that array. More specifically, you use the new command to allocate memory for the array.

Following the constructor is another special method called the *destructor*. This method is called when the object is destroyed. Thus, you can place any cleanup code that you want to occur when the object is destroyed. For example, you can add code to close files or remove temporary storage. However, since C# runs on .NET with a garbage collector (a special feature that automatically frees allocated memory when no longer in scope), you do not have to worry about freeing (deleting) any memory you've allocated.

Next are two public methods, which users (or callers of the instance) can call. The first method, size(), looks as you would expect and in this case simply returns the value of the private member variable sz. The next method is a special form of method called an operator (or get/set) method. Notice that there are two sections or cases: get and set.

The method allows you to use the class instance (object) as if it were an array. The method is best understood by way of an example. The following shows how this operator method is used. Notice the lines in bold:

```
Vector v = new Vector(10);
for (int i=0; i < v.size(); ++i)
{
    v[i] = (i * 3);
}
Console.Write("Values of v: ");
for (int j=0; j < v.size(); ++j)
{
    Console.Write(v[j]);
    Console.Write(" ");
}
Console.WriteLine();
```

The code instantiates an instance of the Vector class requesting storage for ten double values. Next, you iterate over the values in the instance using the operator to set the value of each of the ten elements. This results in the set portion of the operator method executing. Next, the code iterates over the values again, this time requesting (getting) the value for each. This results in the get portion of the operator method executing. Neat, eh?

Note   Some programming language books on C# use the term *function* and *method* interchangeably, while other books make a distinction between the two.[3] However, Microsoft uses the term *method*.

I built this example as a single code file, but had I wanted to use modularization, I would have placed the code for the Vector class in its own source (.cs) file. The name of the source file is not required to be the same as the class it contains. Indeed, a source file may contain multiple classes. Still, you would likely choose a meaningful name.

To add a new source file to a Visual Studio C# solution, simply right-click the project name in Solution Explorer, and then choose *Add* ➤ *Add new item* and choose *C# Class* in the tree. At the bottom of the dialog, you can name the file. When ready, click the *Add* button. You can then create any classes you want (or move classes) in the file. You can also use the same namespace to keep all of your classes in the same namespace. However, if you create a new namespace, you must use the using command to use the new namespace. Figure 6-2 shows an example of adding a new class to an existing C# solution.

---

[3]A function returns a value, whereas a method does not.

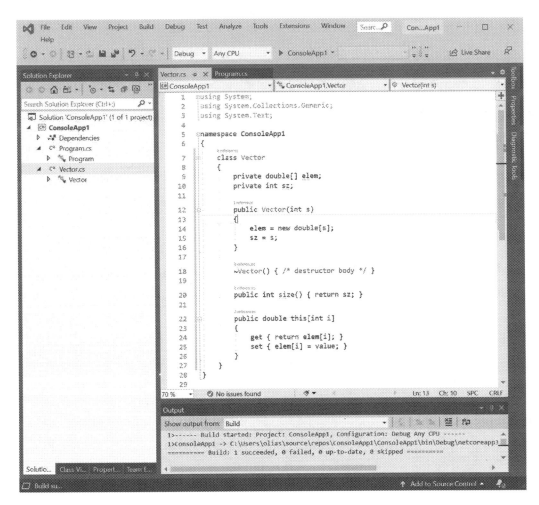

Figure 6-2. *C# solution with a new class*

As you may have surmised, classes are the building blocks for object-oriented programming, and as you learn more about using classes, you can build complex libraries of your own.

Tip   Visual Studio provides a tool called the Class View window that you can use to explore the libraries and classes used in your application.

Curly Braces

Notice that both methods are implemented with a pair of curly braces {} that define the body of the method. Curly braces in C# are used to define a block of code or simply to express a grouping of code. Curly braces are used to define the body of methods, structures, classes, and more. Notice that they are used everywhere, even in the conditional statements (see the if statements).

Tip   Some C# programmers prefer to place the starting curly brace on the same line as the line of code to which it belongs like I did in the example. However, others prefer the open curly brace placed on the next line. You should choose the style you like best.

Comments

One of the most fundamental concepts in any programming language is the ability to annotate your source code with text that not only allows you to make notes among the lines of code but also forms a way to document your source code.[4]

To add comments to your source code, use two slashes, // (no spaces between the slashes). Place them at the start of the line to create a comment for that line, repeating the slashes for each subsequent line. This creates what is known as a *block comment*, as shown. Notice that I used a comment without any text to create whitespace. This helps with readability and is a common practice for block comments.

```
//
// Windows 10 for the IoT Second Edition
//
// Example C# console application rewrite.
//
// Created by Dr. Charles Bell
//
```

---

[4]If you ever hear someone claim, "My code is self-documenting," be cautious when using their code. There is no such thing. Sure, plenty of good programmers can write code that is easy to understand (read), but all fall short of that lofty claim.

You can also use the double slash to add a comment at the end of a line of code. That is, the compiler ignores whatever is written after the double slash to the end of the line. You see an example of this next. Notice that I used the comment symbol (double slash) to comment out a section of code. This can be really handy when testing and debugging, but generally discouraged for final code. That is, don't leave any commented out code in your deliverable (completed) source code. If it's commented out, it's not needed!

```
if (size < max_size) {
  size++;  /* increment the size */
} //else {
  //  return -1;
//}
```

Notice that you also see the use of /* */, which is an alternative C-like mechanism for writing comments. Anything that is included between the symbols becomes a comment and can include multiple lines of code. However, convention seems to favor the // symbol but as you can see you can mix and match however you like. I recommend choosing one or the other with consistency over variety.

Writing good comments and indeed documenting your code well is a bit of an art form, one that I encourage you to practice regularly. Since it is an art rather than a science, keep in mind that your comments should be written to teach others what your code does or is intended to do. As such, you should use comments to describe any preconditions (or constraints) of using the code, limitations of use, errors handled, and a description of how the parameters are used and what data is altered or returned from the code (should it be a method or class member).

How C# Programs Are Structured

Now let's look at how C# programs are structured by examining a slightly different version of the temperature application you saw in Chapter 4. Listing 6-2 shows the code rewritten for C#.

Listing 6-2. Temperature Code Example Rewrite

```
//
// Windows 10 for the IoT Second Edition
//
// Example C# console application rewrite
```

```csharp
//
// Created by Dr. Charles Bell
//
using System;
using System.Collections.Generic;
using System.Linq;
using System.Text;
using System.Threading.Tasks;

namespace temperature_csharp
{
    class Program
    {
        static double convert_temp(char scale, double base_temp)
        {
            if ((scale == 'c') || (scale == 'C'))
            {
                return ((9.0/5.0) * base_temp) + 32.0;
            }
            else
            {
                return (5.0 / 9.0) * (base_temp - 32.0);
            }
        }

        static void Main(string[] args)
        {
            double temp_read = 0.0;
            char scale = 'c';

            Console.WriteLine("Welcome to the temperature conversion
            application.");
            Console.Write("Please choose a starting scale (F) or (C): ");
            scale = Console.ReadKey().KeyChar;
            Console.WriteLine();
            Console.Write("Please enter a temperature: ");
            temp_read = Convert.ToDouble(Console.ReadLine());
```

```csharp
            if ((scale == 'c') || (scale == 'C'))
            {
                Console.WriteLine("Converting value from Celsius to
                Fahrenheit.");
                Console.Write(temp_read);
                Console.Write(" degrees Celsius = ");
                Console.Write(convert_temp(scale, temp_read));
                Console.WriteLine(" degrees Fahrenheit.");
            }
            else if ((scale == 'f') || (scale == 'F'))
            {
                Console.WriteLine("Converting value from Fahrenheit to
                Celsius.");
                Console.Write(temp_read);
                Console.Write(" degrees Fahrenheit = ");
                Console.Write(convert_temp(scale, temp_read));
                Console.WriteLine(" degrees Celsius.");
            }
            else
            {
                Console.Write("ERROR: I'm sorry, I don't understand '");
                Console.Write(scale);
                Console.WriteLine("'.");
            }
        }
    }
}
```

In the example, the only methods created are convert_temp() and Main(), but this is because you are implementing a very simple solution. Had you wanted to model (create a separate class for) temperature, you would still have the one class named Program with the MainPage() method (which is the starting method for the application), but would have added a new class named Temperature, which would contain its own methods for working with temperature. Indeed, this is how one should think when writing C# code—model each distinct concept as a class. Here, you see the sample application named temperature_csharp was implemented with a class with the name Program.

Wow, that's quite a change from the code in the last chapter! While the functionality is exactly the same, the code looks very different from the C++ version. The following describe the C# concepts I have implemented in this example.

The using Keyword

First, you notice a number of lines that begin with using. These are preprocessor directives that tell the compiler you want to "use" a class or a class hierarchy that exists in a particular namespace. The using directive tells the compiler that you are using the namespace System.

```
using System;
```

In the other lines, I have included additional namespaces with multiple names separated by a period. This is how you tell the compiler to use a specific namespace located in libraries of classes often form hierarchies that you can chain together. For example, if you wanted to use the namespace inside the Windows Foundations library named Collections, you would refer to it as follows:

```
using System.Threading.Tasks;
```

This is a very common occurrence in Windows C# applications. In fact, you will use several namespaces in our example project. The following is an example of using the Tasks namespace located in the Threading subclass namespace of the System namespace.

The Main() Method

The MainPage() method is the starting or initial execution for the C# console project. Here, you see the name is preceded by the keyword static (which means its value will not change and indeed cannot change during runtime) followed by a type (in this case void). This tells the C# compiler that this method will not return any value (but you can make methods that return a value).

```
static void Main(string[] args)
```

Next, you see the name, main, followed by a list of parameters enclosed in parentheses. For the MainPage() method, the parameters are fixed and are used to store any command-line arguments provided by the user. In this case, you have the arguments stored in args, which is an array of strings.

A method in C# is used as an organizational mechanism to group functionality and make your programs easier to maintain (methods with hundreds of lines of code are very difficult to maintain), improve comprehensibility, and localize specialized operations in a single location, thereby reducing duplication.

Methods therefore are used in your code to express the concepts of the functionality they provide. Notice how I used the `convert_temp()` method. Here, I declared it as a method that returned a double and takes a character and a `double` as input. As you can see, the body of the method (defined inside the curly braces) uses the character as the scale in the same way as you do in main and uses the double parameter as the target (or base) temperature to convert. Since I made the parameters generic, I can use only one variable.

Tip   Method parameters and values passed must match on type and order when called.

Notice also that I placed it in the line of code that prints the value to the screen. This is a very common practice in C# (and other programming languages). That is, you use the method to perform some operation, and rather than store the result in a variable, you use it directly in the statements (code).

Variables and Types

No program would be very interesting if you did not use variables to store values for calculations. As you saw earlier, variables are declared with a type and once defined with a specific type cannot be changed. Since C# is strongly typed, the compiler ensures that anywhere you use the variable, it obeys its type, for example, that the operation on the variable is valid for the type. Thus, every variable must have a type assigned.

There are a number of simple types that the C# language supports. They are the basic building blocks for more complex types. Each type consumes a small segment of memory which defines not only how much space you have to store a value but also the range of values possible.[5]

---

[5]For a complete list, see `https://docs.microsoft.com/en-us/cpp/cpp/data-type-ranges?redirectedfrom=MSDN&view=vs-2019`.

For example, an integer consumes 4 bytes, and you can store values in the range –2,147,483,648 to 2,147,483,647. In this case, the integer variable is signed (the highest bit is used to indicate positive or negative values). An unsigned integer can store values in the range 0 to 4,294,967,295.

You can declare a variable by specifying its type first and then an identifier. The following shows a number of variables using a variety of types:

```
int num_fish = 0;                   // number of fish caught
double max_length = 0.0;            // length of the longest fish in feet
char[] fisherman = new char[25];    // name of the fisherman
```

Notice also that I have demonstrated how to assign a value to the variable in the declaration. The assignment operator is the equal sign. All assignments must obey the type rules. That is, I cannot assign a floating-point number (e.g., 17.55) to an integer value. Table 6-1 shows a list of the commonly used built-in types you will use in your applications.

***Table 6-1.** Commonly Used Types in C#*

Symbol	Size in Bytes	Range
bool	1	False or true
char	1	−128 to 127
string	User-defined	−128 to 127 per character
sbyte	1	−128 to 127
byte	1	0-255
short	2	−32,768 to 32,767
ushort	2	0 to 65,535
int	4	−2,147,483,648 to 2,147,483,647
uint	4	0 to 4,294,967,295
long	4	−2,147,483,648 to 2,147,483,647
ulong	4	0 to 4,294,967,295
float	4	3.4E +/− 38 (7 digits)
decimal	8	(−7.9 x 1028 to 7.9 x 1028) / (100 to 28)
double	8	1.7E +/− 308 (15 digits)

It is always a good practice to initialize your variables when you declare them. It can save you from some nasty surprises if you use the variable before it is given a value (although the compiler will complain about this).

Arithmetic

You can perform a number of mathematical operations in C#, including the usual primitives, but also logical operations and operations used to compare values. Rather than discuss these in detail, I provide a quick reference in Table 6-2 that shows the operation and an example of how to use the operation.

Table 6-2. *Arithmetic, Logical, and Comparison Operators in C#*

Type	Operator	Description	Example
Arithmetic	+	Addition	int_var + 1
	-	Subtraction	int_var - 1
	*	Multiplication	int_var * 2
	/	Division	int_var / 3
	%	Modulus	int_var % 4
	-	Unary subtraction	-int_var
	+	Unary addition	+int_var
Logical	&	Bitwise and	var1&var2
	|	Bitwise or	var1|var2
	^	Bitwise exclusive	var1^var2
	~	Bitwise compliment	~var1
	&&	Logical and	var1&&var2
	||	Logical or	var1||var2

(*continued*)

Table 6-2. (*continued*)

Type	Operator	Description	Example
Comparison	==	Equal	expr1==expr2
	!=	Not equal	expr1!=expr2
	<	Less than	expr1<expr2
	>	Greater than	expr1>expr2
	<=	Less than or equal	expr1<=expr2
	>=	Greater than or equal	expr1>=expr2

Bitwise operations produce a result on the values performed on each bit. Logical operators (and, or) produce a value that is either true or false and are often used with expressions or conditions.

Now that you understand variables and types, the operations permitted on them, and expressions, let's look at how you can use them in flow control statements.

Flow Control Statements

Flow control statements change the execution of the program. They can be conditionals that cause one section of code to execute vs. another (also called gates). These conditionals use expressions that, when evaluated, restrict execution to only those cases where the expression is true. There are special constructs that allow you to repeat a block of code (loops) as well as functions to switch context to perform some special operations. You've already seen how functions work, so let's look at conditional and loop statements.

Conditionals

Conditional statements allow you to direct execution of your programs to sections (blocks) of code based on the evaluation of one or more expressions. There are two types of conditional statements in C#—the if statement and the switch statement.

You have seen the if statement in action in our example code. In the example, you can have one or more (optional) else phrases that you execute once the expression for the if conditions evaluate to false. You can chain if/else statements to encompass multiple conditions where the code executed depends on the evaluation of several conditions. The following shows the general structure of the if statement:

```
if (expr1) {
  // execute only if expr1 is true
} else if ((expr2) || (expr3)) {
  // execute only if expr1 is false *and* either expr2 or expr3 is true
} else {
  // execute if both sets of if conditions evaluate to false
}
```

While you can chain the statement as much as you want, use some care here because the more else/if sections you have, the harder it becomes to understand, maintain, and avoid logic errors in your expressions.

If you have a situation where you want to execute code based on one of several values for a variable or expression that returns a value (such as a method or a calculation), you can use the switch statement. The following shows the structure of the switch statement:

```
switch (eval) {
  case <value1> :
     // do this if eval == value1
     break;
  case <value2> :
     // do this if eval == value2
     break;
  default :
     // do this if eval != any case value
     break;  // Not needed, but good form
 }
```

The case values must match the type of the thing you are evaluating. That is, case values must be the same type as eval. Notice the break statement. This is used to halt evaluation of the code once the case value is found. Otherwise, each successive case value will be compared. Finally, there is a default section for code you want to execute should eval fail to match any of the values.

> **Tip**   Code style varies greatly in how to space/separate these statements. For example, some indent the case statements; some do not.

Loops

Loops are used to control the repetitive execution of a block of code. There are three forms of loops that have slightly different behavior. All loops use conditional statements to determine whether to repeat execution or not. That is, they repeat as long as the condition is true. The three types of loops are while, do, and for. I explain each with an example.

The while loop has its condition at the "top" or start of the block of code. Thus, while loops only execute the body if and only if the condition evaluates to true on the first pass. The following illustrates the syntax for a while loop. This form of loop is best used when you need to execute code only if some expression(s) evaluates to true, for example, iterating through a collection of things whose number of elements is unknown (loop until you run out of things in the collection).

```
while (expression) {
    // do something here
}
```

The do loop places the condition at the "bottom" of the statement which permits the body of the loop to execute at least once. The following illustrates the do loop. This form of loop is handy for cases where you want to execute code that, depending on the results of that execution, may require repetition, for example, repeatedly asking the user for input that matches one or more known values, repeating the question if the answer doesn't match.

```
do {
  // do something here - always done once
} while (expression);
```

The for loop is sometimes called a counting loop because of its unique form. The for loop allows you to define a counting variable, a condition to evaluate, and an operation on the counting variable. More specifically, the for loop allows you to define stepping code for a precise number of operations. The following illustrates the

235

structure of the `for` loop. This form of loop is best used for a number of iterations for a known number (either at runtime or as a constant) and commonly used to step through memory, count, and so forth.

```
for (<init> ; <expression> ; <increment>) {
// do something
}
```

The `<init>` section or counting variable declaration is executed once and only once. The `<expression>` is evaluated on every pass. The `<increment>` code is executed every pass except the last. The following is an example `for` loop:

```
for (int i; i < 10; i++) {
   // do something here
}
```

Now let's look at some commonly used data structures.

Basic Data Structures

What you have learned so far about C# will allow you to create applications that do simple to moderately complex operations. However, when you start needing to operate on data—either from the user or from sensors and similar sources—you need a way to organize and store data and operations on the data in memory. The following introduces three data structures in order of complexity: arrays, structures, and classes.

Arrays allocate a contiguous area of memory for multiple storage of a type. That is, you can store several integers, characters, and so forth, set aside in memory. Arrays also provide an integer index that you can use to quickly access a specific element. The following illustrates how to create an array of integers and iterate through them with a `for` loop. Array indexes start at 0.

```
int[] num_array = {0,1,2,3,4,5,6,7,8,9};  // an array of 10 integers
for (int i=0; i < 10; ++i) {
  Console.Write(num_array[i]);
  Console.Write(" ");
}
Console.Writeline();
```

You can also define multiple dimensional arrays (arrays of arrays). Arrays can be used with any type or data structure.

If you have a number of data items that you want to group together, you can use a special data structure called, amazingly, `struct`. A `struct` is formed as follows:

```
struct <name> {
  // one or more declarations go here
};
```

You can add whatever declarations you want inside the `struct` body (defined by the curly braces). The following shows a crude example. Notice that you can use the structure in an array:

```
struct address {
  int street_num;
  string street_name;
  string city;
  string state;
  string zip;
};

address[] address_book = new address[100];
```

Arrays and structures can increase the power of your programs by allowing you to work with more complex data types.

I hope that this short crash course in C# has explained enough about the sample programs shown so far that you now know how they work. This crash course also forms the basis for understanding the other C# examples in this book.

OK, now it's time to see some of these fundamental elements of C# in action. Let's look at the blink an LED application you saw in Chapter 3, only this time you're going to write it for Windows 10 IoT Core!

Blink an LED, C# Style

OK, let's write some C# code! This project is the same concept as the project from Chapter 3 where you used Python to blink an LED on your Raspberry Pi. Rather than simply duplicate that project, you'll mix it up a bit and make this example a headed application (recall a headed application has a user interface). The user interface presents the user with a greeting, a symbol that changes color in time with the LED, and a button to start and stop the blink timer.

Rather than build the entire application at once by presenting you a bunch of code, we walk through this example in two phases. The first phase builds the basic user interface. The code for the GPIO is added in the second phase. By using this approach, you can test the user interface on your PC, which is really convenient.

Recall that the PC does not support the GPIO libraries (there is no GPIO!), so if you built the entire application, you would have to test it on the device, which can be problematic if there are serious logic errors in your code. This way, you can ensure that the user interface is working correctly and therefore eliminate any possible issues in that code before you deploy it.

Before you get into the code for the user interface, let's look at the components that you will use and then set up the hardware.

Required Components

The following lists the components that you need. All of these are available in the Microsoft Internet of Things Pack for the Raspberry Pi from Adafruit. If you do not have that kit, you can find these components separately on the Adafruit website (`www.adafruit.com`), from SparkFun (`www.sparkfun.com`), or any electronics store that carries electronic components.

- 560 ohm 5% 1/4W resistor (green, blue, brown stripes[6])
- Diffused 10mm red LED (or similar)
- Breadboard (mini, half, or full sized)
- (2) male-to-female jumper wires

You may notice that this is the same set of components you used in Chapter 3.

Set Up the Hardware

Begin by placing the breadboard next to your Raspberry Pi and power the Raspberry Pi off, orienting the Raspberry Pi with the label facing you (GPIO pins in the upper left).

---

[6]https://en.wikipedia.org/wiki/Electronic_color_code

Next, take one of the jumper wires and connect the female connector to pin 6 on the GPIO. The pins are numbered left to right starting with the lower-left pin. Thus, the left two pins are 1 and 2 with pin 1 below pin 2. Connect the other wire to pin 7 on the GPIO.

Tip   The only component that is polarized is the LED. This longer side is the positive side.

Next, plug the resistor into the breadboard with each pin on one side of the center groove. You can choose whichever area you want on the breadboard. Next, connect the LED so that the long leg is plugged into the same row as the resistor and the other pin on another row. Finally, connect the wire from pin 6 to the same row as the negative side of the LED and the wire from pin 7 to the row with the resistor. Figure 6-3 shows how all of the components are wired together. Be sure to study this drawing and double-check your connections prior to powering on your Raspberry Pi. Once you're satisfied everything is connected correctly, you're ready to power on the Raspberry Pi and write the code.

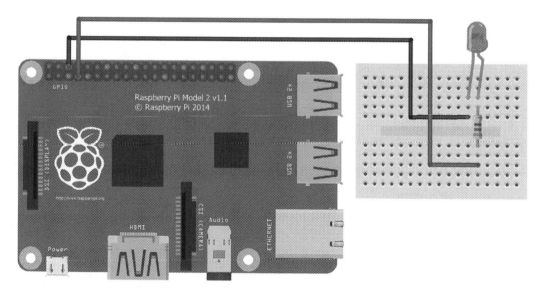

Figure 6-3.   *Wiring the LED to a Raspberry Pi*

Since you are building a headed application, you'll also need a keyboard, mouse, and monitor connected to the Raspberry Pi.

OK, now that the hardware is set up, it's time to start writing the code.

Write the Code: User Interface

Begin by opening a new project template. Choose the *C# Blank App (Universal Windows)* template in the list. This template creates a new solution with all of the source files and resources you need for a UWP headed application. Figure 6-4 shows the project template you need. Use the project name BlinkCSharpStyle.

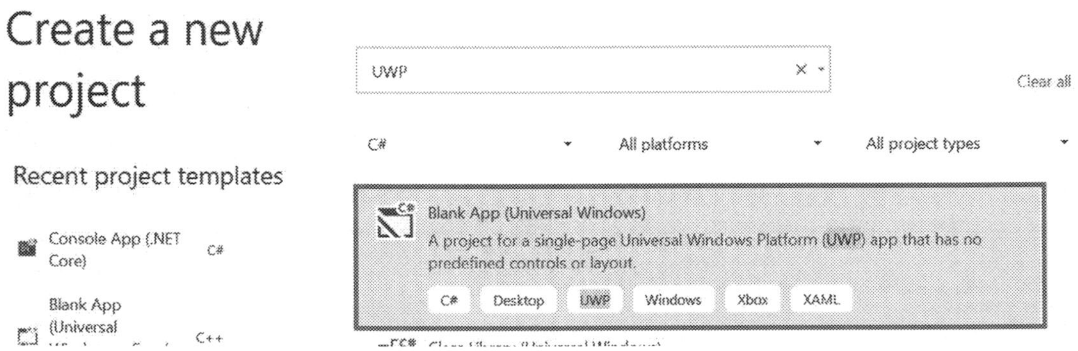

Figure 6-4. *New project dialog: blank C# application*

You will be asked to choose the Windows 10 IoT Core version you want to use, and then Visual Studio will open with the project configured. Notice that there are a number of files created, and the layout of the project is similar to what you've seen in previous examples.

Let's begin our example by adding the XAML code in the MainPage.xaml file. Recall, we need only click the file in the project list to display the GUI and code. You may need to resize the code portion to see it all. Listing 6-3 shows the bare XAML code placed in the file by default. I've added a note that shows where to add new code.

Listing 6-3. Bare XAML Code (MainPage.xaml)

```
<Page
    x:Class="BlinkCSharpStyle.MainPage"
    xmlns="http://schemas.microsoft.com/winfx/2006/xaml/presentation"
    xmlns:x="http://schemas.microsoft.com/winfx/2006/xaml"
    xmlns:local="using:BlinkCSharpStyle"
    xmlns:d="http://schemas.microsoft.com/expression/blend/2008"

    mc:Ignorable="d"
    Background="{ThemeResource ApplicationPageBackgroundThemeBrush}">
```

```
    <Grid>
-----> OUR CODE GOES HERE <----
    </Grid>
</Page>
```

Recall that the XAML file is used to define a user interface in a platform-independent way using an XML-like language. In this project, I demonstrate the more basic controls: a text box, a button, and an ellipse (circle) placed inside a special control called a stacked panel. The stacked panel allows you to arrange the controls in a vertical "stack," making it easier to position them. As you can see in the listing, you want to place your XAML user interface items in the `<Grid></Grid>` section.

In this example, you want a text box at the top and a circle (ellipse) to represent the LED that you will use to turn on (change to green) and off (change to gray) to correspond with the hardware on/off code that you will add later. You also need a button to toggle the blink operation on and off. Finally, you'll add another text box to allow you to communicate with the user about the state of the GPIO code (that you'll add later).

Now let's add the code. Since the stacked panel is a container, all of the controls are placed inside it. Listing 6-4 shows the code you want to add (shown in bold).

Listing 6-4. Adding XAML Code for the User Interface: MainPage.xaml

```
<Page
    x:Class="BlinkCSharpStyle.MainPage"
    xmlns="http://schemas.microsoft.com/winfx/2006/xaml/presentation"
    xmlns:x="http://schemas.microsoft.com/winfx/2006/xaml"
    xmlns:local="using:BlinkCSharpStyle"
    xmlns:d="http://schemas.microsoft.com/expression/blend/2008"

    mc:Ignorable="d"
    Background="{ThemeResource ApplicationPageBackgroundThemeBrush}">

    <Grid Background="{ThemeResource ApplicationPageBackgroundThemeBrush}">
        <StackPanel Width="400" Height="400">
            <TextBlock x:Name="title" Height="60" TextWrapping="NoWrap"
                    Text="Hello, Blinky C# Style!" FontSize="28"
                    Foreground="Blue"
                    Margin="10" HorizontalAlignment="Center"/>
```

```xaml
            <Ellipse x:Name="led_indicator" Fill="LightGray"
            Stroke="Gray"   Width="75"
                    Height="75" Margin="10" HorizontalAlignment="Center"/>
            <Button x:Name="start_stop_button" Content="Start" Width="75"
            ClickMode="Press"
                    Click="start_stop_button_Click" Height="50"
                    FontSize="24"
                    Margin="10" HorizontalAlignment="Center"/>
            <TextBlock x:Name="status" Height="60" TextWrapping="NoWrap"
                    Text="Status" FontSize="28" Foreground="Blue"
                    Margin="10" HorizontalAlignment="Center"/>
        </StackPanel>
    </Grid>
</Page>
```

Notice the button control. Here, you have an event that you want to associate with the button named `start_stop_button_Click`, which you assigned via the `Click` attribute. That is, when the user clicks it, a method named `start_stop_button_Click()` will be called.

XAML provides a great way to define a simple, easy user interface with the XML-like syntax. However, it also provides a mechanism to associate code with the controls. The code is placed in another file called a source-behind file named `MainPage.xaml.cs`. You will place all of the source code for the application in this file.

If you were typing this code in by hand, you will notice a nifty feature of Visual Studio—a context-sensitive help called IntelliSense that automatically completes the code you're typing and provides drop-down lists of choices. For example, when you type in the button control and type *Click=*, a drop-down box will appear, allowing you to create the event handler (a part of the code that connects to the XML). In fact, it creates the code in the `MainPage.xaml.cs` file for you. If you copy and pasted the code, you will not get this option and would have to type in the code manually.

Let's look at the code for the button control implemented in the class that you created in the source code file (`MainPage.xaml.cs`). Listing 6-5 shows the code you need to add in bold. You place everything in the class named `BlinkCPPStyle` (the application).

Listing 6-5. Adding the Base Code: MainPage.xaml.cs

```
//
// Windows 10 for the IoT Second Edition
//
// Blink C# Style Example
//
// Created by Dr. Charles Bell
//
using System;
using Windows.UI.Xaml;
using Windows.UI.Xaml.Controls;
using Windows.UI.Xaml.Media;
// Add using clause for GPIO
using Windows.Devices.Gpio;

namespace BlinkCSharpStyle
{
    public sealed partial class MainPage : Page
    {
        // Create brushes for painting contols
        private SolidColorBrush greenBrush = new SolidColorBrush(Windows.
        UI.Colors.Green);
        private SolidColorBrush grayBrush = new SolidColorBrush(Windows.
        UI.Colors.Gray);

        // Add a variable to control button
        private Boolean blinking = false;

        public MainPage()
        {
            this.InitializeComponent();
            // Add code to initialize the controls
            this.led_indicator.Fill = grayBrush;
        }
```

```
    private void start_stop_button_Click(object sender, RoutedEventArgs e)
    {
        this.blinking = !this.blinking;
        if (this.blinking)
        {
            this.start_stop_button.Content = "Stop";
            this.led_indicator.Fill = greenBrush;
        }
        else
        {
            this.start_stop_button.Content = "Start";
            this.led_indicator.Fill = grayBrush;
        }
    }
}
}
```

OK, there are a few extra bits here that may not be very obvious why they're here. You want to paint the LED control green and gray for on and off. To do that, you need a reference to the green and gray brush resources. Thus, I create a new object from the Windows user interface colors namespace.

You also add the code for the button click event, start_stop_button_Click(), as well as a boolean member variable that you use to trigger the LED timer. You add code that inverts the blinking variable (toggles between false and true), and depending on the value, you turn the LED indicator control green (meaning the LED is on) or gray (meaning the LED is off). You also change the label of the button to correspond with the operation. That is, if the button is labeled Start, the LED indicator is off, and when clicked, the label changes to Stop and the LED indicator is turned on.

That's it! You've finished the user interface. Go ahead and build the solution, correcting any errors that may appear. Once compiled, you're ready to test it.

Test and Execute: User Interface Only

That was easy, wasn't it? Better still, since this is a universal app, you can run this code on your PC. To do so, choose debug and x86 (or x64) from the platform box and press

Ctrl+F5. Figure 6-5 shows an excerpt of the output (just the control itself). Note that you may see the application run with a black background depending on your Windows settings. The figures are shown with a white background for clarity.

Hello, Blinky C# Style!

Start

Status

Figure 6-5. *The user interface: timer off*

Figure 6-6 shows what happens when you click the button. Cool, eh?

Hello, Blinky C# Style!

Stop

Status

Figure 6-6. *The user interface: timer on*

You may be wondering where the blink part is. Well, you haven't implemented it yet. You will do that in the next phase.

Add the GPIO Code

Now, let's add the code to work with the GPIO header. For this phase, you cannot run the code on your PC because the GPIO header doesn't exist, but you can add code to check the GPIO header status—hence the extra text box in the interface.

Note   The following is a bit more complicated and requires additional objects. Thus, I walk through the code changes one part at a time. Henceforth, for brevity, I present excerpts of the files that we will be editing.

Now, let's add the variables you need for the timer in the source file (`MainPage.xaml.cs`). Listing 6-6 shows the code you need to add. Here, you add an instance of the `DispatchTimer` as well as several private variables for working with the pin in the GPIO library. I show the changes in context with the new lines in bold.

Listing 6-6. Adding the Timer and GPIO Variables: MainPage.xaml.cs

```
...
        private SolidColorBrush greenBrush = new SolidColorBrush(
        Windows.UI.Colors.Green);
        private SolidColorBrush grayBrush = new SolidColorBrush(
        Windows.UI.Colors.Gray);

        // Add a Dispatch Timer
        private DispatcherTimer blinkTimer;
        // Add variables for the GPIO
        private const int LED_PIN = 4;
        private GpioPin pin;
        private GpioPinValue pinValue;

        // Add a variable to control button
        private Boolean blinking = false;

...
```

The private variables store the pin value, a constant set to GPIO 4 (hardware pin #7), and a variable to store the pin variable result.

Next, you need a new method to initialize the GPIO—InitGPIO(). You add a new private method to the class and complete it with code to control the GPIO as shown in Listing 6-7. It is placed in the class declaration after the start_stop_button_Click() method. As you can see, there is a lot going on here. Ellipses shown indicate portions of the code omitted for brevity.

Listing 6-7. Adding the InitGPIO() Code: MainPage.xaml.cs

```
namespace BlinkCSharpStyle
{
...
        private void start_stop_button_Click(object sender, RoutedEventArgs e)
        {
...
        }

        private void InitGPIO()
        {
            var gpio_ctrl = GpioController.GetDefault();

            // Check GPIO state
            if (gpio_ctrl == null)
            {
                this.pin = null;
                this.status.Text = "ERROR: No GPIO controller found!";
                return;
            }
            // Setup the GPIO pin
            this.pin = gpio_ctrl.OpenPin(LED_PIN);
            // Check to see that pin is Ok
            if (pin == null)
            {
                this.status.Text = "ERROR: Can't get pin!";
                return;
            }
```

```
        this.pin.SetDriveMode(GpioPinDriveMode.Output);
        this.pinValue = GpioPinValue.Low; // turn off
        this.pin.Write(this.pinValue);
        this.status.Text = "Good to go!";
    }
  }
}
```

The code first creates an instance of the default GPIO controller class. Next, you check to see if that instance is null, which indicates the GPIO header cannot be initiated, and if so you change the label of the status text and return. Otherwise, you open the GPIO pin defined earlier. If that value is null, you print the message that you cannot get the pin. Otherwise, you set up the pin for output mode, and then turn off the bin and state all is well in the status label text.

Next, you add a new method to handle the event fired from the DispatchTimer named BlinkTimer_Tick(). The timer fires (or calls) this method on the interval you specify (see Listing 6-8 for the changes to the MainPage() method). You can place this code immediately after the InitGPIO() method in the class.

Listing 6-8. Adding the BlinkTimer_Tick() Code: MainPage.xaml.cs

```
private void BlinkTimer_Tick(object sender, object e)
{
    // If pin is on, turn it off
    if (this.pinValue == GpioPinValue.High)
    {
        this.led_indicator.Fill = grayBrush;
        this.pinValue = GpioPinValue.Low;
    }
    // else turn it on
    else
    {
        this.led_indicator.Fill = greenBrush;
        this.pinValue = GpioPinValue.High;
    }
    this.pin.Write(this.pinValue);
}
```

In this method, you check the value of the pin. Here is where the real operation of the code happens. In this code, if the pin is set to high (on), you set it to low (off) and paint the LED control gray. Otherwise, you set the pin to high (on) and paint the LED control green. Cool, eh?

Note   You could change this color to match the color of your LED if you wanted. Just remember to change the brush accordingly in the source file.

Next, you need to add code to the start_stop_button_Click() method to start and stop the timer. Listing 6-9 shows the changes in bold.

Listing 6-9. Adding the Timer Control Code: MainPage.xaml.cs

```
private void start_stop_button_Click(object sender, RoutedEventArgs e)
{
    this.blinking = !this.blinking;
    if (this.blinking)
    {
        this.start_stop_button.Content = "Stop";
        this.led_indicator.Fill = greenBrush;
        this.blinkTimer.Start();
    }
    else
    {
        this.start_stop_button.Content = "Start";
        this.led_indicator.Fill = grayBrush;
        this.blinkTimer.Stop();
        this.pinValue = GpioPinValue.Low;
        this.pin.Write(this.pinValue);
    }
}
```

You see here a few things going on. First, notice that you invert the blinking variable to toggle blinking on and off. You then add a call to the blinkTimer instance of the DispatchTimer to start the timer if the user presses the button when it is labeled Start. Notice you also set the label of the button to Stop so that, when clicked again, the code turns off the timer and set the pin value to low (off). This is an extra measure to ensure that if the button is clicked when the timer is between tick events, the LED is turned off. Try removing it and you'll see.

Finally, you must add a few lines of code to the MainPage() method to get everything started when you launch the application. Listing 6-10 shows the code modifications with changes in bold.

Listing 6-10. Adding the Timer and GPIO Initialization Code: MainPage.xaml.cs

```
public MainPage()
{
    this.InitializeComponent();
    // Add code to initialize the controls
    this.led_indicator.Fill = grayBrush;
    // Add code to setup timer
    this.blinkTimer = new DispatcherTimer();
    this.blinkTimer.Interval = TimeSpan.FromMilliseconds(1000);
    this.blinkTimer.Tick += BlinkTimer_Tick;
    this.blinkTimer.Stop();
    // Initialize GPIO
    InitGPIO();
}
```

Notice you add code to create a new instance of the DispatchTimer class, set the interval for the tick event to 1 second (1000 milliseconds), and add the BlinkTimer_ Tick() method to the new instance (this is how you assign a method reference to an existing event handle). Next, you stop the timer and finally call the method that you wrote to initialize the GPIO.

That's it! Now, let's build the solution and check for errors. You should see something like the following in the output window:

```
1>------ Build started: Project: BlinkCSharpStyle, Configuration: Debug x86
------
1>  BlinkCSharpStyle -> C:\Users\olias\source\repos\BlinkCSharpStyle\
BlinkCSharpStyle\bin\x86\Debug\BlinkCSharpStyle.exe
========== Build: 1 succeeded, 0 failed, 0 up-to-date, 0 skipped ==========
```

OK, now you're ready to deploy the application to your device. Go ahead and set up everything, make the connections on the breadboard, and power on your device.

Deploy and Execute: Completed Application

Once your code compiles, you're ready to deploy the application to your Raspberry Pi (or another device). Recall from Chapter 4, we have a different deployment method for C# applications. If you want to install your application, you can do so from Visual Studio with only a few steps. However, if you want to debug your application, you must use a slightly different process. Let's look at each of these options starting with a normal deployment.

Before you attempt to deploy your application, make sure the remote debugger is stopped. We do not need the remote debugger running to deploy C# applications. This is because we will be using a different authentication mode.

Note   Make sure the remote debugger is stopped on your device before attempting to deploy a C# application.

Deploying Your Application (No Debugging)

Deploying your application with a C# project template is easy. Simply select the build mode, platform, and then device in Visual Studio. In this case, we want to select the *Release* or *Debug* build, the *ARM* platform, and select the *Remote Machine* entry in the device drop-down as shown in Figure 6-7. For this form of deployment, it doesn't matter which build mode you choose.

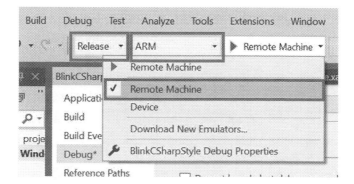

Figure 6-7. *Select the build, platform, and device (Visual Studio)*

When you select the Remote Machine entry, a Remote Connections dialog will appear that permits you to select the device from a list as shown in Figure 6-8. Simply select the device and click Select to select it.

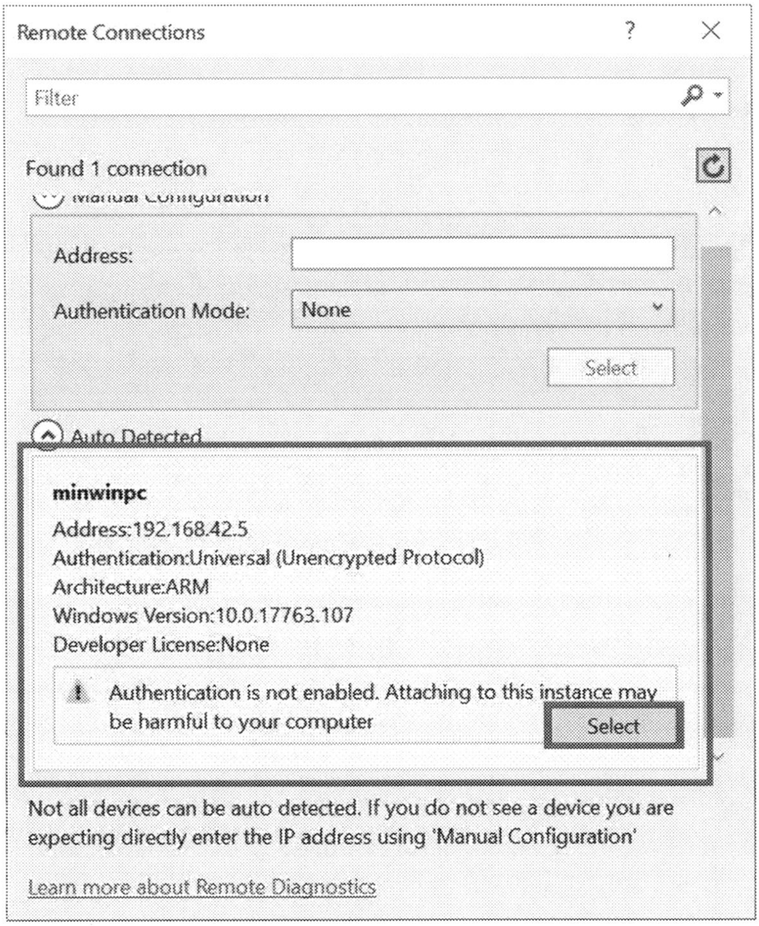

Figure 6-8. *Remote Connections dialog*

OK, we're almost ready to deploy. There is just one more thing to change. I find it helpful to rename the package name using the package manifest. This is the name that will be displayed in the Device Portal. By default, it uses a generated name, but I find renaming it to the same as the project name helpful for easier tracking. Thus, this is an optional step.

If you look at the left side of Visual Studio, you will see one file in the tree view named `Package.appxmanifest`. Double-click that file to open it. As you will see, it is a set of options you can set for your package. The one we want is on the last tab named Packaging. Click that and change the Package name, as shown in Figure 6-9.

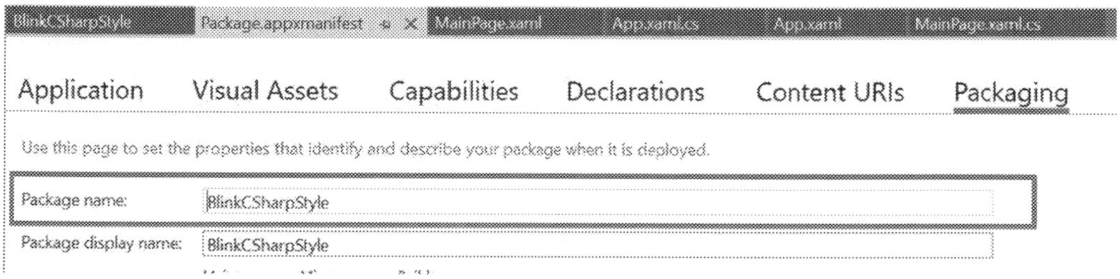

Figure 6-9. *Set the package name (package manifest)*

Now all that is left is to deploy the application. Simply click Build ➤ Deploy Solution, and the deploy will begin. If this is the first time you're deploying the application (or any application) to your device, the deployment could take a while.

Once complete, Visual Studio will indicate the build and deploy was successful similar to the output shown as follows:

```
Creating a new clean layout...
Copying files: Total 7 mb to layout...
Checking whether required frameworks are installed...
Registering the application to run from layout...
Deployment complete (0:00:22.489). Full package name:
"BlinkCSharpStyle_1.0.0.0_arm__mvq1akfegk7g6"
========== Build: 0 succeeded, 0 failed, 1 up-to-date, 0 skipped ==========
========== Deploy: 1 succeeded, 0 failed, 0 skipped ==========
```

Now, you can go to the Device Portal and observe your application in the Apps manager as shown in Figure 6-10. Recall, you can start it by selecting *Start* from the *Actions* drop-down box. Go ahead and do that now. You should see your application start, and you can experiment with it.

Figure 6-10. *Controlling the application in the Apps manager (Device Portal)*

After a few moments, you will see the application start on your device. If you have all of the wiring complete, you can click the button and observe the LED blinking until you stop it. Cool, eh?

If the LED is not blinking, double-check your wiring and ensure that you chose pin 4 in the code and pin 7 on the pin header (recall that the pin is named GPIO 4, but it is pin #7 on the header for the Raspberry Pi).

When you're done, you can stop it from the Apps manager in the Device Portal by using the *Actions* drop-down box and selecting *Stop*. It may take a few seconds for the status to change from *Running* to *Stopped*.

If you want to start the application again, you do not need to redeploy it. The application is installed on the device. To start it, go back to the Device Portal, click *Apps*, then *Apps manager*, and then click the *Actions* drop-down box for the application and click *Start*.

Should you want to uninstall the application, go back to the Device Portal, click *Apps*, then *Apps manager*, and then click the *Actions* drop-down box for the application and click *Uninstall*.

And that's it! Congratulations, you've just written your first C# application that uses the GPIO header to power some electronics!

Now, let's see how to deploy and debug our application.

Deploy and Debug

If you want to debug your application, you can, but once again it is not the same process we saw with the C++ application. Specifically, we do not need to use the remote debugger. Instead, we can make the same selections from the debug toolbar, selecting the *Debug* build, the *ARM* platform, and the *Remote Machine* from the device drop-down. If you have previously completed the Remote Connections dialog, you will not see that again. If you do see the dialog, make sure to select your device as described earlier and click the *Select* button.

Before you start the debugger, let's check a few settings. Most of these should be set correctly, but I've seen a couple of situations where the debug settings have changed. If they are not set correctly, you could have difficulty starting the debug session or may receive a series of strange, cryptic messages explaining either the application is already running or the debug symbols cannot be loaded.[7]

To open the debug settings, right-click the project (not the solution) and choose *Properties*. Then, select the *Debug* page. The principal settings we need to check include the following. Figure 6-11 shows the correct settings and their values. Be sure your project has the correct settings (the device name should be the name of your device).

- *Target device*: Remote Machine

- *Remote Machine*: The name of your device

- (Optional) *Uninstall and then re-install*: Checked

- (Optional) *Deploy optional packages*: Checked

- *Application process*: Managed Only

- *Background task process*: Managed Only

---

[7]Can you guess how I discovered this? Yes, I've been there...

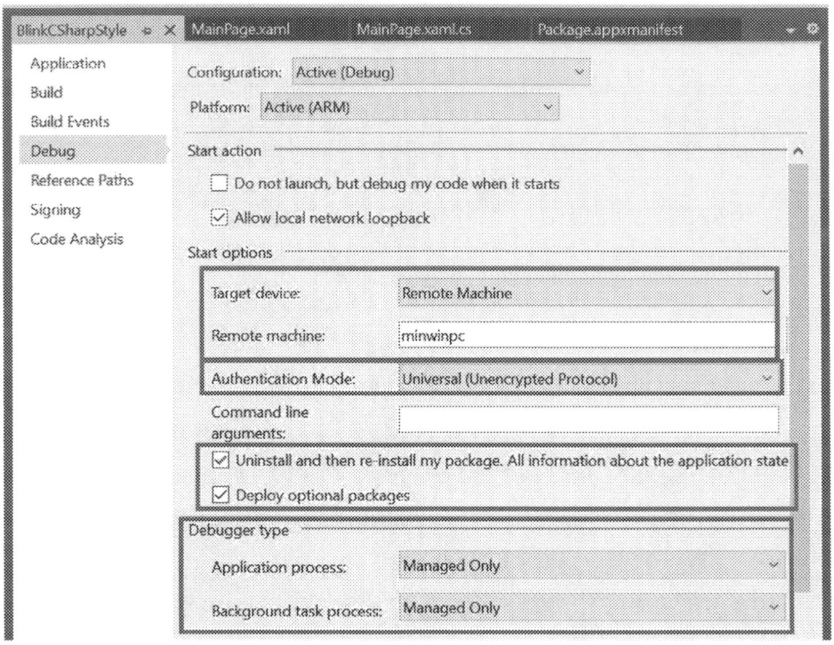

Figure 6-11. *Project debug settings*

Notice there are two optional settings. Checking these will ensure your application will always be deployed (thus overwriting previous installations) and all optional packages are included, which may avoid certain dependency issues if you have installed and uninstalled other applications on the device.

Tip If you want to reset the Remote Connection, remove the IP or device name in the *Remote Machine* text box.

Once you make any changes in this dialog, be sure to save your solution. Building or deploying does not automatically save these settings.

OK, now we are ready to debug our application. I find it helpful to place a breakpoint in the code so that when the application starts, I can step through it. Later, you may want to set a breakpoint somewhere in your code to observe certain variables or check for logic errors.

Recall, we set a breakpoint by clicking the left side of the code window in Visual Studio. A breakpoint is indicated with a red dot on the left side of the code window as demonstrated with an arrow in Figure 6-12.

```
33    public MainPage()
34    {
35        this.InitializeComponent();
36        // Add code to initialize the controls
37        this.led_indicator.Fill = grayBrush;
38        // Add code to setup timer
39        this.blinkTimer = new DispatcherTimer();
40        this.blinkTimer.Interval = TimeSpan.FromMilliseconds(1000);
41        this.blinkTimer.Tick += BlinkTimer_Tick;
42        this.blinkTimer.Stop();
43        // Initalize GPIO
44        InitGPIO();
45    }
```

Figure 6-12. *Setting a breakpoint (Visual Studio)*

Now we can start the debugging session. Simply click *Debug ➤ Start Debugging* or click *F5*. You should see a lengthy set of messages in the output window under the deployment option, but when done, you will see the application running on your device, and it will also be displayed in the Device Portal under the Apps manager. You should also see the interactive debugger start in Visual Studio as shown in Figure 6-13.

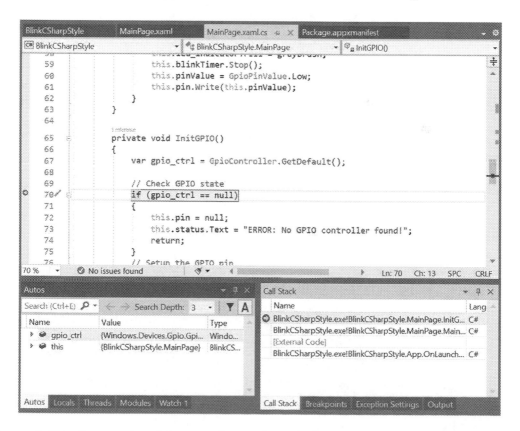

Figure 6-13. *Interactive debugger (Visual Studio)*

Now, you can step through your application using the debug menu. For example, you can step over method calls by clicking *F10* or step into methods with *F11*. Should you want to continue execution, you can click *Continue* in the debug toolbar. Or, you can stop the debugger using the *Stop* button or the *Debug* menu. Figure 6-14 shows what stepping through your code may look like. Yes, this means we can run the application on the device and debug it on our PC. Excellent!

```
Package.appxmanifest        MainPage.xaml        MainPage.xaml.cs  ⊕ ✕   BlinkCSharpStyle

  BlinkCSharpStyle                              ▼   ⚙ BlinkCSharpStyle.MainPage                         ▼

  83                    }
  84                        this.pin.SetDriveMode(GpioPinDriveMode.Output);
  85                        this.pinValue = GpioPinValue.Low; // turn off
  86                        this.pin.Write(this.pinValue);
  87                        this.status.Text = "Good to go!";
  88                    }
                    1 reference
  89                    private void BlinkTimer_Tick(object sender, object e)
  90                    {
  91                        // If pin is on, turn it off
● 92                        if (this.pinValue == GpioPinValue.High)
  93                        {
  94                            this.led_indicator.Fill = grayBrush;
  95                            this.pinValue = GpioPinValue.Low;
  96                        }
  97                        // else turn it on
  98                        else
  99                        {
 100                            this.led_indicator.Fill = greenBrush;
 101                            this.pinValue = GpioPinValue.High;
 102                        }
 103                        this.pin.Write(this.pinValue);
 104                    }
 105
 106                }
 107            }
```

Figure 6-14. *Stepping through code with interactive debugger (Visual Studio)*

Go ahead and experiment with the application until you are satisfied it is working correctly. When you want to stop it, go to the Device Portal, and click *Apps* ➤ *Apps manager* and use the *Actions* drop-down to stop the application.

If you get errors during the deployment, go back and check all of your settings to ensure you have everything correct. See the *C# Application Deployment Troubleshooting* section for more details.

Note   If the app deployed successfully but doesn't show in the drop-down list, try disconnecting and reconnecting. If that doesn't work, try rebooting your device.

C# Application Deployment Troubleshooting

If you're like me, things sometimes go wonky and just don't work, or they present you with an interesting but nearly indecipherable error message. I present a couple of these incidents you may encounter along with actions you can take to prevent or correct them.

Application Already/Not Running

If you get an error dialog saying the application is already or not running, you should double-check your debug settings for the project as described earlier. This can happen if you've deployed the application numerous times or have previous versions installed. Clearing the debug properties selections usually fixes the problem, but if that does not, change the package name or increment the version number, uninstall the application from the device, and restart your deployment.

Missing Framework

If you get an error during deployment that the `Microsoft.NET.Native...` is missing, your deployment will fail, and no amount of fiddling will correct this. Worse, the error message isn't very helpful as shown in the following:

```
DEP0800: The required framework "C:\Program Files (x86)\Microsoft
SDKs\UWPNuGetPackages\runtime.win10-arm.microsoft.net.native.
sharedlibrary\2.2.7-rel-28605-00\build\..\tools\SharedLibrary\ret\Native\
Microsoft.NET.Native.Framework.2.2.appx" failed to install. [0x80073CF9]
AppX Deployment operation failed with error 0x80070002 from API
IsCurrentProfileSpecial
```

What this is trying to tell you is that your application package is missing one or more of the .NET dependencies and that redistributable library is not already on the device.

This can be caused by an incomplete Visual Studio installation. If you see this message, you should first check for updates by launching the Visual Studio Installer and install all updates. You may also have to update the software on your device. You can do this on the Device Portal.

If that does not fix the problem, you can download the offline build tools for Visual Studio from `https://visualstudio.microsoft.com/downloads/`. Scroll down to the Visual Studio tools section and click the *Download* button as shown in Figure 6-15 or visit `https://visualstudio.microsoft.com/thank-you-downloading-visual-studio/?sku=BuildTools&rel=16`.

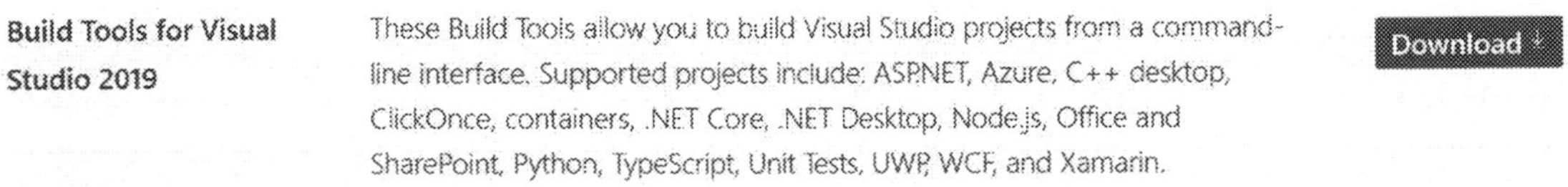

Build Tools for Visual Studio 2019   These Build Tools allow you to build Visual Studio projects from a command-line interface. Supported projects include: ASP.NET, Azure, C++ desktop, ClickOnce, containers, .NET Core, .NET Desktop, Node.js, Office and SharePoint, Python, TypeScript, Unit Tests, UWP, WCF, and Xamarin.   `Download ⌄`

Figure 6-15. *Downloading the Visual Studio build tools*

This is an installer that installs a number of libraries including the development libraries for .NET. However, you should ensure you also update this installation so that its version matches the version of Visual Studio installed. If the versions do not match, the problem may not be fixed.

Failure to Unregister Application

If you have installed the application many times and attempted to uninstall it, or it crashed and became corrupt, or you accidentally (maybe intentionally) removed the application files, you can see an error message during deployment like the following:

```
DEP0900: Failed to unregister application "BlinkCSharpStyle_1.0.0.0_arm__
mvq1akfegk7g6". [0x80073CFA] AppX Deployment operation failed with error
0x80070002 from API IsCurrentProfileSpecial
```

What this means is either the application is still registered on the device or some of its files are present, but it is not registered. To fix this, open an SSH terminal to the device and remove the application directory and all of its files like shown as follows. In this example, I had a derelict package on my device left over from a previous session,[8] and it caused my deployment of a newer application with the same package name to fail. In this case, the package was named BlinkCPPStyle_mvq1akfegk7.

```
[192.168.42.13]: PS C:\Data\Users\Administrator\AppData\Local\Packages>
rmdir /s BlinkCSharpStyle_1.0.0.0_arm__mvq1akfegk7g6
BlinkCSharpStyle_1.0.0.0_arm__mvq1akfegk7g6, Are you sure (Y/N)? Y.
```

Once you've removed the folder, you can deploy your application.

Deploy Succeeds, but Nothing Happens

It is possible you can deploy your application, but it either doesn't start or doesn't show up in your Apps manager on the Device Portal.

If Visual Studio reports the deployment succeeded, but the application doesn't start, that's OK and normal if you are doing a straight deployment. If you are deploying and debugging expecting the application to start, check the project properties as described earlier and correct any settings that may have changed. You can also attempt to uninstall the application using the App manager before retrying the deployment. If that still doesn't work, try removing all applications you've built previously.

If the application doesn't show up in the Apps manager, be sure to double-check all of the project settings in Visual Studio as described earlier, and the device is connected and reachable from the Device Portal. Try using the clean and build solution menu options again before attempting to deploy the application.

Tip   As a challenge, you can modify this project to add a *Close* or an *Exit* button to stop the application.

Summary

If you are learning how to work with Windows 10 IoT Core and don't know how to program with C#, learning C# can be a challenge. While there are many examples on

---

[8]From the first edition of this book.

the Internet you can use, very few are documented in such a way as to provide enough information for someone new to C# to understand or much less get started or even compile and deploy the sample!

This chapter has provided a crash course in C# that covers the basics of the things that you encounter when examining most of the smaller example projects. You discovered the basic syntax and constructs of a Visual C# application, including a walk-through of building a real C# application that blinks an LED. Through the course of that example, you learned a little about XAML, including how to wire events to controls, and even a little about how to use the dispatcher timer.

In the next chapter, you'll discover another programming language named Python. You implement the same example project you saw in this chapter but without the user interface. If you want to see how to work with the GPIO in Python, read on!

## CHAPTER 7

Windows 10 IoT Development with Visual Basic

Now that you have a basic understanding of how to use Visual Studio 2019, you can learn more about some of the languages you may encounter when developing your IoT solutions. One of those languages is Visual Basic (sometimes notated as VB)—another robust and powerful object-oriented language that you can use to write managed Windows .NET and UWP applications. Mastering Visual Basic is not as difficult as C++ or C# due to its simplified syntax. However, if you have never seen or worked with any form of Basic programming languages, the code may appear quite different. Fortunately, you need only learn the syntax since Visual Basic can be used to write the same applications as C++ and C#.

This chapter presents a crash course on the basics of Visual Basic programming in Visual Studio including an explanation about some of the most commonly used language features. As such, this chapter provides you with the skills you need to understand the growing number of IoT project examples available on the Internet. The chapter concludes with a walk-through of a Visual Basic example project that shows you how to interact with hardware. Specifically, you will implement the LED project you saw in Chapter 3. Only this time, you'll be writing it as a Visual Basic Windows 10 IoT Core application. So, let's get started!

Tip If you are not interested in using Visual Basic in your IoT solutions, or you already know the basics of Visual Basic programming, feel free to skim through this chapter. I recommend working through the example project at the end of the chapter, especially if you've not written IoT applications.

© Charles Bell 2021
C. Bell, *Windows 10 for the Internet of Things*, https://doi.org/10.1007/978-1-4842-6609-0_7

Getting Started

The Visual Basic .Net language has been around since the first introduction of the .NET Framework (pronounced "dot net") introduced in 2002. Visual Basic .Net is derived from an earlier product of the same name that originated in 1991 that combined the Basic programming language with a visual form designer, hence the "visual" in the name. The latest version of Visual Basic is version 8.0 and is sometimes called simply Visual Basic or VB.Net or simply VB.

Tip   I use "Visual Basic" to refer to Visual Basic .Net in this chapter because the previous product it replaced is no longer supported.

You may be thinking that Visual Basic and .NET may restrict the types of applications you can write, but that also is not true. You can use Visual Basic to write a host of applications from Windows 10 IoT Core to desktop to web applications and beyond. As you will see, you can also write Visual Basic console applications like you did in Chapter 5 with C++ and Chapter 6 with C#.

If you have ever worked with the Basic programming language or the Visual Basic for Applications (VBA) that is part of the macro language in Microsoft Office, Visual Basic will seem familiar to you. You may think Visual Basic is a "toy" language or too, well, basic for anything of significant complexity, but that is not the case. Visual Basic supports object-oriented and all of the resources and classes needed to write complex Windows applications. What sets Visual Basic apart is its simplified syntax.

Should you require more in-depth knowledge of Visual Basic, there are a number of excellent books on the topic. Here is a list of a few of my favorites. While some are a little dated (they cover Visual Studio 2015 or 2017), they provide excellent resources for the language:

- *Beginning Visual Basic 2015* by Bryan Newsome (Wrox, 2015)

- *Microsoft Visual Basic 2017 for Windows, Web, and Database Applications: Comprehensive* by Shelly Cashman (Microsoft Press, 2017)

- *Visual Basic 2015 in 24 Hours* by James Foxall (Sams Teach Yourself, 2015)

Another excellent resource is Microsoft's documentation on MSDN. The following are some excellent resources for learning Visual Basic .Net:

- Getting Started with Visual Basic .Net (`https://docs.microsoft.com/en-us/dotnet/visual-basic/getting-started/`)

- Visual Basic .Net Programming Guide (`https://docs.microsoft.com/en-us/dotnet/visual-basic/programming-guide/`)

Now that you know some of the origins and references for Visual Basic, let's learn about the syntax and basic language features for creating applications.

Visual Basic Crash Course

Now let's learn some of the basic concepts of Visual Basic programming. Let's begin with the building blocks of the language, such as classes, methods, variables, and basic control structures, and then move into the more complex concepts of data structures and libraries.

While the material may seem to come at you in a rush (hence the crash part), this crash course on Visual Basic covers only the most fundamental knowledge of the language and how to use it in Visual Studio. It is intended to get you started writing Visual Basic Windows 10 IoT Core applications. If you find you want to write more complex applications than the examples in this book, I encourage you to acquire one or more of the resources listed earlier to learn more about the intriguing power of Visual Basic programming.

Visual Basic Fundamentals

There are a number of basic concepts about the Visual Basic programming language that you need to know in order to get started. In this section, I describe some of the fundamental concepts used in Visual Basic, including how the code is organized, how libraries are used, namespaces, and how to document your code.

Visual Basic is a case-sensitive language, so you must take care when typing the names of methods or classes in libraries. Fortunately, Visual Studio's IntelliSense feature recognizes mistyped case letters and autocorrects the names. For example, if you type `BLINKTimer` for a variable that is declared as `blinkTimer`, Visual Studio corrects it for you. Once you get used to this feature, it is very hard to live without it.

Classes

The first thing you may notice is that Visual Basic is an object-oriented language and that every program you write is written as a class. However, simple console applications do not have to use a class and instead use a module definition, which resembles a class (it's very similar). Just think object-oriented when you see the Class keyword and code module when you see the Module keyword. We will see an example of both in this chapter.

UWP applications are implemented using a class definition. A class is more than a simple data structure. You use classes to model concepts that include data and operations on the data. A class can contain private and public definitions (called *members*) and any number of operations (called *methods*) that operate on the data and give the class meaning.

Tip   Source files in Visual Basic have a file extension of .vb.

You can use classes to break your programs down into more manageable chunks. That is, you can place a class you've implemented in its own .vb file and refer to it in any of the code provided there aren't namespace issues, and even then you simply use the namespace you want.

Let's look at a simple class named Vector implemented in Visual Basic. This class manages a list of double variables hiding the data from the caller while providing rudimentary operations for using the class. Listing 7-1 shows how such a class could be constructed.

Listing 7-1. Vector Class in Visual Basic

```
Public Class Vector
    Private elem() As Double
    Private sz As Integer

    Public Sub New(s As Integer)
        ReDim elem(s)
        sz = s
    End Sub
```

```
    Public Function size() As Integer
        Return sz
    End Function

    Public Function getValue(i As Integer) As Double
        Return elem(i)
    End Function

    Public Sub setValue(i As Integer, value As Double)
        elem(i) = value
    End Sub
End Class
```

This is a basic class that is declared with the keyword `Class` followed by a name. The name can be any identifier you want to use, but the convention is to use an initial capital letter for the name. All classes are defined using a common indentation that defines the body or structure of the class.

By convention, you list the member variables (also called *attributes*) indented from the class definition. In this example, you see two member variables that are declared as private. You make them private to hide information from the caller. That is, only member methods inside the class itself can access and modify private member variables. Note that derivatives (classes built from other classes—sometimes called a *child class*) can also access protected member variables.

Notice we don't have curly braces. Instead, we use the `End <keyword>` syntax to define a code block. For example, we use `End Class` to end the class, `End Sub`, and `End Function` to define code blocks for methods (`Sub`) and functions.

Next, you see a special method named `New`. This is called the *constructor*. The constructor is a method that is called when the class is instantiated (used). You call the code that defines the attributes and methods a class, and when executed, you call the resulting object an instance of the class.

In this case, the constructor takes a single integer parameter that is used to define the size of the private member variable that stores the array of double values. Notice how this code is used to dynamically define that array. More specifically, you use the `ReDim` command to reallocate memory for the array.

Next are three public methods, which users (or callers of the instance) can call. The first method, size(), looks as you would expect and in this case simply returns the value of the private member variable sz. The next methods are the get and set operations for storing a double in the array (setValue) and retrieving a specific double from the array using an index (getValue).

Let's see the class in action. The following shows a main method for a typical Visual Basic console application that declares the vector, initializes it with values, then prints out the values. I placed the class in its own file (vector.vb).

```
Imports System

Module Program
    Sub Main(args As String())
        Dim v As New Vector(10)
        For i As Integer = 1 To 10
            v.setValue(i, i * 3)
        Next
        Console.Write("Values of v: ")
        For i As Integer = 1 To 10
            Console.Write(v.getValue(i))
            Console.Write(" ")
        Next
        Console.WriteLine()
    End Sub
End Module
```

Notice that it is easy to read what the code is doing. Neat, eh?

To add a new source file to a Visual Studio Visual Basic solution, simply right-click the project name in Solution Explorer, and then choose *Add ➤ New item...* and click the Code section on the tree and then choose *Class* in the List. At the bottom of the dialog, you can name the file. When ready, click the *Add* button. You can then create any classes you want (or move classes) in the file. You can also use the same namespace to keep all of your classes in the same namespace. However, if you create a new namespace, you must use the using command to use the new namespace. Figure 7-1 shows an example of adding a new class to an existing Visual Basic solution.

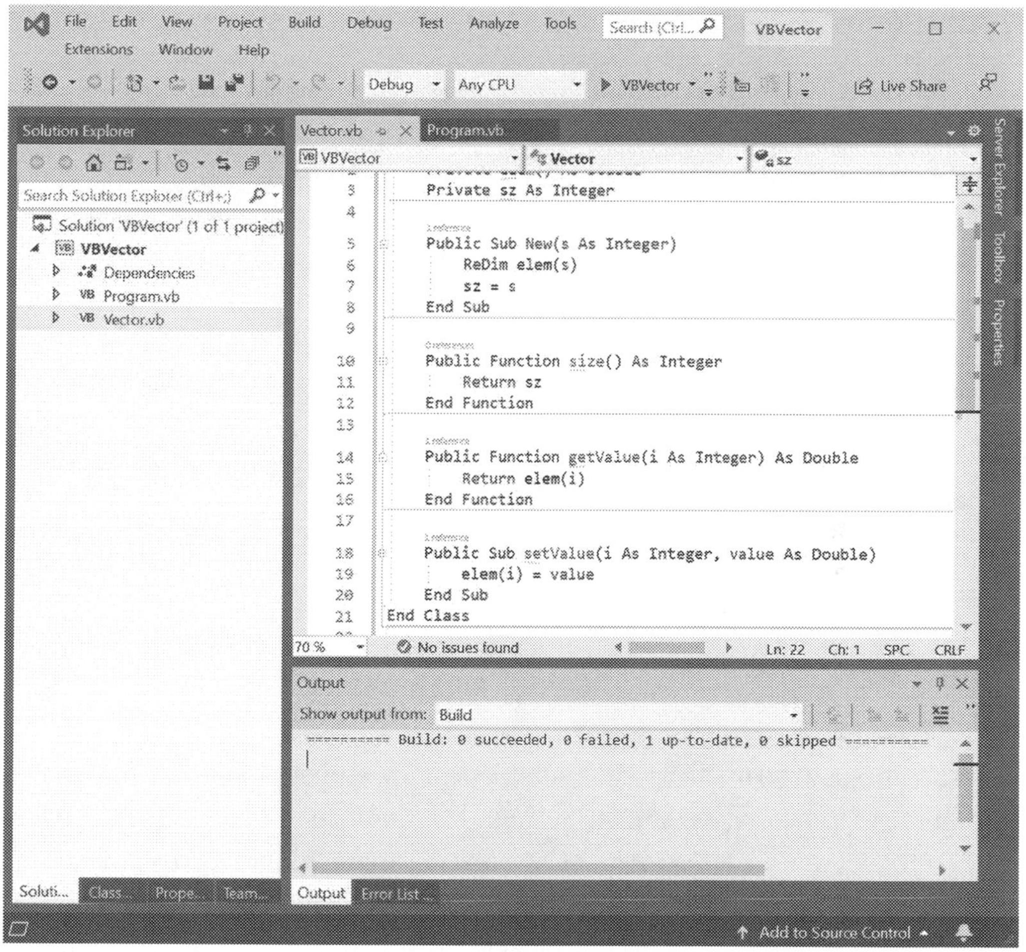
File Edit View Project Build Debug Test Analyze Tools
Extensions Window Help
VBVector
Debug
Any CPU
Live Share
Solution Explorer
Solution 'VBVector' (1 of 1 project)
VBVector
Dependencies
Program.vb
Vector.vb
Vector.vb
Program.vb
Private sz As Integer
Public Sub New(s As Integer)
ReDim elem(s)
sz = s
End Sub
Public Function size() As Integer
Return sz
End Function
Public Function getValue(i As Integer) As Double
Return elem(i)
End Function
Public Sub setValue(i As Integer, value As Double)
elem(i) = value
End Sub
End Class
No issues found
Output
Show output from: Build
========== Build: 0 succeeded, 0 failed, 1 up-to-date, 0 skipped ==========
Error List
Add to Source Control

Figure 7-1. *Visual Basic solution with a new class*

As you may have surmised, classes are the building blocks for object-oriented programming, and as you learn more about using classes, you can build complex libraries of your own.

Tip Visual Studio provides a tool called the Class View window that you can use to explore the libraries and classes used in your application.

Comments

One of the most fundamental concepts in any programming language is the ability to annotate your source code with text that not only allows you to make notes among the lines of code but also forms a way to document your source code.[1]

To add comments to your source code, use a single quote ('). Place them at the start of the line to create a comment for that line, repeating the quote for each subsequent line. This creates what is known as a *block comment*, as shown. Notice that I used a comment without any text to create whitespace. This helps with readability and is a common practice for block comments.

```
'

' Windows 10 for the IoT Second Edition
'

' Example Visual Basic console application rewrite.
'

' Created by Dr. Charles Bell
'
```

You can also use the quote to add a comment at the end of a line of code. That is, the compiler ignores whatever is written after the quote to the end of the line. You see an example of this next. Notice that I used the comment symbol (single quote) to comment out a section of code. This can be really handy when testing and debugging, but generally discouraged for final code. That is, don't leave any commented out code in your deliverable (completed) source code. If it's commented out, it's not needed!

```
Function testComments(i As Integer) As Integer
    If i < 10 Then
        i = i + 1    ' Increment the value
    'Else
    '      Return -1
    End If
End Function
```

---

[1]If you ever hear someone claim, "My code is self-documenting," be cautious when using their code. There is no such thing. Sure, plenty of good programmers can write code that is easy to understand (read), but all fall short of that lofty claim.

Writing good comments and indeed documenting your code well is a bit of an art form, one that I encourage you to practice regularly. Since it is an art rather than a science, keep in mind that your comments should be written to teach others what your code does or is intended to do. As such, you should use comments to describe any preconditions (or constraints) of using the code, limitations of use, errors handled, and a description of how the parameters are used and what data is altered or returned from the code (should it be a method or class member).

How Visual Basic Programs Are Structured

Now let's look at how Visual Basic programs are structured by examining a slightly different version of the temperature application you saw in Chapter 4. Listing 7-2 shows the code rewritten for Visual Basic.

Listing 7-2. Temperature Code Example Rewrite

```vb
'
' Windows 10 for the IoT Second Edition
'
' Example Visual Basic console application rewrite
'
' Created by Dr. Charles Bell
'
Imports System

Module Program
    Public Function convertTemp(scale As Char, baseTemp As Double) As Double
        If scale = "c" Or scale = "C" Then
            Return ((9.0 / 5.0) * baseTemp) + 32.0
        Else
            Return (5.0 / 9.0) * (baseTemp - 32.0)
        End If
    End Function

    Sub Main(args As String())
        Dim tempRead As Double = 0.0
        Dim scale As Char = "c"
```

```vb
        Console.WriteLine("Welcome to the temperature conversion
        application.")
        Console.Write("Please choose a starting scale (F) or (C): ")
        scale = Console.ReadKey().KeyChar
        Console.WriteLine()
        Console.Write("Please enter a temperature: ")
        tempRead = Convert.ToDouble(Console.ReadLine())
        If scale = "c" Or scale = "C" Then
            Console.WriteLine("Converting value from Celsius to
            Fahrenheit.")
            Console.Write(tempRead)
            Console.Write(" degrees Celsius = ")
            Console.Write(convertTemp(scale, tempRead))
            Console.WriteLine(" degrees Fahrenheit.")
        ElseIf scale = "f" Or scale = "F" Then
            Console.WriteLine("Converting value from Fahrenheit to Celsius.")
            Console.Write(tempRead)
            Console.Write(" degrees Fahrenheit = ")
            Console.Write(convertTemp(scale, tempRead))
            Console.WriteLine(" degrees Celsius.")
        Else
            Console.Write("ERROR: I'm sorry, I don't understand '")
            Console.Write(scale)
            Console.WriteLine("'.")
        End If
    End Sub
End Module
```

In the example, the only methods created are convertTemp() and Main(), but this is because you are implementing a very simple solution. Here, you see the sample application named temperatureVB was implemented with a module with the name Program.

Wow, that's quite a change from the code in the last chapter! While the functionality is exactly the same, the code looks very different from the C++ and C# versions. The following describe the Visual Basic concepts I have implemented in this example.

The Imports Keyword

First, you notice a number of lines that begin with `Imports`. These are preprocessor directives that tell the compiler you want to "use" a class or a class hierarchy that exists in a particular namespace. The `Imports` directive tells the compiler that you are using the namespace `System`.

```
Imports System
```

In the other lines, I have included additional namespaces with multiple names separated by a period. This is how you tell the compiler to use a specific namespace located in libraries of classes often form hierarchies that you can chain together. For example, if you wanted to use the namespace inside the Windows Foundations device library named `Gpio`, you would refer to it as follows:

```
Imports Windows.Devices.Gpio
```

This is a very common occurrence in Windows Visual Basic applications. In fact, you will use several namespaces in our example project.

The Main() Method

The `Main()` method is the starting or initial execution for the Visual Basic console project. Here, you see the name is preceded by the keyword Sub as follows:

```
Sub Main(args As String())
```

Next, you see the name, main, followed by a list of parameters enclosed in parentheses. For the `Main()` method, the parameters are fixed and are used to store any command-line arguments provided by the user. In this case, you have the arguments stored in `args`, which is a string.

A method in Visual Basic is used as an organizational mechanism to group functionality and make your programs easier to maintain (methods with hundreds of lines of code are very difficult to maintain), improve comprehensibility, and localize specialized operations in a single location, thereby reducing duplication.

Methods and functions therefore are used in your code to express the concepts of the functionality they provide. Notice how I used the `convertTemp()` method. Here, I declared it as a method that returned a double and takes a character and a `double` as input. As you can see, the body of the method (defined inside the `Function ➤ End Function`) uses the character as the scale in the same way as you do in main and uses the double parameter as the target (or base) temperature to convert. Since I made the parameters generic, I can use only one variable.

Tip Method parameters and values passed must match on type and order when called.

Notice also that I placed it in the line of code that prints the value to the screen. This is a very common practice in Visual Basic (and other programming languages). That is, you use the method to perform some operation, and rather than store the result in a variable, you use it directly in the statements (code).

Variables and Types

No program would be very interesting if you did not use variables to store values for calculations. As you saw earlier, variables are declared with a type (the s <type> syntax) and once defined with a specific type cannot be changed. Since Visual Basic is strongly typed, the compiler ensures that anywhere you use the variable, it obeys its type, for example, that the operation on the variable is valid for the type. Thus, every variable must have a type assigned.

There are a number of simple types that the Visual Basic language supports. They are the basic building blocks for more complex types. Each type consumes a small segment of memory which defines not only how much space you have to store a value but also the range of values possible.[2]

For example, an integer consumes 4 bytes, and you can store values in the range –2,147,483,648 to 2,147,483,647. In this case, the integer variable is signed (the highest bit is used to indicate positive or negative values). An unsigned integer can store values in the range 0 to 4,294,967,295.

You can declare a variable by specifying the keyword Dim, followed by an identifier (name), then its type using the as keyword, and optionally assign it a value with the = operator. The following shows a number of variables using a variety of types:

---

[2]For a complete list, see https://docs.microsoft.com/en-us/cpp/cpp/data-type-ranges?redirectedfrom=MSDN&view=vs-2019.

```
Dim numFish as Integer = 0          ' number of fish caught
Dim maxLength as Double = 0.0;       ' length of the longest fish in feet
Dim fisherman(25) as char           ' name of the fisherman
```

Notice also that I have demonstrated how to assign a value to the variable in the declaration. The assignment operator is the equal sign. All assignments must obey the type rules. That is, I cannot assign a floating-point number (e.g., 17.55) to an integer value. Table 7-1 shows a list of the commonly used built-in types you will use in your applications.

Table 7-1. *Commonly Used Types in Visual Basic*

Symbol	Size in Bytes	Range
Boolean	1	False or true
Char	1	−128 to 127
String	User-defined	−128 to 127 per character
SByte	1	−128 to 127
Byte	1	0–255
Short	2	−32,768 to 32,767
UShort	2	0 to 65,535
Integer	4	−2,147,483,648 to 2,147,483,647
UInteger	4	0 to 4,294,967,295
Long	4	−2,147,483,648 to 2,147,483,647
ULong	4	0 to 4,294,967,295
Single	4	3.4E +/− 38 (7 digits)
Decimal	8	(−7.9 x 1028 to 7.9 x 1028) / (100 to 28)
Double	8	1.7E +/− 308 (15 digits)

It is always a good practice to initialize your variables when you declare them. It can save you from some nasty surprises if you use the variable before it is given a value (although the compiler will complain about this).

Arithmetic

You can perform a number of mathematical operations in Visual Basic, including the usual primitives, but also logical operations and operations used to compare values. Rather than discuss these in detail, I provide a quick reference in Table 7-2 that shows the operation and example of how to use the operation.

Table 7-2. *Arithmetic, Logical, and Comparison Operators in Visual Basic*

Type	Operator	Description	Example
Arithmetic	+	Addition	`intVar + 1`
	-	Subtraction	`intVar - 1`
	*	Multiplication	`intVar * 2`
	/	Division	`intVar / 3`
	Mod	Modulus	`intVar % 4`
	^	Exponent	`intVar ^ 2`
	\	Integer division	`floatVar \ 3`
	+=	Unary addition	`intVar += 1`
	-=	Unary subtraction	`intVar -= 1`
Logical	And	Bitwise and	`var1&var2`
	Or	Bitwise or	`var1|var2`
	Xor	Bitwise exclusive	`var1^var2`
	And	Logical and	`var1&&var2`
	Or	Logical or	`var1||var2`
Comparison	=	Equal	`expr1==expr2`
	<>	Not equal	`expr1!=expr2`
	<	Less than	`expr1<expr2`
	>	Greater than	`expr1>expr2`
	<=	Less than or equal	`expr1<=expr2`
	>=	Greater than or equal	`expr1>=expr2`

Bitwise operations produce a result on the values performed on each bit. Logical operators (and, or) produce a value that is either true or false and are often used with expressions or conditions.

Now that you understand variables and types, the operations permitted on them, and expressions, let's look at how you can use them in flow control statements.

Flow Control Statements

Flow control statements change the execution of the program. They can be conditionals that cause one section of code to execute vs. another (also called gates). These conditionals use expressions that, when evaluated, restrict execution to only those cases where the expression is true. There are special constructs that allow you to repeat a block of code (loops) as well as functions to switch context to perform some special operations. You've already seen how functions work, so let's look at conditional and loop statements.

Conditionals

Conditional statements allow you to direct execution of your programs to sections (blocks) of code based on the evaluation of one or more expressions. There are two types of conditional statements in Visual Basic—the If...Then statement and the Select...Case statement.

You have seen the If statement in action in our example code. In the example, you can have one or more (optional) Else phrases that you execute once the expression for the If conditions evaluates to false. You can chain If/Else statements to encompass multiple conditions where the code executed depends on the evaluation of several conditions. The following shows the general structure of the If statement:

```
If expr1 Then
  ' execute only if expr1 is true
ElseIf expr2 Or Then
  ' execute only if expr1 is false *and* either expr2 or expr3 is true
Else
  ' execute if both sets of if conditions evaluate to false
End If
```

While you can chain the statement as much as you want, use some care here because the more ElseIf sections you have, the harder it becomes to understand, maintain, and avoid logic errors in your expressions.

If you have a situation where you want to execute code based on one of several values for a variable or expression that returns a value (such as a method or a calculation), you can use the Select...Case statement. The following shows the structure of the Select...Case statement as an example from the temperature project earlier with the If...Then statement rewritten as a Select...Case statement:

```
Select Case scale
    Case "c"
    Case "C"
        Console.WriteLine("Converting value from Celsius to Fahrenheit.")
        Console.Write(tempRead)
        Console.Write(" degrees Celsius = ")
        Console.Write(convertTemp(scale, tempRead))
        Console.WriteLine(" degrees Fahrenheit.")
    Case "f"
    Case "F"
        Console.WriteLine("Converting value from Fahrenheit to Celsius.")
        Console.Write(tempRead)
        Console.Write(" degrees Fahrenheit = ")
        Console.Write(convertTemp(scale, tempRead))
        Console.WriteLine(" degrees Celsius.")
    Case Else
        Console.Write("ERROR: I'm sorry, I don't understand '")
        Console.Write(scale)
        Console.WriteLine("'.")
End Select
```

The case values must match the type of the thing you are evaluating. That is, case values must be the same type as scale. Notice I placed Case statements "stacked" so that multiple cases can be directed to the same code block. That is a simple way of achieving that goal. Finally, there is a Case Else section for code you want to execute should scale fail to match any of the values.

Loops

Loops are used to control the repetitive execution of a block of code. There are three forms of loops that have slightly different behavior. All loops use conditional statements to determine whether to repeat execution or not. That is, they repeat as long as the condition is true. The three types of loops are While, Do, and For. I explain each with an example.

The While loop has its condition at the "top" or start of the block of code. Thus, While loops only execute the body (between the While and End While statements) if and only if the condition evaluates to true on the first pass. This form of loop is best used when you need to execute code only if some expression(s) evaluates to true, for example, iterating through a collection of things whose number of elements is unknown (loop until you run out of things in the collection).

The While loop in Visual Basic permits two additional directives. While developers try to structure their code to avoid such constructs, you can control the flow of the While loop using the Exit While and Continue While clauses.

The Exit While does what you expect—it will immediately stop executing any more statements and return control to the next statement after the End While, which is analogous with the break statement from other languages. The Continue While is interesting because it also stops executing any more statements in the loop but instead returns to the While condition (the top of the loop). As you view more Visual Basic examples, you may encounter these clauses.

The following illustrates the syntax for a While loop:

```
Console.WriteLine("While Loop:")
count = 10
While count > 0
    Console.Write(count)
    If count = 1 Then
        Exit While
    End If
    Console.Write(", ")
    count -= 1
End While
Console.WriteLine()
```

The Do loop normally places the condition at the "bottom" of the statement which permits the body of the loop to execute at least once. However, the Do loop in Visual Basic has a second form that permits you to add the condition at the "top" of the loop which performs the same as the While loop.

The Do loop in Visual Basic permits two additional directives. While developers try to structure their code to avoid such constructs, you can control the flow of the Do loop using the Exit Do and Continue Do clauses.

The Exit Do does what you expect—it will immediately stop executing any more statements and return control to the next statement after the End Do, which is analogous with the break statement from other languages. The Continue Do is interesting because it also stops executing any more statements in the loop but instead returns to the While condition (the top or bottom of the loop). As you view more Visual Basic examples, you may encounter these clauses.

This form of loop is handy for cases where you want to execute code that, depending on the results of that execution, may require repetition, for example, repeatedly asking the user for input that matches one or more known values, repeating the question if the answer doesn't match.

The following illustrates the Do loop:

```
Console.WriteLine("Do Loop 'bottom' Condition:")
count = 10
Do
    Console.Write(count)
    If count = 1 Then
        Exit Do
    End If
    Console.Write(", ")
    count -= 1
Loop While count > 0

Console.WriteLine("Do Loop 'top' Condition:")
count = 10
Do While count > 0
    Console.Write(count)
    If count = 1 Then
        Exit Do
    End If
```

```
    Console.Write(", ")
    count -= 1
Loop
```

The For loop has two forms and is sometimes called a counting loop because of its unique way of executing a loop for a specific number of iterations or through the members of a container.

The first form of the For loop allows you to define a counting variable, a condition to evaluate, and an operation on the counting variable. Instead of an "End" clause, the For loop is defined by Next at the end of the body of the loop. More specifically, the For loop allows you to define stepping code for a precise number of operations, a <start> TO <stop> clause and optional Step <increment> clause that you can use to count by a constant such as counting by 2s, 3s, and so on. This form of loop is best used for a number of iterations for a known number (either at runtime or as a constant) and commonly used to step through memory, count, and so forth.

The For counting loop in Visual Basic permits two additional directives. While developers try to structure their code to avoid such constructs, you can control the flow of the For loop using the Exit For and Continue For clauses.

The Exit For does what you expect—it will immediately stop executing any more statements and return control to the next statement after the Next, which is analogous with the break statement from other languages. The Continue For is interesting because it also stops executing any more statements in the loop but instead returns to the count evaluation. As you view more Visual Basic examples, you may encounter these clauses.

The following illustrates the structure of the For counting loop:

```
Dim v As New Vector(10)
For i As Integer = 1 To 10
    v.setValue(i, i * 3)
Next
```

The second form is the For Each loop that iterates through a collection or container and has a slightly different syntax. Instead of a counting variable and optional step clause, we use a variable that contains the type of the things in the collection. For example, if we have a structure (more on those in the next section), we can use the For Each loop to iterate through a list (array) of books.

The following shows an example of using the For Each loop:

```
Public Structure bookInfo
    Public title As String
    Public cost As Double
End Structure
...
Dim books(10) As bookInfo
For Each book As bookInfo In books
    If book.title.StartsWith("S") Then
        Exit For
    End If
    Console.WriteLine(book.title)
Next
```

Tip   For more detailed information about the loops in Visual Basic, see
https://docs.microsoft.com/en-us/dotnet/visual-basic/
programming-guide/language-features/control-flow/loop-
structures.

Now let's look at some commonly used data structures.

Basic Data Structures

What you have learned so far about Visual Basic will allow you to create applications
that do simple to moderately complex operations. However, when you start needing to
operate on data—either from the user or from sensors and similar sources—you need
a way to organize and store data and operations on the data in memory. The following
introduces three data structures in order of complexity: arrays, structures, and classes.

Arrays allocate a contiguous area of memory for multiple storage of a type. That
is, you can store several integers, characters, and so forth, set aside in memory. Arrays
also provide an integer index that you can use to quickly access a specific element. The
following illustrates how to create an array of integers and iterate through them with a
for loop. Array indexes start at *0*.

```
Dim numArray = New Integer() {0, 1, 2, 3, 4, 5, 6, 7, 8, 9}  ' an array Of
10 integers
Console.Write("Values of v: ")
For i As Integer = 0 To 9
    Console.Write(numArray(i))
    Console.Write(" ")
Next
Console.WriteLine()
```

You can also define multiple dimensional arrays (arrays of arrays). Arrays can be used with any type or data structure.

If you have a number of data items that you want to group together, you can use a special data structure called, amazingly, Structure. A Structure is formed as follows:

```
Private Structure <name>
  ' one or more declarations go here
End Structure
```

You can add whatever declarations you want inside the Structure body (before the End Structure statement). The following shows a crude example. Notice that you can use the structure in an array:

```
Private Structure address
    Public streetNum As Integer
    Public streetName As String
    Public city As String
    Public state As String
    Public zip As String
End Structure
...
Dim addressBook(100) As address
```

Arrays and structures can increase the power of your programs by allowing you to work with more complex data types.

Wow! That was a wild ride, wasn't it? I hope that this short crash course in Visual Basic has explained enough about the sample programs shown so far that you now know how they work. This crash course also forms the basis for understanding the other Visual Basic examples in this book.

OK, now it's time to see some of these fundamental elements of Visual Basic in action. Let's look at the blink an LED application you saw in Chapter 3, only this time you're going to write it for Windows 10 IoT Core!

Tip   For a complete, online reference guide to Visual Basic, see the Official Microsoft documentation at `https://docs.microsoft.com/en-us/dotnet/visual-basic`.

Blink an LED, Visual Basic Style

OK, let's write some Visual Basic code! This project is the same concept as the project from Chapter 3 where you used Python to blink an LED on your Raspberry Pi. Rather than simply duplicate that project, you'll mix it up a bit and make this example a headed application (recall a headed application has a user interface). The user interface presents the user with a greeting, a symbol that changes color in time with the LED, and a button to start and stop the blink timer.

Rather than build the entire application at once by presenting you a bunch of code, we walk through this example in two phases. The first phase builds the basic user interface. The code for the GPIO is added in the second phase. By using this approach, you can test the user interface on your PC, which is really convenient.

Recall that the PC does not support the GPIO libraries (there is no GPIO!), so if you built the entire application, you would have to test it on the device, which can be problematic if there are serious logic errors in your code. This way, you can ensure that the user interface is working correctly and therefore eliminate any possible issues in that code before you deploy it.

Before you get into the code for the user interface, let's look at the components that you will use, and then set up the hardware.

Required Components

The following lists the components that you need. All of these are available in the Microsoft Internet of Things Pack for the Raspberry Pi from Adafruit. If you do not have that kit, you can find these components separately on the Adafruit website

(www.adafruit.com), from SparkFun (www.sparkfun.com), or any electronics store that
carries electronic components.

- 560 ohm 5% 1/4W resistor (green, blue, brown stripes[3])

- Diffused 10mm red LED (or similar)

- Breadboard (mini, half, or full sized)

- (2) male-to-female jumper wires

You may notice that this is the same set of components you used in Chapter 3.

Set Up the Hardware

Begin by placing the breadboard next to your Raspberry Pi and power the Raspberry Pi
off, orienting the Raspberry Pi with the label facing you (GPIO pins in the upper left).
Next, take one of the jumper wires and connect the female connector to pin 6 on the
GPIO. The pins are numbered left to right starting with the lower-left pin. Thus, the left
two pins are 1 and 2 with pin 1 below pin 2. Connect the other wire to pin 7 on the GPIO.
Figure 7-2 shows how all of the components are wired together. Be sure to study this
drawing and double-check your connections prior to powering on your Raspberry Pi.

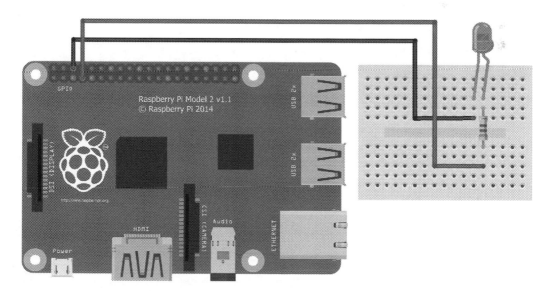

Figure 7-2. *Wiring the LED to a Raspberry Pi*

---

[3]https://en.wikipedia.org/wiki/Electronic_color_code

Tip   The only component that is polarized is the LED. This longer side is the positive side.

Next, plug the resistor into the breadboard with each pin on one side of the center groove. You can choose whichever area you want on the breadboard. Next, connect the LED so that the long leg is plugged into the same row as the resistor and the other pin on another row. Finally, connect the wire from pin 6 to the same row as the negative side of the LED and the wire from pin 7 to the row with the resistor. Once you're satisfied everything is connected correctly, you're ready to power on the Raspberry Pi and write the code.

Since you are building a headed application, you'll also need a keyboard, mouse, and monitor connected to the Raspberry Pi.

OK, now that the hardware is set up, it's time to start writing the code.

Write the Code: User Interface

Begin by opening a new project template. Choose the *Visual Basic Blank App (Universal Windows)* template in the list. This template creates a new solution with all of the source files and resources you need for a UWP headed application. Figure 7-3 shows the project template you need. Use the project name `BlinkVBStyle`.

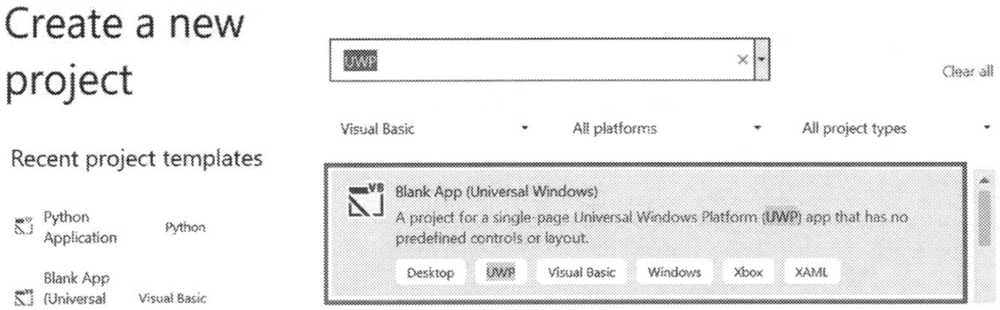

Figure 7-3.   New project dialog: blank Visual Basic application

You will be asked to choose the Windows 10 IoT Core version you want to use, and then Visual Studio will open with the project configured. Notice that there are a number of files created and the layout of the project is similar to what you've seen in previous examples.

Let's begin our example by adding the XAML code in the `MainPage.xaml` file. Recall, we need only click the file in the project list to display the GUI and code. You may need to resize the code portion to see it all. Listing 7-3 shows the bare XAML code placed in the file by default. I've added a note that shows where to add new code.

Listing 7-3. Bare XAML Code (MainPage.xaml)

```
<Page
    x:Class="BlinkVBStyle.MainPage"
    xmlns="http://schemas.microsoft.com/winfx/2006/xaml/presentation"
    xmlns:x="http://schemas.microsoft.com/winfx/2006/xaml"
    xmlns:local="using:BlinkVBStyle"
    xmlns:d="http://schemas.microsoft.com/expression/blend/2008"

    mc:Ignorable="d"
    Background="{ThemeResource ApplicationPageBackgroundThemeBrush}">

    <Grid>
-----> OUR CODE GOES HERE <----
    </Grid>
</Page>
```

Recall that the XAML file is used to define a user interface in a platform-independent way using an XML-like language. In this project, I demonstrate the more basic controls: a text box, a button, and an ellipse (circle) placed inside a special control called a stacked panel. The stacked panel allows you to arrange the controls in a vertical "stack," making it easier to position them. As you can see in the listing, you want to place your XAML user interface items in the `<Grid></Grid>` section.

In this example, you want a text box at the top and a circle (ellipse) to represent the LED that you will use to turn on (change to green) and off (change to gray) to correspond with the hardware on/off code that you will add later. You also need a button to toggle the blink operation on and off. Finally, you'll add another text box to allow you to communicate with the user about the state of the GPIO code (that you'll add later).

Now let's add the code. Since the stacked panel is a container, all of the controls are placed inside it. Listing 7-4 shows the code you want to add (shown in bold).

Listing 7-4. Adding XAML Code for the User Interface: MainPage.xaml

```
<Page
    x:Class="BlinkVBStyle.MainPage"
    xmlns="http://schemas.microsoft.com/winfx/2006/xaml/presentation"
    xmlns:x="http://schemas.microsoft.com/winfx/2006/xaml"
    xmlns:local="using:BlinkVBStyle"
    xmlns:d="http://schemas.microsoft.com/expression/blend/2008"

    mc:Ignorable="d"
    Background="{ThemeResource ApplicationPageBackgroundThemeBrush}">

    <Grid Background="{ThemeResource ApplicationPageBackgroundThemeBrush}">
        <StackPanel Width="400" Height="400">
            <TextBlock x:Name="title" Height="60" TextWrapping="NoWrap"
                    Text="Hello, Blinky Visual Basic Style!" FontSize="28"
                    Foreground="Blue"
                    Margin="10" HorizontalAlignment="Center"/>
            <Ellipse x:Name="led_indicator" Fill="LightGray"
            Stroke="Gray"  Width="75"
                    Height="75" Margin="10" HorizontalAlignment="Center"/>
            <Button x:Name="start_stop_button" Content="Start" Width="75"
            ClickMode="Press"
                    Click="start_stop_button_Click" Height="50"
                    FontSize="24"
                    Margin="10" HorizontalAlignment="Center"/>
            <TextBlock x:Name="status" Height="60" TextWrapping="NoWrap"
                    Text="Status" FontSize="28" Foreground="Blue"
                    Margin="10" HorizontalAlignment="Center"/>
        </StackPanel>
    </Grid>
</Page>
```

Notice the button control. Here, you have an event that you want to associate with the button named start_stop_button_Click, which you assigned via the Click attribute. That is, when the user clicks it, a method named start_stop_button_Click() will be called.

XAML provides a great way to define a simple, easy user interface with the XML-like syntax. However, it also provides a mechanism to associate code with the controls. The code is placed in another file called a source-behind file named `MainPage.xaml.vb`. You will place all of the source code for the application in this file.

If you were typing this code in by hand, you will notice a nifty feature of Visual Studio—a context-sensitive help called IntelliSense that automatically completes the code you're typing and provides drop-down lists of choices. For example, when you type in the button control and type *Click=*, a drop-down box will appear, allowing you to create the event handler (a part of the code that connects to the XML). In fact, it creates the code in the `MainPage.xaml.vb` file for you. If you copy and pasted the code, you will not get this option and would have to type in the code manually.

Let's look at the code for the button control implemented in the class that you created in the source code file (`MainPage.xaml.vb`). Listing 7-5 shows the code you need to add in bold. You place everything in the class named `BlinkVBStyle` (the application).

Listing 7-5. Adding the Base Code: MainPage.xaml.vb

```
'
' Windows 10 for the IoT Second Edition
'
' Blink Visual Basic Style Example
'
' Created by Dr. Charles Bell
'
Imports Windows.Devices.Gpio

Public NotInheritable Class MainPage
    Inherits Page

    ' Create brushes for painting contols
    Private greenBrush As SolidColorBrush = New SolidColorBrush(Windows.
    UI.Colors.Green)
    Private grayBrush As SolidColorBrush = New SolidColorBrush(Windows.
    UI.Colors.Gray)

    ' Add a variable to control button
    Private blinking As Boolean = False
```

```vb
Public Sub New()
    Me.InitializeComponent()
    ' Add code to initialize the controls
    led_indicator.Fill = grayBrush
End Sub

Private Sub start_stop_button_Click(sender As Object, e As
RoutedEventArgs)
    blinking = Not blinking
    If (blinking) Then
        start_stop_button.Content = "Stop"
        led_indicator.Fill = greenBrush
    Else
        start_stop_button.Content = "Start"
        led_indicator.Fill = grayBrush
    End If
End Sub
End Class
```

Note   Notice the New() method. This is the constructor for a Visual Basic class. In other languages, this method would have the same name as the class.

OK, there are a few extra bits here that may not be very obvious why they're here. You want to paint the LED control green and gray for on and off. To do that, you need a reference to the green and gray brush resources. Thus, I create a new object from the Windows user interface colors namespace.

You also add the code for the button click event, start_stop_button_Click(), as well as a boolean member variable that you use to trigger the LED timer. You add code that inverts the blinking variable (toggles between false and true), and depending on the value, you turn the LED indicator control green (meaning the LED is on) or gray (meaning the LED is off). You also change the label of the button to correspond with the operation. That is, if the button is labeled Start, the LED indicator is off, and when clicked, the label changes to Stop and the LED indicator is turned on.

That's it! You've finished the user interface. Go ahead and build the solution correcting any errors that may appear. Once compiled, you're ready to test it.

Test and Execute: User Interface Only

That was easy, wasn't it? Better still, since this is a universal app, you can run this code on your PC. To do so, choose debug and x86 (or x64) from the platform box and press *Ctrl+F5*. Figure 7-4 shows an excerpt of the output (just the control itself). Note that you may see the application run with a black background depending on your Windows settings. The figures are shown with a white background for clarity.

Figure 7-4. *The user interface: timer off*

Figure 7-5 shows what happens when you click the button. Cool, eh?

Figure 7-5. *The user interface: timer on*

You may be wondering where the blink part is. Well, you haven't implemented it yet. You will do that in the next phase.

Add the GPIO Code

Now, let's add the code to work with the GPIO header. For this phase, you cannot run the code on your PC because the GPIO header doesn't exist, but you can add code to check the GPIO header status—hence the extra text box in the interface.

Note   The following is a bit more complicated and requires additional objects. Thus, I walk through the code changes one part at a time. Henceforth, for brevity, I present excerpts of the files that we will be editing.

Now, let's add the variables you need for the timer in the source file (`MainPage.xaml.vb`). Listing 7-6 shows the code you need to add. Here, you add an instance of the `DispatchTimer` as well as several private variables for working with the pin in the GPIO library. I show the changes in context with the new lines in bold.

Listing 7-6. Adding the Timer and GPIO Variables: MainPage.xaml.vb

```
...
    ' Create brushes for painting contols
    Private greenBrush As SolidColorBrush = New SolidColorBrush(Windows.
    UI.Colors.Green)
    Private grayBrush As SolidColorBrush = New SolidColorBrush(Windows.
    UI.Colors.Gray)

    ' Add a Dispatch Timer
    Private blinkTimer As DispatcherTimer

    ' Add variables for the GPIO
    Private Const LED_PIN As Integer = 4
    Private pin As GpioPin
    Private pinValue As GpioPinValue

    ' Add a variable to control button
    Private blinking As Boolean = False
...
```

The private variables store the pin value, a constant set to GPIO 4 (hardware pin #7), and a variable to store the pin variable result.

Next, you need a new method to initialize the GPIO—InitGPIO(). You add a new private method to the class and complete it with code to control the GPIO as shown in Listing 7-7. It is placed in the class declaration after the start_stop_button_Click() method. As you can see, there is a lot going on here. Ellipses shown indicate portions of the code omitted for brevity.

Listing 7-7. Adding the InitGPIO() Code: MainPage.xaml.vb

```vb
Public NotInheritable Class MainPage
    Inherits Page
...

    Private Sub InitGPIO()
        Dim gpio_ctrl = GpioController.GetDefault()

        ' Check GPIO state
        If (gpio_ctrl Is Nothing) Then
            pin = Nothing
            status.Text = "ERROR: No GPIO controller found!"
            Return
        End If
        ' Setup the GPIO pin
        pin = gpio_ctrl.OpenPin(LED_PIN)
        ' Check to see that pin Is Ok
        If (pin Is Nothing) Then
            status.Text = "ERROR: Can't get pin!"
            Return
        End If
        pin.SetDriveMode(GpioPinDriveMode.Output)
        pinValue = GpioPinValue.Low    ' turn off
        pin.Write(pinValue)
        status.Text = "Good to go!"
    End Sub

End Class
```

The code first creates an instance of the default GPIO controller class. Next, you check to see if that instance is null, which indicates the GPIO header cannot be initiated, and if so you change the label of the status text and return. Otherwise, you open the GPIO pin defined earlier. If that value is null, you print the message that you cannot get the pin. Otherwise, you set up the pin for output mode, and then turn off the bin and state all is well in the status label text.

Next, you add a new method to handle the event fired from the `DispatchTimer` named `BlinkTimer_Tick()`. The timer fires (or call) this method on the interval you specify (see Listing 7-8 for the changes to the `New()` method). You can place this code immediately after the `InitGPIO()` method in the class.

Listing 7-8. Adding the BlinkTimer_Tick() Code: MainPage.xaml.vb

```
...
    Private Sub BlinkTimer_Tick(sender As Object, e As Object)
        ' If pin Is on, turn it off
        If (pinValue = GpioPinValue.High) Then
            led_indicator.Fill = grayBrush
            pinValue = GpioPinValue.Low
        Else
            ' else turn it on
            led_indicator.Fill = greenBrush
            pinValue = GpioPinValue.High
        End If
        pin.Write(pinValue)
    End Sub
End Class
```

In this method, you check the value of the pin. Here is where the real operation of the code happens. In this code, if the pin is set to high (on), you set it to low (off) and paint the LED control gray. Otherwise, you set the pin to high (on) and paint the LED control green. Cool, eh?

Note   You could change this color to match the color of your LED if you wanted. Just remember to change the brush accordingly in the source file.

Next, you need to add code to the start_stop_button_Click() method to start and stop the timer. Listing 7-9 shows the changes in bold.

Listing 7-9. Adding the Timer Control Code: MainPage.xaml.vb

```vb
Private Sub start_stop_button_Click(sender As Object, e As RoutedEventArgs)
    blinking = Not blinking
    If (blinking) Then
        start_stop_button.Content = "Stop"
        led_indicator.Fill = greenBrush
        blinkTimer.Start()
    Else
        start_stop_button.Content = "Start"
        led_indicator.Fill = grayBrush
        blinkTimer.Stop()
        pinValue = GpioPinValue.Low
        pin.Write(pinValue)
    End If
End Sub
```

You see here a few things going on. First, notice that you invert the blinking variable to toggle blinking on and off. You then add a call to the blinkTimer instance of the DispatchTimer to start the timer if the user presses the button when it is labeled Start. Notice you also set the label of the button to Stop so that, when clicked again, the code turns off the timer and sets the pin value to low (off). This is an extra measure to ensure that if the button is clicked when the timer is between tick events, the LED is turned off. Try removing it and you'll see.

Finally, you must add a few lines of code to the New() method to get everything started when you launch the application. Listing 7-10 shows the code modifications with changes in bold.

Listing 7-10. Adding the Timer and GPIO Initialization Code: MainPage.xaml.vb

```
Public Sub New()
    Me.InitializeComponent()
    ' Add code to initialize the controls
    led_indicator.Fill = grayBrush
    ' Add code to setup timer
    blinkTimer = New DispatcherTimer()
    blinkTimer.Interval = TimeSpan.FromMilliseconds(1000)
    AddHandler blinkTimer.Tick, AddressOf BlinkTimer_Tick
    blinkTimer.Stop()
    ' Initialize GPIO
    InitGPIO()
End Sub
```

Notice you add code to create a new instance of the `DispatchTimer` class, set the interval for the tick event to 1 second (1000 milliseconds), and add the `BlinkTimer_Tick()` method to the new instance (this is how you assign a method reference to an existing event handle). Next, you stop the timer and finally call the method that you wrote to initialize the GPIO.

That's it! Now, let's build the solution and check for errors. You should see something like the following in the output window:

```
1>------ Build started: Project: BlinkVBStyle, Configuration: Debug ARM
------
1>  BlinkVBStyle -> C:\Users\olias\source\repos\BlinkVBStyle\BlinkVBStyle\
bin\ARM\Debug\BlinkVBStyle.exe
========== Build: 1 succeeded, 0 failed, 0 up-to-date, 0 skipped ==========
```

OK, now you're ready to deploy the application to your device. Go ahead and set up everything, make the connections on the breadboard, and power on your device.

Deploy and Execute: Completed Application

Once your code compiles, you're ready to deploy the application to your Raspberry Pi (or another device). Recall from Chapter 4, we have a different deployment method for Visual Basic applications. If you want to install your application, you can do so from

Visual Studio with only a few steps. However, if you want to debug your application, you must use a slightly different process. Let's look at each of these options starting with a normal deployment.

Before you attempt to deploy your application, make sure the remote debugger is stopped. We do not need the remote debugger running to deploy Visual Basic applications. This is because we will be using a different authentication mode.

Note   Make sure the remote debugger is stopped on your device before attempting to deploy a Visual Basic application.

Deploying Your Application (No Debugging)

Deploying your application with a Visual Basic project template is easy. Simply select the build mode, platform, and then device in Visual Studio. In this case, we want to select the *Release* or *Debug* build, the *ARM* platform, and select the *Remote Machine* entry in the device drop-down as shown in Figure 7-6. For this form of deployment, it doesn't matter which build mode you choose.

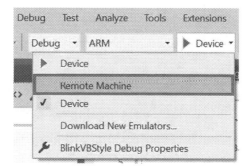

Figure 7-6.   *Select the build, platform, and device (Visual Studio)*

When you select the Remote Machine entry, a Remote Connections dialog will appear that permits you to select the device from a list as shown in Figure 7-7. Simply select the device and click Select to select it.

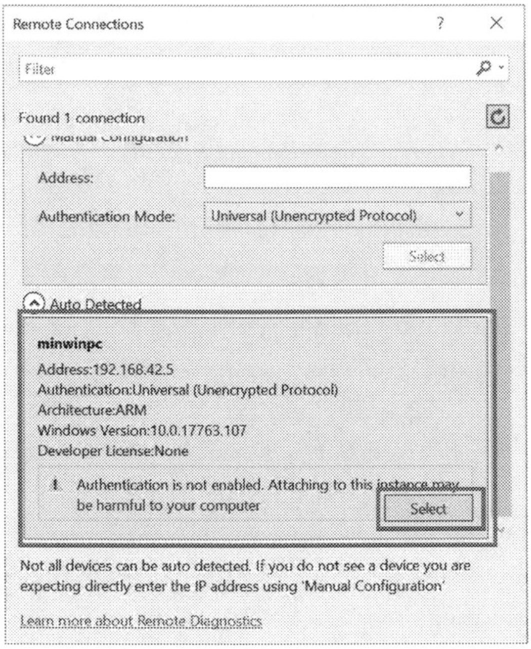

Figure 7-7. *Remote Connections dialog*

OK, we're almost ready to deploy. There is just one more thing to change. I find it helpful to rename the package name using the package manifest. This is the name that will be displayed in the Device Portal. By default, it uses a generated name, but I find renaming it to the same as the project name helpful for easier tracking. Thus, this is an optional step.

If you look at the left side of Visual Studio, you will see one file in the tree view named `Package.appxmanifest`. Double-click that file to open it. As you will see, it is a set of options you can set for your package. The one we want is on the last tab named Packaging. Click that and change the Package name, as shown in Figure 7-8.

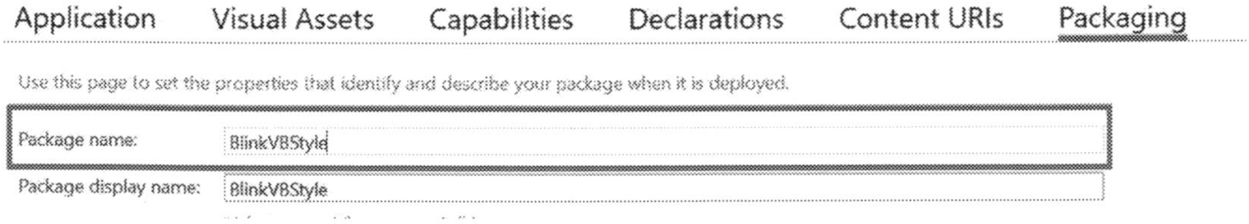

Figure 7-8. *Set the package name (package manifest)*

Now all that is left is to deploy the application. Simply click Build ➤ Deploy Solution, and the deploy will begin. If this is the first time you're deploying the application (or any application) to your device, the deployment could take a while.

Once complete, Visual Studio will indicate the build and deploy was successful similar to the output shown as follows:

```
Creating a new clean layout...
Copying files: Total 7 mb to layout...
Checking whether required frameworks are installed...
Registering the application to run from layout...
Deployment complete (0:00:22.489). Full package name:
"BlinkVBStyle_1.0.0.0_arm__vmq2bagagk5h4"
========== Build: 0 succeeded, 0 failed, 1 up-to-date, 0 skipped ==========
========== Deploy: 1 succeeded, 0 failed, 0 skipped ==========
```

Now, you can go to the Device Portal and observe your application in the Apps manager as shown in Figure 7-9. Recall, you can start it by selecting *Start* from the *Actions* drop-down box. Go ahead and do that now. You should see your application start, and you can experiment with it.

App Name	App Type	Startup	Status	
BlinkCSharpStyle	Foreground	○	Stopped	Actions
BlinkVBStyle	Foreground	○	Stopped	Actions
Connect	Foreground	○	Stopped	Actions Start Uninstall Details Actions
IOTCoreDefaultApplication	Foreground	◉	Running	
IoTUAPOOBE	Foreground	○	Stopped	
IoTOnboardingTask	Background	⬤	Running	Actions

Figure 7-9. *Controlling the application in the Apps manager (Device Portal)*

After a few moments, you will see the application start on your device. If you have all of the wiring complete, you can click the button and observe the LED blinking until you stop it. Cool, eh?

If the LED is not blinking, double-check your wiring and ensure that you chose pin 4 in the code and pin 7 on the pin header (recall that the pin is named GPIO 4, but it is pin #7 on the header for the Raspberry Pi).

When you're done, you can stop it from the Apps manager in the Device Portal by using the *Actions* drop-down box and selecting *Stop*. It may take a few seconds for the status to change from *Running* to *Stopped*.

If you want to start the application again, you do not need to redeploy it. The application is installed on the device. To start it, go back to the Device Portal, click *Apps*, then *Apps manager*, and then click the *Actions* drop-down box for the application and click *Start*.

Should you want to uninstall the application, go back to the Device Portal, click *Apps*, then *Apps manager*, and then click the *Actions* drop-down box for the application and click *Uninstall*.

And that's it! Congratulations, you've just written your first Visual Basic application that uses the GPIO header to power some electronics!

Now, let's see how to deploy and debug our application.

Deploy and Debug

If you want to debug your application, you can, but once again it is not the same process we saw with the C++ application. Specifically, we do not need to use the remote debugger. Instead, we can make the same selections from the debug toolbar, selecting the *Debug* build, the *ARM* platform, and the *Remote Machine* from the device drop-down. If you have previously completed the Remote Connections dialog, you will not see that again. If you do see the dialog, make sure to select your device as described earlier and click the *Select* button.

Before you start the debugger, let's check a few settings. Most of these should be set correctly, but I've seen a couple of situations where the debug settings have changed. If they are not set correctly, you could have difficulty starting the debug session or may receive a series of strange, cryptic messages explaining either the application is already running or the debug symbols cannot be loaded.[4]

To open the debug settings, right-click the project (not the solution) and choose *Properties*. Then, select the *Debug* page. The principal settings we need to check include the following. Figure 7-10 shows the correct settings and their values. Be sure your project has the correct settings (the device name should be the name of your device).

- *Target device*: Remote Machine

---

[4]Can you guess how I discovered this? Yes, I've been there....

- *Remote Machine*: The name of your device

- (Optional) *Uninstall and then re-install*: Checked

- (Optional) *Deploy optional packages*: Checked

- *Application process*: Managed Only

- *Background task process*: Managed Only

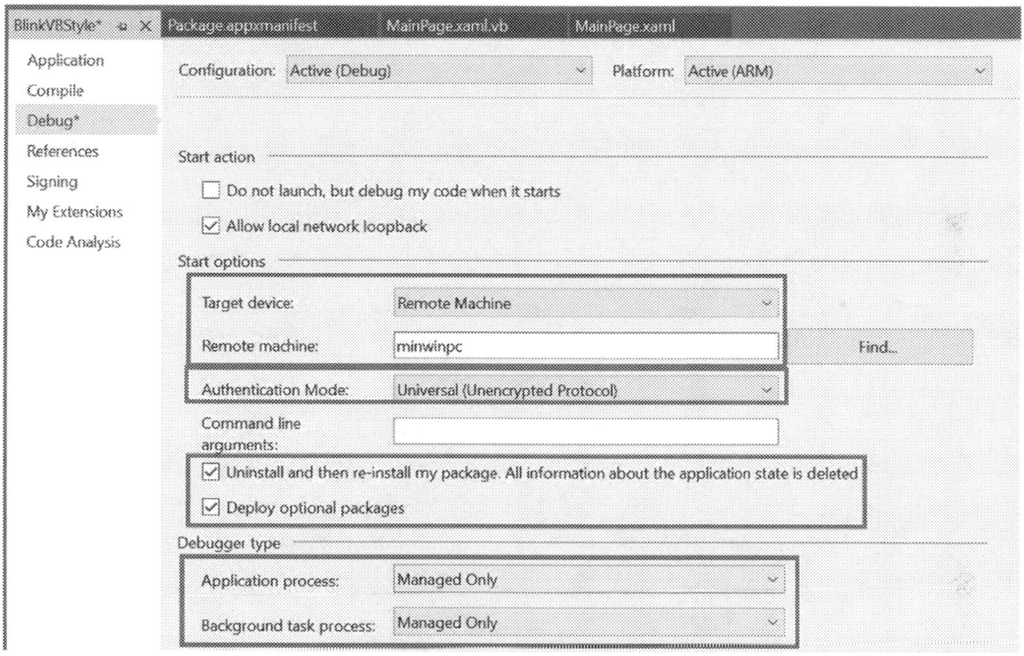

Figure 7-10. *Project debug settings*

Notice there are two optional settings. Checking these will ensure your application will always be deployed (thus overwriting previous installations) and all optional packages are included, which may avoid certain dependency issues if you have installed and uninstalled other applications on the device.

Tip   If you want to reset the Remote Connection, remove the IP or device name in the *Remote Machine* text box.

Once you make any changes in this dialog, be sure to save your solution. Building or deploying does not automatically save these settings.

OK, now we are ready to debug our application. I find it helpful to place a breakpoint in the code so that when the application starts, I can step through it. Later, you may want to set a breakpoint somewhere in your code to observe certain variables or check for logic errors.

Recall, we set a breakpoint by clicking the left side of the code window in Visual Studio. A breakpoint is indicated with a red dot on the left side of the code window as demonstrated with an arrow in Figure 7-11.

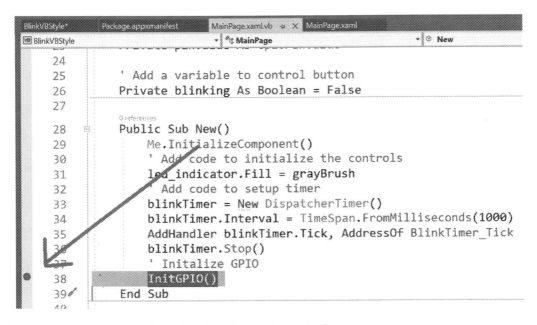

Figure 7-11. *Setting a breakpoint (Visual Studio)*

Now we can start the debugging session. Simply click *Debug ➤ Start Debugging* or click *F5*. You should see a lengthy set of messages in the output window under the deployment option, but when done, you will see the application running on your device, and it will also be displayed in the Device Portal under the Apps manager. You should also see the interactive debugger start in Visual Studio as shown in Figure 7-12.

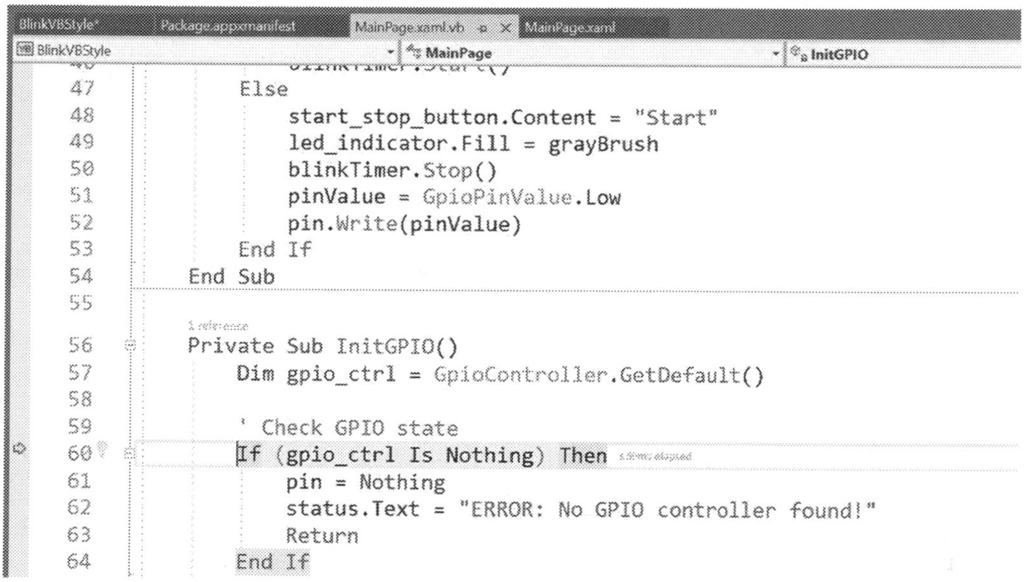

```
BlinkVBStyle*          Package.appxmanifest        MainPage.xaml.vb  ↦  ×   MainPage.xaml
BlinkVBStyle                              ▾  MainPage                                          ▾  InitGPIO

47              Else
48                  start_stop_button.Content = "Start"
49                  led_indicator.Fill = grayBrush
50                  blinkTimer.Stop()
51                  pinValue = GpioPinValue.Low
52                  pin.Write(pinValue)
53              End If
54          End Sub
55
          1 reference
56          Private Sub InitGPIO()
57              Dim gpio_ctrl = GpioController.GetDefault()
58
59              ' Check GPIO state
60          If (gpio_ctrl Is Nothing) Then  5.99ms elapsed
61                  pin = Nothing
62                  status.Text = "ERROR: No GPIO controller found!"
63                  Return
64          End If
```

Figure 7-12. *Interactive debugger (Visual Studio)*

Now, you can step through your application using the debug menu. For example, you can step over method calls by clicking *F10* or step into methods with *F11*. Should you want to continue execution, you can click *Continue* in the debug toolbar. Or, you can stop the debugger using the *Stop* button or the *Debug* menu. Figure 7-13 shows what stepping through your code may look like. Yes, this means we can run the application on the device and debug it on our PC. Excellent!

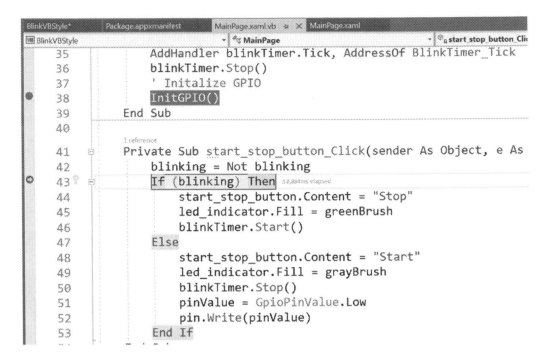

Figure 7-13. Stepping through code with interactive debugger (Visual Studio)

Go ahead and experiment with the application until you are satisfied it is working correctly. When you want to stop it, go to the Device Portal, and click *Apps* ➤ *Apps manager* and use the *Actions* drop-down to stop the application.

If you get errors during the deployment, go back and check all of your settings to ensure you have everything correct. See the *Visual Basic Application Deployment Troubleshooting* section for more details.

Note   If the app deployed successfully but doesn't show in the drop-down list, try disconnecting and reconnecting. If that doesn't work, try rebooting your device.

Visual Basic Application Deployment Troubleshooting

If you're like me, things sometimes go wonky and just don't work, or they present you with an interesting but nearly indecipherable error message. I present a couple of these incidents you may encounter along with actions you can take to prevent or correct them.

Application Already/Not Running

If you get an error dialog saying the application is already or not running, you should double-check your debug settings for the project as described earlier. This can happen if you've deployed the application numerous times or have previous versions installed. Clearing the debug properties selections usually fixes the problem, but if that does not, change the package name or increment the version number, uninstall the application from the device, and restart your deployment.

Failure to Unregister Application

If you have installed the application many times and attempted to uninstall it, or it crashed and became corrupt, or you accidentally (maybe intentionally) removed the application files, you can see an error message during deployment like the following:

```
DEP0900: Failed to unregister application
"BlinkVBStyle_1.0.0.0_arm__vmq2bagagk5h4". [0x80073CFA] AppX Deployment
operation failed with error 0x80070002 from API IsCurrentProfileSpecial
```

What this means is either the application is still registered on the device or some of its files are present, but it is not registered. To fix this, open an SSH terminal to the device and remove the application directory and all of its files like shown as follows. In this example, I had a derelict package on my device left over from a previous session,[5] and it caused my deployment of a newer application with the same package name to fail. In this case, the package was named BlinkVBStyle_ vmq2bagagk5h4.

```
[192.168.42.13]: PS C:\Data\Users\Administrator\AppData\Local\Packages>
rmdir /s BlinkVBStyle_1.0.0.0_arm__vmq2bagagk5h4
BlinkVBStyle_1.0.0.0_arm__vmq2bagagk5h4, Are you sure (Y/N)? Y.
```

Once you've removed the folder, you can deploy your application.

Deploy Succeeds, but Nothing Happens

It is possible you can deploy your application, but it either doesn't start or doesn't show up in your Apps manager on the Device Portal.

---

[5]From the first edition of this book.

If Visual Studio reports the deployment succeeded, but the application doesn't start, that's OK and normal if you are doing a straight deployment. If you are deploying and debugging expecting the application to start, check the project properties as described earlier and correct any settings that may have changed. You can also attempt to uninstall the application using the App manager before retrying the deployment. If that still doesn't work, try removing all applications you've built previously.

If the application doesn't show up in the Apps manager, be sure to double-check all of the project settings in Visual Studio as described earlier, and the device is connected and reachable from the Device Portal. Try using the clean and build solution menu options again before attempting to deploy the application.

Tip   As a challenge, you can modify this project to add a *Close* or an *Exit* button to stop the application.

Summary

If you are learning how to work with Windows 10 IoT Core and don't know how to program with Visual Basic, learning Visual Basic can be fun given its easy-to-understand syntax. While there are many examples on the Internet that you can use, most are not UWP projects and thus can be a challenge to understand how they translate to UWP applications. As you have seen, the UWP application template for Visual Basic follows the same pattern and uses the same libraries as the C# example from the last chapter. Only, the syntax is less complex.

This chapter has provided a crash course in Visual Basic that covers the basics of the things you encounter when examining most of the smaller example projects. You discovered the basic syntax and constructs of a Visual Basic application, including a walk-through of building a real Visual Basic application that blinks an LED. Through that example, you learned how to work with headless applications, including how to manage a startup background application.

The next chapter takes a short detour in your exploration of Windows 10 IoT Core projects. You are introduced to the basics of working with electronics. Like the programming crash course, the chapter provides a short introduction to working with electronics. A mastery of electronics in general is not required for the projects you explore, but if you've never worked with electronic components before, the next chapter will prepare you for the more advanced projects in Chapters 10–14.

## CHAPTER 8

Electronics for Beginners

If you're new to the IoT or have never worked with electronics, you may be wondering how you're going to get your ideas for an IoT solution realized. The projects in this book walk you through how to connect the various components used, and thus you can complete them without a lot of additional information or specialized skills.

However, if something goes wrong or you want to create projects on your own, you may need a bit more information than "plug this end in here." More specifically, you need to know enough about how the components work in order to successfully complete your project—whether that is completing the examples in this book or examples found elsewhere on the Internet.

Rather than attempt to present a comprehensive tutorial on electronics, which would take several volumes, this chapter presents an overview of electronics for those who want to work with the types of electronic components commonly found in IoT projects. I include an overview of some of the basics, descriptions of common components, and a look at sensors. If you are new to electronics, this chapter gives you the extra boost you need to understand the components used in the projects in this book.

However, if you have experience with electronics either at the hobbyist or enthusiast level or have experience or formal training in electronics, you may want to skim this chapter or read the sections with topics that interest you.

SELF-PACED ELECTRONICS TRAINING

If you find you need or want to learn more about electronics, especially the types of electronics you need for an IoT solution, check out the set of electronics books by Charles Platt. I've found these books to be very well written, opening the door for many to learn electronics without having to spend years learning the tedious (but no less important) theory and mathematics of electronics. I recommend the following books for anyone wanting to learn more about electronics:

© Charles Bell 2021
C. Bell, *Windows 10 for the Internet of Things*, https://doi.org/10.1007/978-1-4842-6609-0_8

- *Make: Electronics, Second Edition* by Charles Platt (O'Reilly, 2015)

- *Make: More Electronics* by Charles Platt (O'Reilly, 2014)

- *Encyclopedia of Electronic Components* by Charles Platt (O'Reilly, 2012)

Chaney Electronics sells companion kits that contain all of the parts you need to complete the experiments in the *Make: Electronics* and *Make: More Electronics* books (www.electronickitsbychaneyelectronics.com/products.asp?dept=192). The books together with the kits make for an excellent self-paced learning experience.

Let's begin with a look at the basics of electronics. Once again, this is in no way a tutorial that covers all there is to know, but it gets you to the point where the projects make sense in how they connect and use components.

The Basics

This section presents a short overview of some of the most common tools and techniques you need to use when working with electronics. As you will see, you only need the most basic of tools, and the skills or techniques are not difficult to learn. However, before you get into those, let's discuss the most fundamental concept you must understand when working with electronics—power!

Powering Your Electronics

Electricity[1] is briefly defined as the flow of electric charge and, when used, provides power for your electronics—a common light bulb, a ceiling fan, a high-definition television, a tablet. Whether you are powering your electronics with batteries or a power supply, you are initiating a circuit where electrons flow in specific patterns. There are two forms (or kinds) of power that you will use. Your home is powered by alternating current, and your electronics are powered by direct current.

The term alternating current (AC) is used to describe the flow of charged particles that changes direction periodically at a specific rate (or cycle), reversing the voltage along with the current. Thus, AC systems are designed to work with a specific range of cycles as well as voltage. Typically, AC systems use higher voltages than direct current systems.

---

[1]See https://learn.sparkfun.com/tutorials/what-is-electricity.

The term direct current (DC) is used to describe the flow of charged particles that do not change direction and thus always flow in a specific "direction." Most electronics systems are powered with DC voltages and are typically at lower voltages than AC systems; for example, IoT projects typically run on lower direct current (DC) voltages in the range 3.3V–24V.

Tip   For more information about AC and DC current and the differences, see `https://learn.sparkfun.com/tutorials/alternating-current-ac-vs-direct-current-dc`.

Since DC flows in a single direction, components that operate on DC have a positive side and a negative side, where current flows from positive to negative. The orientation of these sides—one to positive and one to negative—is called *polarity*. Some components, such as resistors, can operate in either "direction," but you should always be sure to connect your components according to its polarity. Most components are clearly marked, but those that are not have a well-known arrangement. (For example, the positive pole (side) of an LED is the longer of the two legs; it is called an *anode*. The negative and shorter leg is called the *cathode*).

Despite the lower voltages, you mustn't think that they are completely harmless or safe. Incorrectly wiring electronics (reversing polarity) or shorting (connecting positive and negative together) can damage your electronics and in some cases cause overheating, which, in extreme cases, causes electronics to catch fire.

I had a lesson in just how real this scenario can be a couple of years ago. I was changing the batteries in my smoke detectors. I took the old batteries out and placed them in my pocket. I had forgotten I had a small penknife in the same pocket. One of the 9V batteries shorted on the knife, and within about ten minutes, the battery heated to an alarming temperature. It wasn't enough to burn, but had I left something like that unattended, it could have been bad.

That's a scary thought, isn't it? Consider it an admonishment as well as a warning; you should never relax your safe handling practices even for lower voltage projects.

Finally, DC components are often rated for a specific voltage range. Recall from the discussion on the various low-cost computing boards and GPIO headers, some boards operate at 5V, whereas others operate at 3.3V (or less). Fortunately, there are several ways you can adapt components that work at different voltages—by using other components!

> **Note**   I have deliberately kept the discussion on power simple. There is far more to electrical current—even DC—than what I've described here. As long as you understand these basics, you'll be able to work with the projects in this book and more.

Now let's take a look at some of the tools you need to work on your IoT projects.

Tools

The vast majority of tools you need to construct your IoT projects are common hand tools (screwdrivers, small wrenches, pliers, etc.). For larger projects or for creating enclosures, you may need additional tools, such as power tools, but I concentrate only on those tools needed for building the projects. The following is a list of tools that I recommend to get you started:

- Breadboard
- Breadboard wires (also called jumpers)
- Electrostatic discharge (ESD) safe tweezers
- Helping hands or printed circuit board (PCB) holder
- Multimeter
- Needle-nose pliers
- Screwdrivers: assorted sizes (micro, small)
- Solder
- Soldering iron
- Solder remover (solder sucker)
- Tool case, roll, or box for storage
- Wire strippers

However, you cannot go wrong if you prefer to buy a complete electronics toolset, such as those from SparkFun (`www.sparkfun.com/categories/47`) or Adafruit (`www.adafruit.com/categories/83`). You can often find electronics kits at major brand electronics stores and home improvement centers. If you are fortunate enough to live near a Fry's Electronics, you can find just about any electronics tool made. Most electronics kits have all the hand tools that you need. Some even come with a multimeter, but more often you have to buy it separately.

Most of the tools in the list do not need any explanation except to say you should purchase the best tools that your budget permits. The following paragraphs describe some of the tools that are used for special tasks, such as stripping wires, soldering, and measuring voltage and current.

Multimeter

A *multimeter* is one of those tools that you need when building IoT solutions. You also need it to do almost any electrical repair on your circuits. There are many different multimeters available with prices ranging from inexpensive, basic units to complex, feature-rich, incredibly expensive units. For most IoT projects, including most IoT kits, a basic unit is all that you need. However, if you plan to build more than one IoT solution or want to assemble your own electronics, you may want to invest a bit more in a more sophisticated multimeter. Figure 8-1 shows a basic digital multimeter (costing about $10) on the left and a professional multimeter from BK Precision on the right.

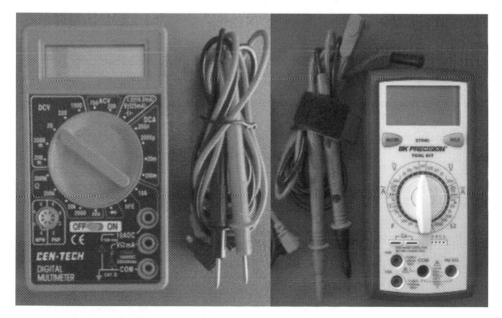

Figure 8-1. *Digital multimeters*

Notice that the better meter has more granular settings and more features. Again, you probably won't need more than the basic unit. You need to measure voltage, current, and resistance at a minimum. Whichever meter you buy, make sure that it has modes for measuring AC and DC voltage, continuity testing (with an audible alert), and checking resistance. I explain how to use a multimeter in a later section.

Tip   Most multimeters including the inexpensive ones come with a small instruction booklet that shows you how to measure voltage, resistance, and other functions of the unit.

Soldering Iron

A *soldering iron* is not required for any of the projects in this book because you use a breadboard to lay out and connect the components. However, if you plan to build a simple IoT solution where you need to solder wires, or maybe a few connectors, a basic soldering iron from an electronics store such as Radio Shack is all you need. On the other hand, if you plan to assemble your own electronics, you may want to consider getting a good, professional soldering iron, such as a Hakko. The professional models include features that allow you to set the temperature of the wand, have a wider array of tips available, and tend to last a lot longer. Figure 8-2 shows a well-used entry-level Radio Shack. Figure 8-3 shows a professional model Hakko soldering iron.

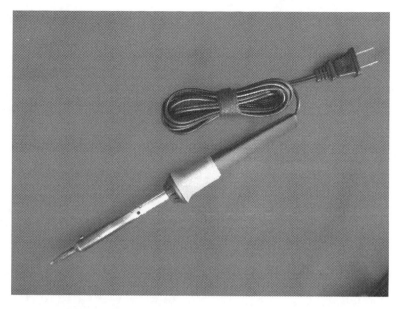

Figure 8-2. *Entry-level soldering iron*

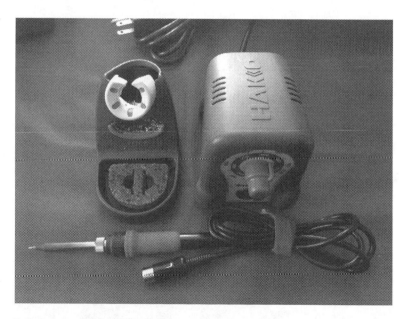

Figure 8-3. *Professional soldering iron*

Tip   For best results, choose a solder with a low lead content in the 37%–40% range. If you use a professional soldering iron, adjust the temperature to match the melting point of the solder (listed on the label).

DO I NEED TO LEARN TO SOLDER?

If you do not know how to solder or it has been a while since you've used a soldering iron, you may want to check out the book *Learn to Solder* by Brian Jepson, Tyler Moskowite, and Gregory Hayes (O'Reilly Media, 2012) or Google how-to videos on soldering. Or you could buy the SparkFun WeevilEye – Beginner Soldering Kit from SparkFun (`www.sparkfun.com/products/10723`), which comes with a soldering iron, wire cutters, supplies, and more—everything you need to learn how to solder. Cool.

Wire Strippers

There are several types of *wire strippers*. In fact, there are probably a dozen or more designs out there. But there really are two kinds: ones that only grip and cut the insulation as you pull it off the wire and those that grip, cut, and remove the insulation. The first type is more common and, with some practice, does just fine for most small jobs (like repairing a broken wire); but the second type makes a larger job—such as wiring electronics from bare wire (no prefab connectors)—much faster. As you can imagine, the first type is considerably cheaper. Figure 8-4 shows both types of wire strippers. Either is a good choice.

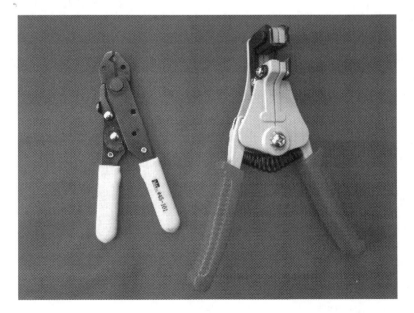

Figure 8-4. *Wire strippers*

Helping Hands

There is one other tool that you may want to get, especially if you need to do any soldering; it's called *helping hands* or a *third-hand tool*. Most have a pair of alligator clips to hold wires, printed circuit boards, or components while you solder. Figure 8-5 shows an excellent example from Adafruit (`www.adafruit.com/products/291`).

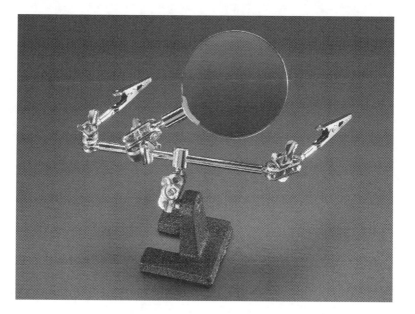

Figure 8-5. *Helping hands tool (courtesy of* `www.adafruit.com`*)*

Now let's take a look at some of the skills you are likely to need when working with advanced IoT projects.

ESD IS THE ENEMY

You should take care to make sure that your body, your workspace, and your project are grounded to avoid electrostatic discharge (ESD). ESD can damage your electronics— permanently. The best way to avoid this is to use a grounding strap that loops around your wrist and attaches to an anti-static mat like those available from `www.uline.com/ BL_7403/Anti-Static-Table-Mats`.

Using a Multimeter

The electrical skills needed for IoT projects can vary from plugging in wires on a breadboard—as you saw with the projects so far—to needing to solder components together or to printed circuit boards (PCBs). Regardless of whether you need to solder the electronics, you need to be able to use a basic multimeter to measure resistance and check voltage and current.

A multimeter is a very useful and essential tool for any electronics hobbyist and downright required for any enthusiast of worth. A typical multimeter has a digital display[2] (typically an LCD or similar numeric display), a dial, and two or more posts or ports for plugging in test leads with probe ends. Most multimeters have ports for lower current (that you will use most) and ports for higher current. Test leads use red for positive and black for negative (ground). The ground port is where you plug in the black test lead and is often marked either with a dash or COM for common. Which of the other ports you use depends on what you are testing.

One thing to note on the dial is that there are many settings (with some values repeated) or those that look similar. For example, you see a set of values (sometimes called a *scale*) for ohms, one or two sets of values for amperage, and one or two sets of values for volts. The DC voltage is indicated by a V with a solid and dashed line over it, whereas the AC voltage is indicated by a V with a wavy line over it. Amperage ranges are marked in the same manner. Figure 8-6 shows a close-up of a multimeter dial labeled with the sets of values that I mentioned.

---

[2]Older multimeters have an analog gauge. You can still find them if you want a bit of old school feel.

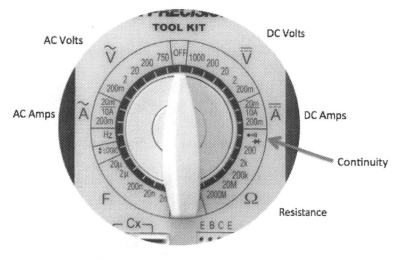

Figure 8-6. *Multimeter dial (typical)*

Tip   When not in use, be sure to turn your multimeter dial to off or one of the
voltage ranges if it has a separate off button.

There is a lot you can do with a multimeter. You can check voltage, measure
resistance, and even check continuity. Most basic multimeters do all of these functions;
however, some multimeters have a great many more features, such as testing capacitors,
and the ability to test AC and DC.

Let's see how you can use a multimeter to perform the most common tasks you
need for IoT projects: testing continuity, measuring voltage in a DC circuit, measuring
resistance, and measuring current.

Testing Continuity

You test for continuity to determine if there is a path for the charged particles to flow.
That is, your wires and components are connected properly; for example, you may want
to check to ensure that a wire has been spliced correctly.

To test for continuity, turn your multimeter dial to the position marked with an
audible symbol, bell, or triangle with an arrow through it. Plug the black test lead into
the *COM* port and the red test lead in the port marked with Hz VΩ or similar. Now you
can touch the probe end of the test leads together to hear an audible tone or beep.
Some multimeters don't have an audible tone but instead may display "1" or similar to

indicate continuity. Check your manual for how your multimeter indicates continuity. Figure 8-7 shows how to set a multimeter to check for continuity including which ports to plug in the test leads.

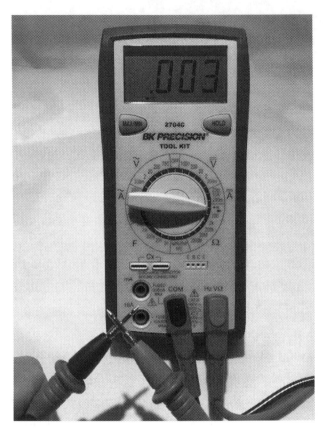

Figure 8-7. *Settings for checking continuity*

In Figure 8-7, I simply touched the probes together to demonstrate how to check for continuity. I like to do this just to ensure that my multimeter is turned on and in the correct setting.[3]

Another excellent use for the continuity test is when diagnosing or discovering how cables are wired. For example, you can use the continuity test to discover which connector is connected on each end of the cable (sometimes called *wire sorting* or *ringing out*, from the old telephone days).

---

[3]Yes, a bit of OCD there. Check, double-check, check again.

Measuring Voltage

Our IoT projects use DC. To measure voltage in the circuit, you use the DC range on the multimeter. The DC range has several stops. This is a scale selection. Choose the scale that closely matches the voltage range you want to test. For example, for our IoT projects, you often measure 3.3V–12V, so you choose 20 on the dial. Next, plug the black test lead into the COM port and the red test lead into the port labeled **Hz VΩ**.

Now you need something to measure! Take any battery you have in the house and touch the black probe to the negative side and the red probe to the positive side. You should see a value appear on the display that is close to the range for the battery; for example, if you used a 1.5V battery, you should see close to 1.5V. It may not be exactly 1.5V–1.6V if the battery is depleted. So now you know how to test batteries for freshness! Figure 8-8 shows how to measure voltage of a battery.

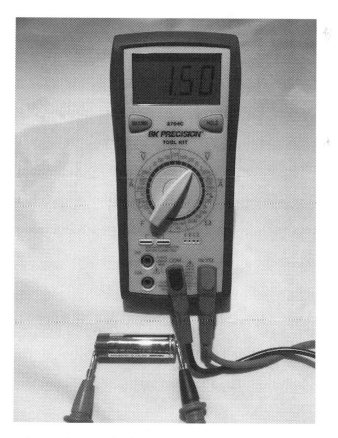

Figure 8-8. *Measuring voltage of a battery*

The readout displays 1.50, which is the correct voltage for this AA battery. If I had reversed the probes—the red one on negative and the black on positive, the display would have read –1.50. This is OK because it shows the current is flowing in the opposite direction of how the probes are oriented.

Note   If you use the wrong probe when measuring voltage in a DC circuit, most multimeters display the voltage as a negative number. Try that with your battery. It won't hurt the multimeter (or the battery)!

You can use this technique to measure voltage in your projects. Just be careful to place the probes on the appropriate positions and try not to cross or short by touching more than one component at a time with a single probe tip.

Measuring Current

Current is measured as amperage (actually milliAmps (mA)). Thus, you use the range marked with an A with a straight and dashed line (not the wavy one, that's AC). You measure current in series. That is, you must place the multimeter in the circuit. This can be a little tricky because you must interrupt the flow of current and put the meter inline.

Let's set up an experiment to measure current. Get your breadboard, an LED, a resistor, and two jumper wires you used in the blink project. Wire everything up the same way except don't complete the circuit for the GPIO4 pin. Instead, you use the multimeter to complete the circuit by touching one probe to the positive 5V pin on the GPIO and the other probe on the resistor. Figure 8-9 shows how to set up the circuit with the multimeter inline.

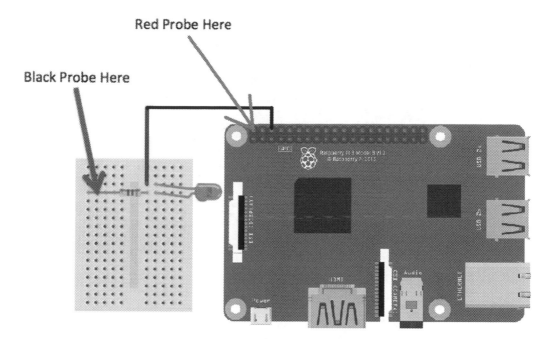

Figure 8-9. *Measuring current - locations*

Before powering on your Raspberry Pi, plug the black test lead into the COM port and the other test leads into the port labeled mA. Some multimeters use the same port for measuring voltage as well as current. Turn the dial on the multimeter to the 200mA setting. Then power on the Raspberry Pi and touch the leads to the places indicated. Be careful to touch only the 5V pin on the header. If you want to err on the side of caution, use the remaining jumper wire and connect it to the 5V pin, and then touch the probe to the other end of the jumper wire. Once the Raspberry Pi is powered on, you should see a value on the multimeter. Figure 8-10 shows how to use a multimeter to measure current in a circuit.

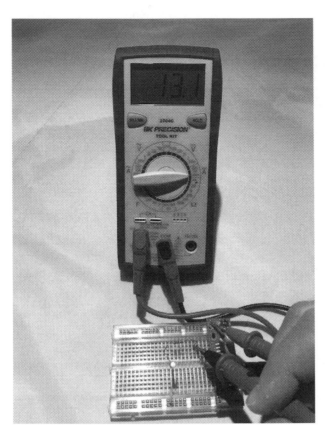

Figure 8-10. *Measuring current with a multimeter*

In Figure 8-10, I am using a breadboard with a breadboard power supply instead of the Raspberry Pi. Whereas the value you see on the multimeter may differ, the demonstration accomplishes the same goal. In this case, I touch the red probe to the positive pole on the power supply and the black probe on the resistor in the same manner as I described.

There is one other tricky thing about measuring current. If you attempt to measure current that is greater than the maximum for the port (e.g., the meter in the photo has a maximum of 20mA on the one port) it would likely blow a fuse in the multimeter. This is not desirable, but at least there is a fuse that you can replace should you make a mistake and choose the wrong port.

Measuring Resistance

Resistance is measured in ohms (Ω). A *resistor* is the most common component that you use to introduce resistance in a circuit. You can test the resistance of the charge through the resistor with your multimeter. To test resistance, choose the ohm scale that is closest to the rating of the resistor. For example, I tested a resistor that I believed about 200 ohms, but since I was not sure, I chose the 2K setting.

Next, plug the black test lead into the COM port and the red test lead into the port labeled **Hz VΩ**. Now, touch a probe to one side of a resistor and the other probe to the other side. It doesn't matter which side you choose—a resistor works in both directions. Notice the readout. The meter reads one of three things: 0.00, 1, or the actual resistor value.

In this case, the meter reads 0.219, meaning this resistor has a value of 220Ω. Recall, I used the 2K scale, which means a resistor of 1K would read 1.0. Since the value is a decimal, I can move the decimal point to the left to get a whole number.

If the multimeter displays another value, such as 0 or 1, it indicates the scale is wrong and you should try a higher scale. This isn't a problem. It just means you need to choose a larger scale. On the other hand, if the display shows 0 or a really small number, you need to choose a lower scale. I like to go one tick of the knob either way when I am testing resistance in an unknown component or circuit.

Figure 8-11 shows an example of measuring resistance for a resistor. The display reads 219. I am testing a resistor rated at 220 ohms. The reason it is 219 instead of 220 is because the resistor I am using is rated at 220 +/– 5%. Thus, the acceptable range for this resistor is 209 ohms to 231 ohms.

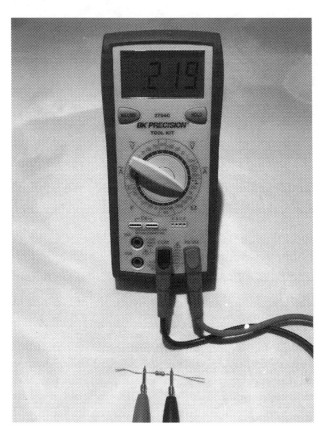

Figure 8-11. *Measuring resistance of a resistor*

Now you know how to test a resistor to discover its rating. As you will see, those rings around the body of the resistor are the primary way you know its rating, but you can always test it if you're unsure, someone has painted over it (hey, it happens), or you're too lazy to look it up.

Electronic Components

Aside from learning how to use a multimeter and possibly learning to solder, you also need to know something about the electronic components available to build your projects. In this section, I describe some common components—listed in alphabetical order by name—that you encounter when building IoT solutions. I also cover breakout boards and logic circuits, which are small circuits built with a set of components that provide a feature or solve a problem. For example, you can get breakout boards for USB host connections, Ethernet modules, logic shifters, real-time clocks, and more.

Button

A *button* (sometimes called a *momentary button*) is a mechanism that makes a connection when pressed. More specifically, a button connects two or more poles together while it is pressed. A common (and perhaps overused) example of a button is a home doorbell. When pressed, it completes a circuit that triggers a chime, bell, tone, or music to play. Some older doorbells continue to sound while the button is pressed.

In IoT projects, you use buttons to trigger events, start and stop actions, and similar operations. A button is a simple form of a switch, but unlike a switch, you must continue to press the button to make the electrical connections. Most buttons have at least two legs (or pins) that are connected when the button is pressed. Some have more than two legs connected in pairs, and some of those can permit multiple connections. Figure 8-12 shows a number of buttons.

Figure 8-12. *Momentary buttons*

There is a special variant of a momentary button called a *latching momentary button*. This version uses a notch or detent to keep the poles connected until it is pushed again. If you've seen a button on a stereo or in your car that remains depressed until pressed again, it is likely a latching momentary button.

There are all kinds of buttons from those that can be used with breadboards (the spacing of the pins allows it to be plugged into a breadboard), can be mounted in a panel, or those made for soldering to printed circuit boards.

Capacitor

A *capacitor* is designed to store charges. As current flows through the capacitor, it accumulates charge and can discharge after the current is disconnected. In this way, it is like a battery, but unlike a battery, a capacitor charges and discharges very fast. You use capacitors for all manner of current storage from blocking current, reducing noise in power supplies, in audio circuits, and more. Figure 8-13 shows a number of capacitors.

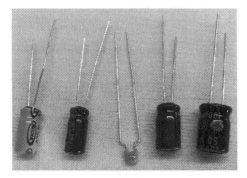

Figure 8-13. *Capacitors*

There are several types of capacitors, but you most often encounter capacitors when building power supplies for IoT projects. Most capacitors have two legs (pins) that are polarized. That is, one is positive and the other negative. Be sure to connect the capacitor with the correct polarity in your circuit.

Diode

A *diode* is designed to allow current to flow in only one direction. Most are marked with an arrow pointing to a line, which indicates the direction of flow. A diode is often used as rectifiers in AC-to-DC converters (devices that convert AC to DC voltage), used in conjunction with other components to suppress voltage spikes, or to protect components from reversed voltage. It is often used to protect against current flowing into a device.

Most diodes are shaped like a small cylinder, are usually black with silver writing, and have two legs. They look a little like resistors. You use a special variant called a *Zener diode* in power supplies to help regulate voltages. Figure 8-14 shows a number of Zener diodes.

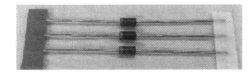

Figure 8-14. *Diodes*

Fuse

A fuse is designed to protect a device (actually the entire circuit) from current greater than what the components can safely operate. Fuses are placed inline on the positive pole. When too much current flows through the fuse, the internal parts trigger a break in the flow of current.

Some fuses use a special wire inside that melts or breaks (thereby rendering it useless but protecting your equipment), while other fuses use a mechanism that operates like a switch (many of these are resettable). When this happens, you say the fuse has "blown" or "tripped." Fuses are rated at a certain current in amperage, indicating the maximum amps that the fuse permits to flow without tripping.

Fuses come in many shapes and varieties. They work with either AC or DC voltage. The fuses that you use are of the disposable variety. Figure 8-15 shows an example of two fuses: an automotive-style blade fuse on the left and a glass cartridge fuse on the right.

Figure 8-15. *Fuses*

If you are familiar with your home's electrical panel that contains the circuit breakers, they are resettable fuses. So, the next time one of them goes "click" and the lights go out, you can say, "Hey, a fuse has tripped!" Better still, now you know why—you have exceeded the maximum rating of the circuit breaker.

This is probably fine in situations where you accidentally left that infrared heater on when you dropped the toast and started the microwave (it happens), but if you are tripping breakers frequently without any load, you should call an electrician to have the circuit checked.

Light-Emitting Diode (LED)

Recall from Chapter 3 that an LED is a special diode that produces light when powered.

As you learned in Chapter 3, an LED has two legs: the longer leg is positive and the shorter is negative. LEDs also have a flat edge that also indicates the negative leg. They come in a variety of sizes ranging from as small as 3mm to 10mm. Figure 8-16 shows an example of some smaller LEDs.

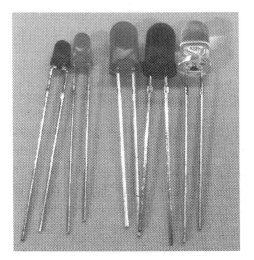

Figure 8-16. *Light-emitting diodes*

Recall you also need to use a resistor with an LED. You need this to help reduce the flow of the circuit to lower the current flowing through the LED. LEDs can be used with lower current (they burn a bit dimmer than normal) but should not be used with higher current.

To determine what size resistor you need, you need to know several things about the LED. This data is available from the manufacturer who provides the data in the form of a data sheet or, in the case of commercially packaged products, lists the data on the package. The data you need includes the maximum voltage, the supply voltage (how many volts are coming to the LED), and the current rating of the LED.

For example, if I have an LED like the one you used in the last chapter, in this case a 5mm red LED, you find on Adafruit's website (`www.adafruit.com/products/297`) that the LED operates at 1.8V–2.2V and 20mA of current. Let's say you want to use this with a 5V supply voltage. You can then take these values and plug them into this formula:[4]

```
R = (Vcc-Vf)/I
```

---

[4]A variant of Ohm's law (`https://en.wikipedia.org/wiki/Ohm's_law`).

Using more descriptive names for the variable, you get the following:

```
Resistor = (Volts_supply - Volts_forward) / Desired_current
```

Plugging in the data, you get this result. Note that you have mA so you must use the correct decimal value (divide by 1000)—in this case, 0.020, and you pick a voltage in the middle.

```
Resistor = (5 - 2.0) / 0.020
         = 3.0 / 0.020
         = 150
```

Thus, you need a resistor of 150 ohms. Cool. Sometimes, the formula produces a value that does not match any existing resistors. In that case, choose one closest to the value but a bit higher. Remember, you want to limit current and thus err on the side of more restrictive than less restrictive. For example, if you found you need a resistor of 95 ohms, you can use one rated at 100 ohms, which is safer than using one rated at 90 ohms.

Also, if you use LEDs in serial or parallel, the formula is a little different. See `https://learn.adafruit.com/all-about-leds` for more information about using LEDs in your projects and calculating the size of resistors to use with LEDs.

Relay

A relay is an interesting component that helps you control higher voltages with lower voltage circuits. For example, suppose you wanted to control a device that is powered by 12V from your Raspberry Pi, which only produces a maximum of 5V. A relay can be used with a 5V circuit to turn on (or relay) power from that higher source. In this example, you would use the Raspberry Pi's output to trigger the relay to switch on the 12V power. Thus, relays are a form of switch. Figure 8-17 shows a typical relay and how the pins are arranged.

Figure 8-17. *Relay*

Relays can take a lot of different forms and typically have slightly different wiring options, such as where the supply voltage is attached and where the trigger voltage attaches as well as whether the initial state is open (no flow) or close (flow) and thus the behavior of how it controls voltage. Some relays come mounted on a PCB with clearly marked terminals that show where to change the switching feature and where everything plugs in. If you want to use relays in your projects, always check the data sheet to make sure that you are wiring it correctly, based on its configuration.

You can also use relays to allow your DC circuit to turn AC appliances on and off like Controllable Four Outlet Power Relay Module version 2 from Adafruit (`www.adafruit.com/product/2935`).

Resistor

A resistor is one of the standard building blocks of electronics. Its job is to impede current and impose a reduction in voltage (which is converted to heat). Its effect, known as resistance, is measured in ohms. A resistor can be used to reduce voltage to other components, limiting frequency response, or protect sensitive components from overvoltage. Figure 8-18 shows a number of resistors.

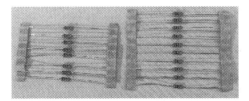

Figure 8-18. *Resistors*

When a resistor is used to pull up voltage (by attaching one end to positive voltage) or pull down voltage (by attaching one end to the ground) (resistors are bidirectional), it eliminates the possibility of the voltage floating in an indeterminate state. Thus, a pull-up resistor ensures that the stable state is positive voltage, and a pull-down resistor ensures that the stable state is zero voltage (ground).

Switch

A switch is designed to control the flow of current between two or more pins. Switches come in all manner of shapes, sizes, and packaging. Some are designed as a simple on/off, while others can be used to change current from one set of pins to another. Like buttons, switches come in a variety of mounting options from PCB (also called a *through hole*) to panel mount for mounting in enclosures. Figure 8-19 shows a variety of switches.

Figure 8-19. *Various switches*

Switches that have only one pole (leg or side) are called *single-pole switches*. Switches that can divert current from one set of poles to another set are called *two-pole switches*. Switches where there is only one secondary connection per pole are called *single-throw switches*. Switches that disconnect from one set of poles and connect to another while maintaining a common input are called *double-throw switches*. These are often combined together and form the switch type (or kind) as follows:

- *SPST*: Single pole, single throw

- *DPST*: Double pole, single throw

- *SPDT*: Single pole, double throw

- *DPDT*: Double pole, double throw

- *3PDT*: Three pole, double throw

There may be other variants that you could encounter. I like to keep it simple: if I have just an on/off situation, I want a single-throw switch. How many poles depends on how many wires or circuits I want to turn on or off at the same time. I use these for double-throw switches when I have an "A" condition and a "B" condition in which I want A on when B is off, and vice-versa. I sometimes use multiple-throw switches when I want A, B, and off situations. I use the center position (throw) as off. You can be very creative with switches!

Transistor

A transistor (a bipolar transistor) is designed to switch current on/off in a cycle or amplify fluctuations in current. Interestingly, transistors used to amplify current replaced vacuum tubes. If you are an audiophile, you likely know a great deal about vacuum tubes. When a resistor operates in switching mode, it behaves similar to a relay, but its "off" position still allows a small amount of current to flow. Transistors are used in audio equipment, signal processing, and switching power supplies. Figure 8-20 shows two varieties of transistors.

Figure 8-20. *Transistors*

Transistors come in all manner of varieties, packaging, and ratings that make it suitable for one solution or another.

Voltage Regulator

A voltage regulator (linear voltage regulator) is designed to keep the flow of current constant. Voltage regulators often appear in electronics when you need to condition or lower current from a source. For example, you want to supply 5V to a circuit, but only have a 9V power supply. Voltage regulators accomplish this (roughly) by taking current in and dissipating the excess current through a heat sink. Thus, voltage regulators have three legs: positive current in, negative, and positive current out. They are typically shaped like those shown in Figure 8-21, but other varieties exist.

Figure 8-21. *Voltage regulators*

The small hole in the plate that extends out of the voltage regulator is where the heat sink is mounted. Voltage regulators are often numbered to match their rating; for example, a LM7805 produces 5V, whereas a LM7833 produces 3.3V.

An example of using a voltage regulator to supply power to a 3.3V circuit on a breadboard is shown in Figure 8-22. This circuit was designed with capacitors to help smooth or condition the power. Notice that the capacitors are rated by uF, which means *microfarad*.

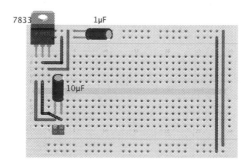

Figure 8-22. *Power supply circuit on a breadboard with voltage regulator*

Breakout Boards and Circuits

Breakout boards are your modular building blocks for IoT solutions. They typically combine several components together to form a function, such as measuring temperature, enabling reading GPS data, communicating via cellular services, and more. Figure 8-23 shows two breakout boards. On the left is an Adafruit AC/DC converter (`www.adafruit.com/products/1083`), and on the right is an Adafruit barometric pressure sensor breakout board (`www.adafruit.com/products/391`).

Figure 8-23. *Breakout boards*

Whenever you design a circuit or IoT solution, you should consider using breakout boards as much as possible because they simplify the use of the components. Take the barometric pressure sensor, for example; Adafruit has designed this board so that all you need to do to use it is to attach power and connect it to your IoT device on its I2C bus. An I2C bus is a fast digital protocol that uses two wires (plus power and ground) to read data from circuits (or devices).

Thus, there is no need to worry about how to connect the sensor to other components to use it—just connect it like any I2C device and start reading data! You use several breakout boards in the projects later in this book.

Using a Breadboard to Build Circuits

If you have been following along with the projects thus far in the book, you have already encountered a breadboard to make a very simple circuit. Recall from Chapter 3 that a breadboard is a tool you use to plug components into to form circuits. Technically, you're using a solderless breadboard. A solder breadboard has the same layout, only it has only through-hole solder points on a PCB.

A breadboard allows you to create prototypes for your circuits or simply temporary circuits without having to spend the time (and cost) to make the printed circuit board. Prototyping is the process of experimenting with a circuit by building and testing your ideas. In fact, once you've got your circuit to work correctly, you can use the breadboard layout to help you design a PCB. Figure 8-24 shows a number of breadboards.

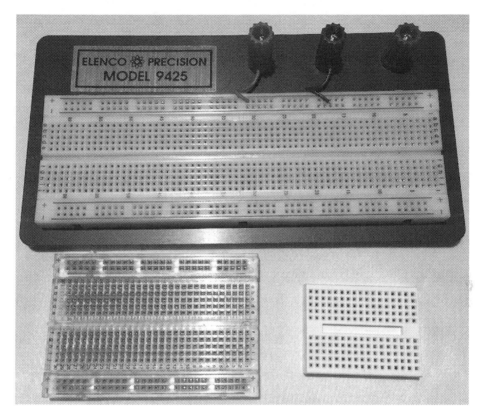

Figure 8-24. *Assorted breadboards*

WHY ARE THEY CALLED BREADBOARDS?

In the grand old days of microelectronics before discrete components became widely available for experimentation, when if you wanted to prototype a circuit, some would use a piece of wood with nails driven into it (sometimes in a grid pattern) where connections were made (called *runs*) by wrapping wire around the nails. Some actually used a breadboard from the kitchen to build their wire wrap prototypes. The name has stuck ever since.

Most breadboards (there are several varieties) have a center groove (called a *ravine*) or a printed line down the center of the board. This signifies the terminal strips that run perpendicular to the channel are not connected. That is, the terminal strip on one side is not connected to the other side. This allows you to plug integrated circuits (IC) or chip that are packaged as two rows of pins. Thus, you can plug the IC into the breadboard with one set of pins on each side of the breadboard. You see this in the following example.

Most breadboards also have one or more sets of power rails that are connected together parallel to the ravine. If there are two sets, the sets are not connected together. The power rails may have a colored reference line, but this is only for reference; you can make either one positive with the other negative. Finally, some breadboards number the terminal strip rows. These are for reference only and have no other meaning. However, they can be handy for making notes in your engineering notebook. Figure 8-25 shows the nomenclature of a breadboard and how the terminal strips and power rails are connected together.

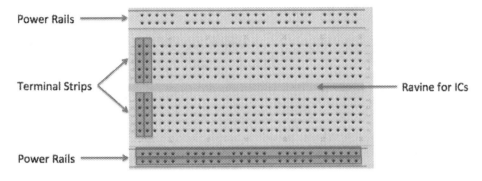

Figure 8-25. *Breadboard layout*

Note   The sets of power rails are not connected together. If you want to have power on both sides of the breadboard, you must use jumpers to connect them.

It is sometimes desirable to test a circuit out separately from code. For example, if you want to make sure that all of your devices are connected together properly, you can use a breadboard power supply to power the circuit. This way, if something goes horribly wrong, you don't risk damaging your IoT device. Most breadboard power supplies are built on a small PCB with a barrel jack for a wall wart power supply, two sets of pins to plug into the power rails on the breadboard, and an off switch (very handy), and some can generate different voltages. Figure 8-26 shows one of my favorite breadboard power supplies from SparkFun (`www.sparkfun.com/products/13157`).

Figure 8-26. *Breadboard power supply*

Should your circuits require more room than what is available on a single breadboard, you can use multiple breadboards by simply jumping the power rails and continuing the circuit. To facilitate this, some breadboards can be connected together using small nubs and slots on the side. Finally, most breadboards also come with an adhesive backing that you can use to mount on a plate or inside an enclosure or similar workspace. If you decide to use the adhesive backing, be forewarned that they cannot be unstuck easily—they stay put quite nicely.

FRITZING: A BREADBOARDING SOFTWARE APPLICATION

The drawings of breadboards in this book were made with a program named Fritzing (`http://fritzing.org/home/`). This open source application allows you to create a digital representation of a circuit on a breadboard. It is really quite handy to use. If you find yourself wanting to design a prototype circuit, using Fritzing can help save you a lot of trial and error. As a bonus, Fritzing allows you to see the same circuit in an electronic schematic or PCB layout view. I recommend downloading and trying this application out.

Now that you know more about how breadboards work, let's discuss the component your IoT solutions employ to collect data: sensors.

What Are Sensors?

A *sensor* is a device that measures phenomena of the physical world. These phenomena can be things you see, like light, gases, water vapor, and so on. They can also be things you feel, like temperature, electricity,[5] water, wind, and so on. Humans have senses

---

[5]Shocking, isn't it?

that act like sensors, allowing you to experience the world around you. However, there are some things your body can't see or feel, such as radiation, radio waves, voltage, and amperage. Upon measuring these phenomena, it's the sensors' job to convey a measurement in the form of either a voltage representation or a number.

There are many forms of sensors. They're typically low-cost devices designed for a single purpose and with a limited capability for processing. Most simple sensors are discrete components; even those that have more sophisticated parts can be treated as separate components. Sensors are either analog or digital and are typically designed to measure only one thing. But an increasing number of sensor modules are designed to measure a set of related phenomena, such as the USB weather board from SparkFun Electronics (`www.sparkfun.com/products/13956`).

The following sections examine how sensors measure data, how to store that data, and examples of some common sensors.

How Sensors Measure

Sensors are electronic devices that generate a voltage based on the unique properties of their chemical and mechanical construction. They don't actually manipulate the phenomena they're designed to measure. Rather, sensors sample some physical variable and turn it into a proportional electric signal (voltage, current, digital, etc.).

For example, a humidity sensor measures the concentration of water (moisture) in the air. Humidity sensors react to these phenomena and generate a voltage that the microcontroller or similar device can then read and use to calculate a value on a scale. A basic, low-cost humidity sensor is the DHT-22 available from most electronics stores.

The DHT-22 is designed to measure temperature as well as humidity. It generates a digital signal on the output (data pin). Although simple to use, it's a bit slow and should be used to track data at a reasonably slow rate (no more frequently than about once every 3 or 4 seconds).

When this sensor generates data, that data is transmitted as a series of high (interpreted as a 1) and low (interpreted as a 0) voltages that the microcontroller can read and use to form a value. In this case, the microcontroller reads a value that is 40 bits in length (40 pulses of high or low voltage)—that is, 5 bytes—from the sensor and places it in a program variable. The first two bytes are the value for humidity, the second two are for temperature, and the fifth byte is the checksum value to ensure an accurate read. Fortunately, all of this hard work is done for you in the form of a special library designed for the DHT-22 and similar sensors.

The DHT-22 produces a digital value. Not all sensors do this; some generate a voltage range instead. These are called *analog sensors*. Let's take a moment to understand the differences. This becomes essential information as you plan and build your sensor nodes.

Analog Sensors

Analog sensors are devices that generate a voltage range, typically between 0 and 5 volts. An analog-to-digital circuit is needed to convert the voltage to a number. But it isn't that simple (is it ever?). Analog sensors work like resistors and, when connected to GPIO pins, often require another resistor to "pull up" or "pull down"[6] the voltage to avoid spurious changes in voltage known as *floating*. This is because voltage flowing through resistors is continuous in both time and amplitude.

Thus, even when the sensor isn't generating a value or measurement, there is still a flow of voltage through the sensor that can cause spurious readings. Your projects require a clear distinction between OFF (zero voltage) and ON (positive voltage). Pull-up and pull-down resistors ensure that you have one of these two states. It's the responsibility of the A/D converter to take the voltage read from the sensor and convert it to a value that can be interpreted as data.

When sampled (when a value is read from a sensor), the voltage read must be interpreted as a value in the range specified for the given sensor. Remember that a value of, say, 2 volts from one analog sensor may not mean the same thing as 2 volts from another analog sensor. Each sensor's data sheet shows you how to interpret these values.

As you can see, working with analog sensors is a lot more complicated than using the DHT-22 digital sensor. With a little practice, you find that most analog sensors aren't difficult to use once you understand how to attach them to a microcontroller and how to interpret their voltage on the scale in which the sensor is calibrated to work.

Digital Sensors

Digital sensors like the DHT-22 are designed to produce a string of bits using serial transmission (one bit at a time). However, some digital sensors produce data via parallel

---

[6]Broadly, the term pull up means to raise voltage to the reference voltage of the circuit (think "positive"), and the term pull down means to lower voltage to zero (think "negative"). See `www.electronics-tutorials.ws/logic/pull-up-resistor.html` for more details.

transmission (one or more bytes[7] at a time). As described previously, the bits are represented as voltage, where high voltage (say, 5 volts) or ON is 1 and low voltage (0 or even –5 volts) or OFF is 0. These sequences of ON and OFF values are called *discrete values* because the sensor is producing one or the other in pulses—it's either ON or OFF.

Digital sensors can be sampled more frequently than analog signals because they generate the data more quickly and because no additional circuitry is needed to read the values (such as A/D converters and logic or software to convert the values to a scale). As a result, digital sensors are generally more accurate and reliable than analog sensors. But the accuracy of a digital sensor is directly proportional to the number of bits it uses for sampling data.

The most common form of digital sensor is the pushbutton or switch. What, a button is a sensor? Why, yes, it's a sensor. Consider for a moment the sensor attached to a window in a home security system. It's a simple switch that is closed when the window is closed and open when the window is open. When the switch is wired into a circuit, the flow of current is constant and unbroken (measuring positive volts using a pull-up resistor) when the window is closed and the switch is closed, but the current is broken (measuring zero volts) when the window and switch are open. This is the most basic of ON and OFF sensors.

Most digital sensors are actually small circuits of several components designed to generate digital data. Unlike analog sensors, reading their data is easy because the values can be used directly without conversion (except to other scales or units of measure). Some may suggest this is more difficult than using analog sensors, but that depends on your point of view. An electronics enthusiast would see working with analog sensors as easier, whereas a programmer would think digital sensors are simpler to use.

Now let's take a look at some of the sensors available and the types of phenomena they measure.

Examples of Sensors

An IoT solution that observes something may use at least one sensor and a means to read and interpret the data. You may be thinking of all manner of useful things you can measure in your home or office or even in your yard or surroundings. You may want to measure the temperature changes in your new sunroom, detect when the mail carrier

---

[7]This depends on the width of the parallel buffer. An 8-bit buffer can communicate 1 byte at a time, a 16-bit buffer can communicate 2 bytes at a time, and so on.

has tossed the latest circular in your mailbox, or perhaps keep a log of how many times your dog uses his doggy door. I hope that by now you can see these are just the tip of the iceberg when it comes to imagining what you can measure.

What types of sensors are available? The following list describes some of the more popular sensors and what they measure. This is just a sampling of what is available. Perusing the catalogs of online electronics vendors like Mouser Electronics (www.mouser.com), SparkFun Electronics (www.sparkfun.com), and Adafruit Industries (www.adafruit.com) reveals many more examples.

- *Accelerometers*: These sensors measure motion or movement of the sensor or whatever it's attached to. They're designed to sense motion (velocity, inclination, vibration, etc.) on several axes. Some include gyroscopic features. Most are digital sensors. A Wii Nunchuck (or WiiChuck) contains a sophisticated accelerometer for tracking movement. Aha, now you know the secret of those funny little thingamabobs that came with your Wii.

- *Audio sensors*: Perhaps this is obvious, but microphones are used to measure sound. Most are analog, but some of the better security and surveillance sensors have digital variants for higher compression of transmitted data.

- *Barcode readers*: These sensors are designed to read barcodes. Most often, barcode readers generate digital data representing the numeric equivalent of a barcode. Such sensors are often used in inventory-tracking systems to track equipment through a plant or during transport. They're plentiful, and many are economically priced, enabling you to incorporate them into your own projects.

- *RFID sensors*: Radio frequency identification uses a passive device (sometimes called an *RFID tag*) to communicate data using radio frequencies through electromagnetic induction. For example, an RFID tag can be a credit card–sized plastic card, a label, or something similar that contains a special antenna, typically in the form of a coil, thin wire, or foil layer that is tuned to a specific frequency. When the tag is placed in close proximity to the reader, the reader emits a radio signal; the tag can use the electromagnet energy to transmit a nonvolatile message embedded in the antenna, in the form of radio signals, which is then converted to an alphanumeric string.[8]

- *Biometric sensors*: A sensor that reads fingerprints, irises, or palm prints contains a special sensor designed to recognize patterns. Given the uniqueness inherent in patterns, such as fingerprints and palm prints, they make excellent components for a secure access system. Most biometric sensors produce a block of digital data that represents the fingerprint or palm print.

- *Capacitive sensors*: A special application of capacitive sensors, pulse sensors are designed to measure your pulse rate and typically use a fingertip for the sensing site. Special devices known as pulse oximeters (called *pulse-ox* by some medical professionals) measure pulse rate with a capacitive sensor and determine the oxygen content of blood with a light sensor. If you own modern electronic devices, you may have encountered touch-sensitive buttons that use special capacitive sensors to detect touch and pressure.

- *Coin sensors*: This is one of the most unusual types of sensors.[9] These devices are like the coin slots on a typical vending machine. Like their commercial equivalent, they can be calibrated to sense when a certain size of coin is inserted. Although not as sophisticated as commercial units that can distinguish fake coins from real ones, coin sensors can be used to add a new dimension to your projects. Imagine a coin-operated Wi-Fi station. Now, that should keep the kids from spending too much time on the Internet!

- *Current sensors*: These are designed to measure voltage and amperage. Some are designed to measure change, whereas others measure load.

- *Flex/force sensors*: Resistance sensors measure flexes in a piece of material or the force or impact of pressure on the sensor. Flex sensors may be useful for measuring torsional effects or as a means

---

[8]http://en.wikipedia.org/wiki/Radio-frequency_identification
[9]www.sparkfun.com/products/11719

to measure finger movements (like in a Nintendo Power Glove). Flex sensor resistance increases when the sensor is flexed.

- *Gas sensors*: There are a great many types of gas sensors. Some measure potentially harmful gases, such as LPG and methane, and other gases, such as hydrogen, oxygen, and so on. Other gas sensors are combined with light sensors to sense smoke or pollutants in the air. The next time you hear that telltale and often annoying low-battery warning beep[10] from your smoke detector, think about what that device contains. Why, it's a sensor node!

- *Light sensors*: Sensors that measure the intensity or lack of light are special types of resistors: light-dependent resistors (LDRs), sometimes called *photo resistors* or *photocells*. Thus, they're analog by nature. If you own a Mac laptop, chances are you've seen a photo resistor in action when your illuminated keyboard turns itself on in low light. Special forms of light sensors can detect other light spectrums, such as infrared (as in older TV remotes).

- *Liquid flow sensors*: These sensors resemble valves and are placed inline in plumbing systems. They measure the flow of liquid as it passes through. Basic flow sensors use a spinning wheel and a magnet to generate a Hall effect (rapid ON/OFF sequences whose frequency equates to how much water has passed).

- *Liquid-level sensors*: A special resistive solid-state device can be used to measure the relative height of a body of water. One example generates low resistance when the water level is high and higher resistance when the level is low.

- *Location sensors*: Modern smartphones have GPS sensors for sensing location, and of course GPS devices use the GPS technology to help you navigate. Fortunately, GPS sensors are available in low-cost forms, enabling you to add location sensing to your project. GPS sensors generate digital data in the form of longitude and latitude, but some can also sense altitude.

---

[10]I for one can never tell which detector is beeping, so I replace the batteries in all of them.

- *Magnetic stripe readers*: These sensors read data from magnetic stripes (like that on a credit card) and return the digital form of the alphanumeric data (the actual strings).

- *Magnetometers*: These sensors measure orientation via the strength of magnetic fields. A compass is a sensor for finding magnetic north. Some magnetometers offer multiple axes to allow even finer detection of magnetic fields.

- *Proximity sensors*: Often thought of as distance sensors, proximity sensors use infrared or sound waves to detect distance or the range to/from an object. Made popular by low-cost robotics kits, the Parallax Ultrasonic Sensor uses sound waves to measure distance by sensing the amount of time between pulse sent and pulse received (the echo). For approximate distance measuring,[11] it's a simple math problem to convert the time to distance. How cool is that?

- *Radiation sensors*: Among the more serious sensors are those that detect radiation. This can also be electromagnetic radiation (there are sensors for that too), but a Geiger counter uses radiation sensors to detect harmful ionizing. In fact, it's possible to build your very own Geiger counter using a sensor and an Arduino (and a few electronic components).

- *Speed sensors*: Similar to flow sensors, simple speed sensors like those found on many bicycles use a magnet and a reed switch to generate a Hall effect. The frequency combined with the circumference of the wheel can be used to calculate speed and, over time, distance traveled. The speed sensor on the wheel and fork provides the data for the monitor on your handlebars.

- *Switches and pushbuttons*: These are the most basic of digital sensors used to detect if something is set (ON) or reset (OFF).

---

[11]Accuracy may depend on environmental variables, such as elevation, temperature, and so on.

- *Tilt switches*: These sensors can detect when a device is tilted one way or another. Although very simple, they can be useful for low-cost motion detection sensors. They are digital and are essentially switches.

- *Touch sensors*: The touch-sensitive membranes formed into keypads, keyboards, pointing devices, and the like are an interesting form of sensor. You can use touch-sensitive devices like these for collecting data from humans.

- *Video sensors*: As mentioned previously, it's possible to obtain very small video sensors that use cameras and circuitry to capture images and transmit them as digital data.

- *Weather sensors*: Sensors for temperature, barometric pressure, rainfall, humidity, wind speed, and so on are all classified as weather sensors. Most generate digital data and can be combined to create comprehensive environmental solutions. Yes, it's possible to build your own weather station from about a dozen inexpensive sensors, an Arduino (or a Raspberry Pi), and a bit of programming to interpret and combine the data.

Summary

Learning how to work with electronics as a hobby or as a means to create an IoT solution does not require a lifetime of study or a change of vocation. Indeed, learning how to work with electronics is all part of the fun of experimenting with the IoT!

As you've seen in this chapter, knowing about the types of components available, the types of sensors, and a bit of key knowledge of how to use a multimeter goes a long way toward becoming proficient with electronics. You also learned about one of the key components of an IoT solution—sensors. You discovered two ways they communicate and a bit of what types of sensors are available.

In the next chapter, you take a look at a couple of special kits for experimenting with IoT solutions. As you will see, the kits provide a set of electronic components as well as several tools and accessories, such as a breadboard and a power supply.

CHAPTER 9

The Adafruit Microsoft IoT Pack for Raspberry Pi

When working with electronics projects like those in this book, it is often the case that you have to acquire a host of components and tools in order to get started. The projects so far in this book have minimized the components needed, and I have listed what you need for each project. However, if you want to expand your inventory of components but have little experience with electronics, you may not know what to buy. Fortunately, some vendors such as Adafruit are packaging electronic components, accessories, and even some tools together in a kit, making it simple to get started—you just buy the kit!

Indeed, some kits package together a number of common components, such as resistors, LEDs, and a breadboard and jumpers—all the things that you need to get started if you already have the basic parts (low-cost computing board, power supply, etc.). At least one kit goes a bit further providing a more complete set of components and accessories including a development board (e.g., a Raspberry Pi), power supply, and more—everything you need to build an IoT solution using the Windows 10 IoT Core.

Tip For a complete list of the components used in this book and sources for purchasing them and related tools, see the appendix.

This chapter explores the Adafruit Microsoft IoT Pack for Raspberry Pi 3—available with or without the Raspberry Pi. The chapter also demonstrates a small project that uses the components in the kit (well, mostly) to read data from a simple sensor. Let's begin with a look at what is in the kit.

© Charles Bell 2021
C. Bell, *Windows 10 for the Internet of Things*, https://doi.org/10.1007/978-1-4842-6609-0_9

Overview

The Microsoft IoT Pack for Raspberry Pi 3 (hence kit) is the result of the collaboration between Microsoft and Adafruit to provide a one-stop shopping solution for those who want to explore IoT solutions using Windows 10. In fact, the kit comes with everything you need to run Windows 10 IoT Core including the Raspberry Pi 3, power supply, and a micro-SD card with the Windows 10 IoT Core image installed. For most, the kit is the best way to get started using Windows 10 and your Raspberry Pi.

The kit comes in two varieties: one with the Raspberry Pi (www.adafruit.com/products/2733) and one without the Raspberry Pi (www.adafruit.com/products/2702) for those who already own a Raspberry Pi 2 or 3.[1]

Adafruit has made several updates to the kit including keeping up with the latest releases of the Windows 10 IoT Core. The kit has been a huge success and sells out regularly. Fortunately, you will not have to wait more than a few weeks for them to restock.

Tip   You may also purchase them from third-party retailers that carry Adafruit products such as Digi-Key (www.digikey.com/products/en?mpart=2733&v=1528). If the kits are out of stock there as well, you can purchase the components separately (or just the ones you need—see the "Are There Alternatives?" section for more details) from Adafruit.

The kit with the Raspberry Pi costs about $114.95, and the kit without the Raspberry Pi costs about $75.00. Clearly, if you already have a Raspberry Pi, you can save some money there. In fact, for those who want to use a different low-cost computer board, you can buy the kit without the Raspberry Pi—except for the micro-SD card with Windows 10 and possibly the power supply, all the other components work with other boards.

Tip   You can also use the kit with Raspbian Linux and Python. Adafruit has a long list of tutorials to explore.

---

[1]While the Raspberry Pi 3 is the latest board, the Raspberry Pi 2 is more than capable for implementing any IoT project.

The kit comes with a number of handy components including prototyping tools and a few sensors. It is even certified for use with Microsoft Azure (Microsoft's cloud computing platform)! Figure 9-1 shows a photo of all the components included in the kit. I discuss all the components in the next section.

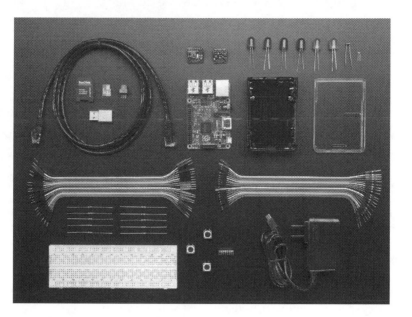

Figure 9-1. *The Adafruit Windows IoT Pack for Raspberry Pi 3 (courtesy of* www.*adafruit.com)*

Components Included

As you can see, there are a lot of pieces in the kit. There are three categories of components: electronic components, accessories for the Raspberry Pi, and sensors.

The electronic components provided in the kit include the following:

- (2) breadboard trim potentiometers
- (5) 10K 5% 1/4W resistors
- (5) 560 ohm 5% 1/4W resistors
- (2) diffused 10mm blue LEDs
- (2) diffused 10mm red LEDs
- (2) diffused 10mm green LEDs

- (1) electrolytic capacitor: 1.0uF

- (3) 12mm tactile switches[2]

The list of accessories in the kit is long. The following includes all the accessories included in the kit. I describe some of these in more detail:

- *Adafruit Raspberry Pi B+ case, smoke base/clear top*: An excellent case to protect your Pi from accidents.

- *Full-sized breadboard*: Plenty of space to spread out your circuits.

- *Premium male-to-male jumper wires, 20 × 6 inches (150mm)*: Used to jump from one port to another on the breadboard. They're extra long and come molded in a ribbon so you can peel off only those you need.

- *Premium female-to-male extension jumper wires, 20 × 6 inches*: Used to jump from male GPIO pins to the breadboard ports. They also come molded in a ribbon.

- *Miniature Wi-Fi module*: A Raspberry Pi–approved Wi-Fi dongle (not needed for the Raspberry Pi 3).

- *5V 2A switching power supply with a 6-foot micro-USB cable*: Meets the Raspberry Pi requirements for power.

- *MCP3008—8-channel 10-bit ADC with SPI interface*: A breakout board that you can use to expand the number of SPI interface channels for larger IoT projects.

- *Ethernet cable, 5-foot*: A nice touch considering the kit has a Wi-Fi dongle—good to have a backup plan!

- *8GB class 10 SD/micro-SD memory card*: Windows 10 IoT Core preloaded!

The sensors included with the kit are an unexpected surprise. They provide what you need to create some interesting IoT solutions. Best of all, they are packaged as breakout boards, making them easy to wire into our circuits. The following lists the sensors included in the kit:

- *1 photocell*: A simple component used to measure light

---

[2]Technically, switches are the simplest of all sensors.

- *Assembled Adafruit BME280 temperature, pressure, and humidity sensor*: Measures temperature, barometric pressure, and humidity

- *Assembled TCS34725 RGB color sensor*: Measures color. Comes with an infrared filter and white LED

Some of the parts in this kit require a bit more explanation. The following sections describe some of the more interesting parts in more detail.

Environmental Sensor: BME280

This sensor is great for all sorts of environmental sensor projects. It features both I2C[3] and SPI[4] interfaces, making this a very versatile breakout board. Figure 9-2 shows an image of the BME280 sensor breakout board.

Figure 9-2. *Environmental sensor: BME280 (courtesy of www.adafruit.com)*

Note   The photos show the breakout boards without the headers (row of pins) soldered. The components included in the kit come with the headers soldered in place.

The sensor on the board has a ±3% accuracy for measuring humidity, barometric pressure with ±1 hPa absolute accuracy, and temperature within ±1.0°C accuracy. In addition, the breakout board includes a 3.3V regulator and level shifting, so you can use

---

[3]https://en.wikipedia.org/wiki/I%C2%B2C
[4]https://en.wikipedia.org/wiki/Serial_Peripheral_Interface_Bus

it with a 3V or 5V connections. For more information about how to use the sensor, see the tutorial at https://learn.adafruit.com/adafruit-bme280-humidity-barometric-pressure-temperature-sensor-breakout.

Color Sensor: TCS34725

If you want to measure light beyond the basic intensity that the photocell sensor provides, such as determining the color of light, you can use this sensor to add that capability to your IoT projects. Adafruit combines a highly accurate color sensor, the TCS34725, and bundles it with other components to make a sensor capable of "seeing" infrared and more. Figure 9-3 shows a photo of the color sensor.

Figure 9-3. *Color sensor: TCS34725 (courtesy of* www.adafruit.com*)*

Like the environmental sensor, it has a 3.3V regulator with level shifting for the I2C pins, so they can be used with 3.3V or 5V. You can find out more about the color sensor at http://learn.adafruit.com/adafruit-color-sensors/overview.

8-Channel 10-Bit ADC with SPI Interface: MCP3008

If you need more analog inputs for your IoT project, Adafruit has provided a nifty integrated circuit in the form of the MCP3008 that you can use to add additional inputs. It uses an SPI interface so you only need to use four pins to connect to the chip. Figure 9-4 shows a photo of the IC.

Figure 9-4. *8-Channel 10-bit ADC with SPI interface (courtesy of* www.adafruit.com*)*

If you would like to see how to use this IC, see `http://learn.adafruit.com/` `reading-a-analog-in-and-controlling-audio-volume-with-the-raspberry-pi`.

Are There Alternatives?

If you are planning to use a board other than the Raspberry Pi, are on a more limited hobby budget, or want only the bare essentials, there are alternatives to the Microsoft IoT Pack from Adafruit. In fact, Adafruit sells another kit that includes almost everything you need for the projects in this book. It doesn't come with sensors, but all the basic bits and bobs are in there, and you can always buy the sensors separately.

The Adafruit Parts Pal comes packaged in a small plastic case with a host of electronic components (`www.adafruit.com/products/2975`). Figure 9-5 shows the Parts Pal kit.

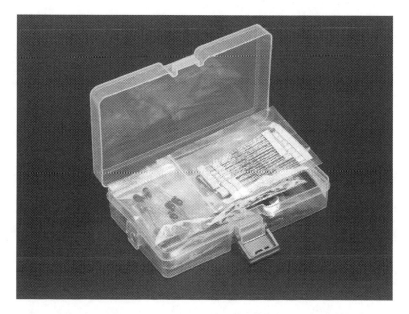

Figure 9-5. *Adafruit Parts Pal (courtesy of* www.adafruit.com*)*

The kit includes the following components: prototyping tools, LEDs, capacitors, resistors, some basic sensors, and more. In fact, there are more components in this kit than the Windows IoT Pack for the Raspberry Pi 3. Better still, the kit costs only $19.95, making it a good deal (and the case is a great bonus).

- (1) storage box with latch
- (1) half-sized breadboard
- (20) jumper wires: male-to-male, 3 inches (75mm)
- (10) jumper wires: male-to-male, 6 inches (150mm)
- (5) 5mm diffused green LEDs
- (5) 5mm diffused red LEDs
- (1) 10mm diffused common-anode RGB LED
- (10) 1.0uF ceramic capacitors
- (10) 0.1uF ceramic capacitors
- (10) 0.01uF ceramic capacitors
- (5) 10uF 50V electrolytic capacitors
- (5) 100uF 16V electrolytic capacitors
- (10) 560 ohm 5% axial resistors
- (10) 1K ohm 5% axial resistors
- (10) 10K ohm 5% axial resistors
- (10) 47K ohm 5% axial resistors
- (5) 1N4001 diodes
- (5) 1N4148 signal diodes
- (5) NPN transistor PN2222 TO-92
- (5) PNP transistor PN2907 TO-92
- (2) 5V 1.5A linear voltage regulator, 7805 TO-220
- (1) 3.3V 800mA linear voltage regulator, LD1117-3.3 TO-220
- (1) TLC555 wide-voltage range, low-power 555 timer

- (1) photocell

- (1) thermistor (breadboard version)

- (1) vibration sensor switch

- (1) 10K breadboard trim potentiometer

- (1) 1K breadboard trim potentiometer

- (1) Piezo buzzer

- (5) 6mm tactile switches

- (3) SPDT slide switches

- (1) 40-pin break-away male header strip

- (1) 40-pin female header strip

The only thing that I feel is missing is the male/female jumpers, but you can buy them separately (www.adafruit.com/product/1954). For only $1.95 more, they're worth adding to your order!

Tip   If you want to save some money and don't need the accessories in the Windows IoT Pack for Raspberry Pi 3, you should consider buying the Adafruit Parts Pal and male/female jumpers. With a cost of about $22–24, they're a great bargain.

Now, let's put our new kit to work with a simple project that uses a very simple sensor.

Example Project: A Simple Sensor

The projects thus far in the book have not used a sensor or read any input other than interacting with the user (which is still a form of sensing). In this project, you see how to write an IoT solution that uses a simple sensor (a pushbutton) and models a real-life solution that uses sensors. No matter which kit you decide to buy, each contains a pushbutton.

While the pushbutton is easy to use, the code is a bit more complicated than the examples from previous chapters. This is not due to the complexity of the problem, rather due to a new concept that you must consider when writing applications without graphical interfaces that use facilities from the UWP libraries. That is, you have a problem dealing with scope of the threads. You'll see this in the following code section.

The solution you're modeling is a subset of a typical traffic light in an urban setting—a pedestrian crosswalk pushbutton. More specifically, you implement a single traffic light for a one-way street with only a single crosswalk button so that you can keep the circuit simple. You can extend the circuit to include two buttons if you would like, and I encourage you to do so once you've mastered the project as written.

So how does this pedestrian crosswalk button work? When a pedestrian presses the crosswalk request button, the traffic light cycles from green to yellow to red, and then the walk signal cycles from DON'T WALK to WALK. A yellow LED is used for WALK and a red LED is used for DON'T WALK. After some time, the walk light flashes, warning the pedestrian that the traffic cycles back to green soon.

Thus, if you watch how the traffic lights work when you signal that you want to cross the street (at least in some US cities), notice that there are several states that the lights go through. I have simplified the states a bit as follows. I use *cycle* to indicate one light is turned off and another is turned on.

1. In the default state, the traffic light is green and the walk light is red.

2. When a pedestrian presses the walk button, the traffic light waits a few seconds and then cycles to yellow.

3. After a few seconds, the traffic light cycles to red.

4. After a few seconds, the walk light cycles to yellow.

5. After a few seconds, the yellow walk light begins to blink.

6. After a few seconds, the walk light cycles to red.

7. After a few seconds, the traffic light cycles to green—returns to state (1).

You learn how to write code to execute these states in a later section, but first let's look at the hardware that you need to build a circuit for the project.

Required Components

The following lists the components you need. You can find these components in either of the kits mentioned previously, or you can purchase the components separately from Adafruit (`www.adafruit.com`), SparkFun (`www.sparkfun.com`), or any electronics store that carries electronic components.

- (1) pushbutton (breadboard pin spacing)

- (2) red LEDs

- (2) yellow LEDs (or blue is OK)

- (1) green LED

- (5) 150 ohm resistors (see upcoming notes)

- Jumper wires

- Breadboard (full size recommended but half size is OK)

- Raspberry Pi 2 or 3

- Power supply

Set Up the Hardware

This project has many more connections than the projects thus far in the book. In order to set up the hardware correctly and make all the connections that you need, it is always best to make a plan for how things should connect.

For example, since you are using five LEDs and a pushbutton, as well as at least one connection to ground, you make seven connections to the GPIO header. Keeping all of those connections straight and planning which GPIO pins to use could be tricky if you didn't have a plan. I like to call my wiring plans "maps" because they map the connections from the GPIO header to the breadboard. Table 9-1 shows the map I designed for this project. Notice I leave a space for you to make any notes as you learn more about the connections and code.

Table 9-1. *Connection Map for Pedestrian Crossing Project*

GPIO	Connection	Function	Notes
4	Resistor for LED	Red traffic light	
5	Resistor for LED	Yellow traffic light	
6	Resistor for LED	Green traffic light	
16	Button	Walk request	
20	Resistor for LED	Red walk light	
21	Resistor for LED	Green walk light	
GND	Breadboard power rail	Ground	

Recall that you must use a resistor when connecting an LED directly to power (in this case, a GPIO pin set to HIGH) because the LED does not operate at 5V. Furthermore, LEDs are not all rated the same. Their power requirements can vary from one manufacturer to another as well as one color to another. That is, a green LED may have different power requirements than a blue LED. Thus, you must check the manufacturer (or vendor) to get the data sheet for the LED and write down the power requirements. Table 9-2 shows a number of different LEDs from Adafruit and SparkFun. I used Ohm's law (see Chapter 8) to figure out the right sized LED.

Table 9-2. *Various LEDs and Resistors for 5V*

Source	LED	Power Requirements	Resistor Needed
Adafruit Microsoft IoT Pack	10mm red	1.85–2.5V, 20mA	150 ohms
	10mm blue	3.0–3.4V, 20mA	100 ohms
	10mm green	2.2–2.5V, 20mA	150 ohms
SparkFun	3mm red	1.9–2.3V, 20mA	180 ohms
	3mm yellow	2.0–2.4V, 20mA	150 ohms
	3mm green	2.0–2.5V, 20mA	150 ohms

If you do not have the resistor listed, you can use the next higher resistor value or the one closest to but greater than the value listed. You can use the next higher value resistor safely because the higher the value of the resistor, the less current is fed to the LED. You should not use a smaller resistor value because too much current damages the LED.

Tip   There are several online LED resistor calculators. I used the one at `http://led.linear1.org/1led.wiz` for this data.

COOL GADGET: GPIO REFERENCE CARD

There is a very cool gadget that helps you sort out the connections with ease. It is called the GPIO Reference Card for Raspberry Pi 2 or 3 and is available from Adafruit (`www.adafruit.com/products/2263`). The following shows how handy it is to use when making connections to the GPIO.

To use this gadget, place it over the GPIO pins on your Raspberry Pi. Now, when you make the connections, you can clearly see which pin to use! How cool is that?

Since you are going to model a traffic light and a walk light, it would be best if you arrange the components so that the three LEDs for the traffic light are grouped together, and likewise the walk light are grouped together. I arranged the components on my breadboard in this way, as shown in Figure 9-6.

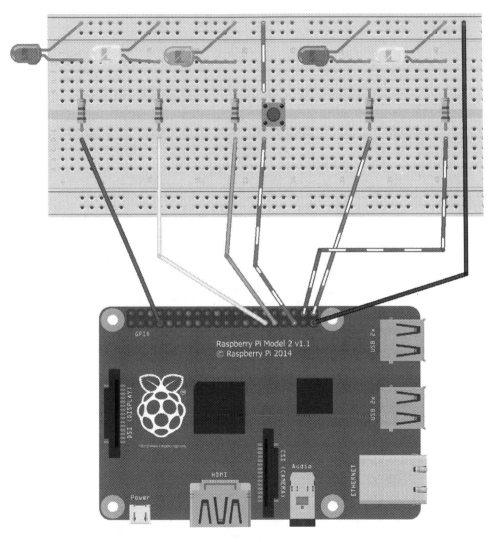

Figure 9-6. *Pedestrian crossing circuit*

Notice how I arranged the LEDs. More specifically, notice that I plugged the negative leg of each LED into the negative side of the power rail on the breadboard. This allows me to make one connection from one of the ground pins on the GPIO to the breadboard rail, which I can then use to plug in the ground side of the components. For example, each positive leg of each LED is plugged into the breadboard so that you can plug the corresponding resistor across the DIP trough and connect those to the appropriate GPIO pin. Recall there are no connections from one side of the trough to the other.

Also notice that I placed the button in the center spanning the DIP trough. One side is connected to the ground and the other to the appropriate GPIO pin. If you are unsure which way to orient the pushbutton, you can use a multimeter to test continuity among the pins. Use the pins where continuity is found (the multimeter beeps) when the pushbutton is pressed but no connection when released.

Note   If you do not have a yellow LED, the blue LED in the kit can be used in its place. Just be sure to use the right resistors as described earlier.

If you are following along with this chapter working on the project, go ahead and make the hardware connections now. Don't power on the board yet, but do double- and triple-check the connections.

Write the Code

The code for this project is a bit more involved than the previous projects. This is partly because of the extra LEDs and it is written in C++, but more so because you want to use a button and need to write code to determine when the sensor indicates the event (the button is pressed). We will use a simple user interface to make things interesting.

Tip   If you are using the same source code from the book website and encounter an error opening the solution or projects regarding the project folders, you can unload the project by right-clicking the project and choosing *Unload Project*. Then, reload the project by right-clicking the project and choosing *Reload Project*. This should resolve any issues with different project folders.

You also need to use the `DispatcherTimer` class to control the light sequence like you did in the last project.[5] More importantly, you will not use a user interface, which simplifies the code a bit (but you can add a user interface if you want).

---

[5]There are other ways to do this, but this uses a technique you've seen previously.

Just like you planned the GPIO connections, you also need to plan the code so that all goes as well as possible on the first attempt. It is a rare case that your code works on the first implementation (unless you're following an example like this one). The following lists the major decisions and features/areas of the code. I explain each in upcoming sections.

- Which project template do you want to use?

- How should the lights work?

- How do you read the button events?

Perhaps the more important question is which language will you use? I've decided to use C++, but you could implement this project in Visual Basic or C#. If you are a big fan of those languages, I encourage you to do just that using the following as a pattern!

Note   The sample source code for the book includes the C# version of this sample application. If you like C#, check it out to see how it differs from the C++ version.

New Project

You may be wondering what project template to use. Since you will not have a user interface, you may be tempted to use a headless, background application. However, you cannot use such a template because you want to use the dispatcher timer, which is only available in the headed project types. You can still use a blank template and can run the application without a user interface, but to get the support for the timer, you need to use a headed application template.

Thus, open a new project and choose the *C++ Blank App (Universal Windows)* project choice from the New Project dialog. Use the name PedestrianCrossing for the project name. You can save the project wherever you like or use the default. Figure 9-7 shows an example of the project template you should choose. Recall, you must also accept the minimal IoT Core version (just accept the default).

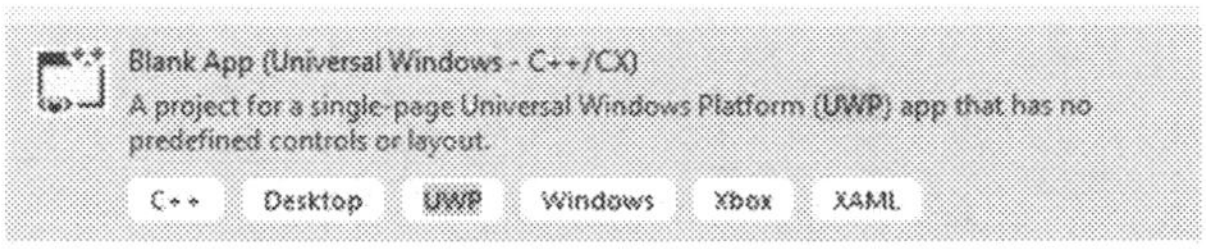

Figure 9-7. *New project template selection*

User Interface

Once the project opens, double-click the MainPage.xaml file. This is where you put all the user interface code for the project. Listing 9-1 shows the code you need to add for the user interface. The lines shown in bold are the ones you need to add (the others are in the template).

Listing 9-1. User Interface Code (MainPage.xaml)

```
<Page
    x:Class="PedestrianCrossing.MainPage"
    xmlns="http://schemas.microsoft.com/winfx/2006/xaml/presentation"
    xmlns:x="http://schemas.microsoft.com/winfx/2006/xaml"
    xmlns:local="using:PedestrianCrossing"
    xmlns:d="http://schemas.microsoft.com/expression/blend/2008"

    mc:Ignorable="d"
    Background="{ThemeResource ApplicationPageBackgroundThemeBrush}"
    Width="1419" Height="971">

    <Grid Background="{ThemeResource ApplicationPageBackgroundThemeBrush}"
    Margin="0,0,935,677">
        <Grid.ColumnDefinitions>
            <ColumnDefinition Width="240"/>
            <ColumnDefinition Width="240"/>
        </Grid.ColumnDefinitions>
        <Grid.RowDefinitions>
            <RowDefinition Height="300"/>
        </Grid.RowDefinitions>
        <StackPanel Grid.Column="0" Margin="10,10,10,10"
        VerticalAlignment="Center">
            <TextBlock x:Name="traffic_title" Height="30"
            TextWrapping="NoWrap"
                    Text="Traffic Light" FontSize="20" Foreground="Blue"
                    Margin="10" HorizontalAlignment="Center"/>
            <Ellipse x:Name="traffic_red" Fill="LightGray"
            Stroke="Gray"   Width="40"
                    Height="40" Margin="2" HorizontalAlignment="Center"/>
```

```xml
                <Ellipse x:Name="traffic_yellow" Fill="LightGray"
                Stroke="Gray"   Width="40"
                        Height="40" Margin="2" HorizontalAlignment="Center"/>
                <Ellipse x:Name="traffic_green" Fill="LightGray"
                Stroke="Gray"   Width="40"
                        Height="40" Margin="2" HorizontalAlignment="Center"/>
            </StackPanel>
            <StackPanel Grid.Column="1" Margin="10,0,10,0"
            VerticalAlignment="Center" Height="182">
                <TextBlock x:Name="pedestrian_title" Height="30"
                TextWrapping="NoWrap"
                        Text="Pededstrian" FontSize="20" Foreground="Blue"
                        Margin="10" HorizontalAlignment="Center"/>
                <Ellipse x:Name="pedestrian_red" Fill="LightGray"
                Stroke="Gray"   Width="40"
                        Height="40" Margin="2" HorizontalAlignment="Center"/>
                <Ellipse x:Name="pedestrian_yellow" Fill="LightGray"
                Stroke="Gray"   Width="40"
                        Height="40" Margin="2" HorizontalAlignment="Center"/>
            </StackPanel>
        </Grid>
</Page>
```

Header File

Next, we will add code to the MainPage.xaml.h file (the header file). Open the file and add the following are the namespaces that you need to include. Go ahead and add those now, as shown here:

```cpp
using namespace Platform;
using namespace Windows::UI;
using namespace Windows::UI::Xaml;
using namespace Windows::UI::Xaml::Media;
using namespace Windows::UI::Core;          // DispatcherTimer
using namespace Windows::Devices::Gpio;
```

Next, you need to add some variables and constants. Let's start with constants for the lights in the user interface. We will create brushes for the traffic and pedestrian lights. Add these to the `private:` section in the class.

```
// Add references for color brushes to paint the led_indicator control
SolidColorBrush^ redFill = ref new SolidColorBrush(Colors::Red);
SolidColorBrush^ yellowFill = ref new SolidColorBrush(Colors::Yellow);
SolidColorBrush^ greenFill = ref new SolidColorBrush(Colors::Green);
SolidColorBrush^ grayFill = ref new SolidColorBrush(Colors::LightGray);
```

Next, we add some constants for the lights. You use these constants with an array for each of the traffic and walk lights. By using constants instead of integers, you make the code easier to read.[6]

```
// Light constants
const int RED = 0;
const int YELLOW = 1;
const int GREEN = 2;
```

Next, let's define the pins for each of our components: the traffic light, button, and the walk light.

```
// Traffic light pins
const int TRAFFIC_PINS[3] = { 4, 5, 6 };

// Button pin
const int BUTTON_PIN = 19;

// Walk light pins
const int WALK_PINS[2] = { 20, 21 };
```

---

[6]Which is always a good choice and worth the effort.

Next, you add constants to describe the states or stages that the lights cycle through. In this case, the values (integers) represent the time sequence in seconds when the lights change. Once again, using constants makes the code easier to read, as you shall see later.

```
const int GREEN_TO_YELLOW = 4;
const int YELLOW_TO_RED = 8;
const int WALK_ON = 12;
const int WALK_WARNING = 22;
const int WALK_OFF = 30;
```

Now you can create the variables to hold the GPIO pin instances for the traffic light, button, and walk light.

```
// Traffic light pin variables
Array<GpioPin^>^ Traffic_light = ref new Array<GpioPin^>(3);
// Walk light pin variables
Array<GpioPin^>^ Walk_light = ref new Array<GpioPin^>(2);

// Button pin variable
GpioPin^ Button;
```

Since you decide to use the `DispatcherTimer` class, you can create the variable for that too.

```
// Add a Dispatcher Timer
DispatcherTimer^ walkTimer;
```

Next, you need a variable to use for counting the seconds that have elapsed since the light sequence was started. This allows us to keep certain LEDs on for a specific time period.

```
// Variable for counting seconds elapsed
int secondsElapsed = 0;
```

Finally, we will add some method signatures for the methods we will need. These include initializing the GPIO, a callback for the timer tick and button click events. Add these after the variables in the same section.

```
void InitGPIO();
void WalkTimer_Tick(Platform::Object^ sender, Platform::Object^ e);
void Button_ValueChanged(GpioPin^ sender, GpioPinValueChangedEventArgs^ e);
```

Source Code

OK, we're done with the header file. Now, open the source code file named `MainPage.xaml.cpp`. We will begin by adding the namespaces we need as shown in the following:

```
using namespace Windows::UI::Core;          // DispatcherTimer
using namespace Windows::Devices::Gpio;      // Add this
using namespace Windows::System::Diagnostics;
```

Next, we will add the `InitGPIO()` method, as well as code, to set up the GPIO. As you discovered in the last project, you use this method to set up the GPIO pins for all the LEDs and buttons. Listing 9-2 shows the complete code for this method.

Listing 9-2. GPIO Initialization Code

```
void MainPage::InitGPIO()
{
    auto gpio = GpioController::GetDefault();
    // Do nothing if there is no GPIO controller
    if (gpio == nullptr) {
        OutputDebugString(L"No GPIO Controller!");
        return;
    }

    // Initialize the GPIO pins
    for (int i = 0; i < 3; i++)
    {
        Traffic_light[i] = gpio->OpenPin(TRAFFIC_PINS[i]);
        Traffic_light[i]->SetDriveMode(GpioPinDriveMode::Output);
    }
    Button = gpio->OpenPin(BUTTON_PIN);
    for (int i = 0; i < 2; i++)
    {
        Walk_light[i] = gpio->OpenPin(WALK_PINS[i]);
        Walk_light[i]->SetDriveMode(GpioPinDriveMode::Output);
    }
    Traffic_light[RED]->Write(GpioPinValue::Low);
    Traffic_light[YELLOW]->Write(GpioPinValue::Low);
    traffic_green->Fill = greenFill;
```

```
Traffic_light[GREEN]->Write(GpioPinValue::High);
pedestrian_red->Fill = redFill;
Walk_light[RED]->Write(GpioPinValue::High);
Walk_light[YELLOW]->Write(GpioPinValue::Low);

// Check if input pull-up resistors are supported
if (Button->IsDriveModeSupported(GpioPinDriveMode::InputPullUp))
    Button->SetDriveMode(GpioPinDriveMode::InputPullUp);
else
    Button->SetCriveMode(GpioPinDriveMode::Input);

// Set a debounce timeout to filter out switch bounce noise from a
button press
std::chrono::milliseconds ms(50);
TimeSpan ts;
ts.Duration = 500000; // 10000000 per second * 0.050
Button->DebounceTimeout = ts;

// Register for the ValueChanged event so our Button_ValueChanged
// function is called when the button is pressed
Button->ValueChanged += ref new TypedEventHandler<GpioPin^, GpioPin
ValueChangedEventArgs^>(this, &MainPage::Button_ValueChanged);
OutputDebugString(L"You're good to go!");
}
```

Notice how I used the arrays to initialize the LEDs. Isn't the code easier to read with the constants? Take a moment to ensure that you understand how the LEDs are set up. That is, the green LED in the traffic light and the red LED in the walk light are on (set to HIGH) at the initial state. This mimics the scenario where the pedestrian approaches a busy street where traffic is flowing (hence the green traffic light).

Notice also we added code for debouncing the button when pressed. Notice you first check to see if the GPIO header has pull-up resistors (it "pulls" the voltage high),[7] and if it does, you use the `GpioPinDriveMode.InputPullUp` mode for the GPIO pin. If it does not have pull-up resistors, you simply set the mode to `GpioPinDriveMode. Input`. Recall you used the `SetDriveMode()` method for the LEDs but set the mode to `GpioPinDriveMode.Output`.

---

[7]See https://learn.sparkfun.com/tutorials/pull-up-resistors.

We also have code to set the circles in the traffic and pedestrian lights in the user interface mockup. I named the walk lights "pedestrian" to help keep them separate and easier to read. You may want to consider renaming `walk_light` to `pedestrian_light` if you would like to keep things a bit more tidy.

Next, I added code to return from the method if the GPIO library cannot be initialized. Recall you used a text label on the user interface to communicate an error. Here, the absence of any LEDs illuminating indicates that something is wrong.

Be sure to add the method call to the `MainPage()` method right after the `InitializeComponent()` call as shown in Listing 9-3. Also, you set the initial value for `secondsElapsed` to 0.

Listing 9-3. The MainPage() Method

```
MainPage::MainPage()
{
    InitializeComponent();
    InitGPIO();
    secondsElapsed = 0;

    // Add code to setup timer
    walkTimer = ref new DispatcherTimer();
    // doing 1/2 second ticks for demo purposes
    std::chrono::milliseconds ms(500);
    TimeSpan ts;
    ts.Duration = 5000000; // 10000000 per second * 0.50
    walkTimer->Interval = ts;
    walkTimer->Tick += ref new EventHandler<Object^>(this,
    &MainPage::WalkTimer_Tick);
    walkTimer->Stop();
}
```

That may seem like a lot of initialization code, and I suppose it is compared to the previous projects, but you use all of these in the rest of the code. Now that you have the basic project code, let's add the code for controlling the LEDs.

Light Sequences

Recall from the description of the project that you want the lights to cycle through several stages based on time. Thus, you use the `DispatcherTimer` class to start a timer that you then count as seconds expired to control the stages. This is done by making a new method to fire whenever the timer event occurs. You named the `DispatcherTimer` class `walkTimer` earlier, so you use `WalkTimer_Tick()` for the event method.

Inside this method, you use the `secondsElapsed` variable to count the seconds (tick events), and when the count reaches the number you assigned to the state changes, you change the light. For example, when the `secondsElapsed` reaches 4 (`GREEN_TO_YELLOW`), you execute state (2) from earlier, where the green traffic light is turned off and the yellow traffic light is turned on. You do the same for the other states. Listing 9-4 shows the code you need to control the light sequences.

Listing 9-4. Light Sequence Code

```
void MainPage::WalkTimer_Tick(Platform::Object^ sender, Platform::Object^ e)
{
    // Change green to yellow
    if (secondsElapsed == GREEN_TO_YELLOW)
    {
        traffic_green->Fill = grayFill;
        Traffic_light[GREEN]->Write(GpioPinValue::Low);
        traffic_yellow->Fill = yellowFill;
        Traffic_light[YELLOW]->Write(GpioPinValue::High);
    }

    else if (secondsElapsed == YELLOW_TO_RED)
    {
        traffic_yellow->Fill = grayFill;
        Traffic_light[YELLOW]->Write(GpioPinValue::Low);
        traffic_red->Fill = redFill;
        Traffic_light[RED]->Write(GpioPinValue::High);
    }
    else if (secondsElapsed == WALK_ON)
    {
        pedestrian_red->Fill = grayFill;
```

```cpp
        Walk_light[RED]->Write(GpioPinValue::Low);
        pedestrian_yellow->Fill = yellowFill;
        Walk_light[YELLOW]->Write(GpioPinValue::High);
    }
    else if ((secondsElapsed >= WALK_WARNING) &&
        (secondsElapsed < WALK_OFF))
    {
        // Blink the walk warning light
        if ((secondsElapsed % 2) == 0)
        {
            pedestrian_yellow->Fill = grayFill;
            Walk_light[YELLOW]->Write(GpioPinValue::Low);
        }
        else
        {
            pedestrian_yellow->Fill = yellowFill;
            Walk_light[YELLOW]->Write(GpioPinValue::High);
        }
    }
    else if (secondsElapsed == WALK_OFF)
    {
        pedestrian_yellow->Fill = grayFill;
        Walk_light[YELLOW]->Write(GpioPinValue::Low);
        pedestrian_red->Fill = redFill;
        Walk_light[RED]->Write(GpioPinValue::High);
        traffic_red->Fill = grayFill;
        Traffic_light[RED]->Write(GpioPinValue::Low);
        traffic_green->Fill = greenFill;
        Traffic_light[GREEN]->Write(GpioPinValue::High);
        secondsElapsed = 0;
        walkTimer->Stop();
        return;
    }
    // increment the counter
    secondsElapsed += 1;
}
```

Notice you used a return in the state where you turn the walk light off. This terminates the code so that the seconds elapsed variable is not incremented. You also reset the `secondsElapsed` variable so that the next pedestrian can initiate the crossing. You also increment the variable each time the event fires otherwise. I leave the user interface code for you to decipher.

Button

We also need to create a method to fire when the button is pressed. You'll name that method `Button_ValueChanged()`. You need to assign the `ValueChanged` attribute of the button to this method. You see that in the last line of code.

Now you can write the code to execute when the button is pressed. This may seem like a simple thing, but there is one aspect you must consider. When a button is pressed, it may not make a connection right away, or it may be the case that there is some hesitation on the user's part. This creates a condition where the event could trigger prematurely. This is called bouncing and can be controlled with a bit more code. In this case, you check the edge attribute of the button to see if the state is on the trailing edge—in other words, that the button has been pressed for a period of time. The condition you use is (`e->Edge == GpioPinEdge::FallingEdge`).

There is one trickier bit. Once the button event fires, you cannot call back to the main code directly. This is because the button event is running simultaneously with the main code. More specifically, there are two threads involved. Thus, you need to make a call to the dispatcher to run a task. In this case, you want to simply turn on the timer. Listing 9-5 shows the method that you use to execute when the button is pressed.

Listing 9-5. Button Value Changed Method

```
void MainPage::Button_ValueChanged(GpioPin^ sender,
GpioPinValueChangedEventArgs^ e)
{
    // Pedestrian has pushed the button. Start timer for going red.
    if (e->Edge == GpioPinEdge::FallingEdge)
    {
        // Start the timer if and only if not in a cycle
        if (secondsElapsed == 0)
        {
            // need to invoke UI updates on the UI thread because this event
```

```cpp
        // handler gets invoked on a separate thread.
        auto task = Dispatcher->RunAsync(CoreDispatcherPriority::Normal,
            ref new DispatchedHandler([&, e]()
        {
            if (e->Edge == GpioPinEdge::FallingEdge)
            {
                walkTimer->Start();
            }
        }));
        }
    }
}
```

Notice the code to make the call to the asynchronous feature of the dispatcher. The code is formatted to read like normal code, but it is actually a method inside the RunAsync() call. This is an advanced feature that you may have to use in some of your projects. For more detailed information about this feature, see the online documentation at https://docs.microsoft.com/en-us/uwp/api/Windows.UI.Core.CoreDispatcher? redirectedfrom=MSDN&view=winrt-19041#Windows_UI_Core_CoreDispatcher_ RunAsync_Windows_UI_Core_CoreDispatcherPriority_Windows_UI_Core_ DispatchedHandler_.

Notice one more thing. I use another conditional to only allow turning on the dispatcher if the secondsElapsed variable is 0. Thus, even if someone pressed the button once the light sequence starts, it is ignored until the variable returns to 0 at the end of the light sequence code. Cool, eh?

Completing the Code

OK, now you have all the pieces of code that you need, but you may not know where each piece goes. Rather than relist the entire code, I include the skeleton of the code in Listing 9-6 to help you put things in the right places. The ellipses are used to represent code omitted.

Listing 9-6. Pedestrian Crossing Code Layout

```
#include "pch.h"
#include "MainPage.xaml.h"

using namespace PedestrianCrossing;

using namespace Platform;
using namespace Windows::Foundation;
using namespace Windows::Foundation::Collections;
using namespace Windows::UI::Xaml;
using namespace Windows::UI::Xaml::Controls;
using namespace Windows::UI::Xaml::Controls::Primitives;
using namespace Windows::UI::Xaml::Data;
using namespace Windows::UI::Xaml::Input;
using namespace Windows::UI::Xaml::Media;
using namespace Windows::UI::Xaml::Navigation;
using namespace Windows::UI::Core;              // DispatcherTimer
using namespace Windows::Devices::Gpio;         // Add this
using namespace Windows::System::Diagnostics;

MainPage::MainPage()
{
...
}

void MainPage::InitGPIO()
{
...
}

// Here you do the lights state change if and only if elapsed_seconds > 0
void MainPage::WalkTimer_Tick(Platform::Object^ sender, Platform::Object^ e)
{
...
}
```

```
void MainPage::Button_ValueChanged(GpioPin^ sender,
GpioPinValueChangedEventArgs^ e)
{
...
}
```

Wow, that is a lot of code! Clearly, this project is larger than the examples so far. If you have not written a project of this size, be sure to check the listing earlier to ensure that you have all the code in the right place. Once you have entered all the code, you should now attempt to compile the code. Correct any errors you find until the code compiles without errors or warnings.

Deploy and Execute

Now it is time to deploy our application! Be sure to fix any compilation errors first. Like you have with other applications, you want to compile the application in debug first (but you can compile in release mode if you'd prefer by choosing *Release* in the drop-down box). While we have seen how to deploy our C++ applications in previous chapters, I will show you a different process here that is similar to how we deploy with C# or Visual Basic.

Specifically, unlike we saw in Chapter 5, we do not need the remote debugger running to deploy our C++ application using this alternate process. Briefly, we will still use the debug settings for the project, but instead of manually configuring the IP address and security protocol, we will use the locate feature to find our device.

Note   Make sure the remote debugger is stopped on your device before attempting to deploy the application using this method. The method in Chapter 5 also works.

We begin by selecting the build mode, platform, and then device in Visual Studio. In this case, we want to select the *Release* or *Debug* build, the *ARM* platform, and select the *Remote Machine* entry in the device drop-down as shown in Figure 9-8. For this form of deployment, it doesn't matter which build mode you choose.

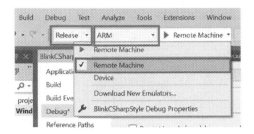

Figure 9-8. *Select the build, platform, and device (Visual Studio)*

Unlike C# and Visual Basic, when you select the *Remote Machine* entry, nothing happens—no Remote Connections dialog will appear. Rather, we will click the drop-down again (the small down arrow to the right of *Remote Machine*) and choose the Debug settings. In this case, the entry will read *PedestrianCrossing Debug Properties*.

In the dialog that appears, click the *Machine Name* drop-down and choose *<Locate...>* as shown in Figure 9-9.

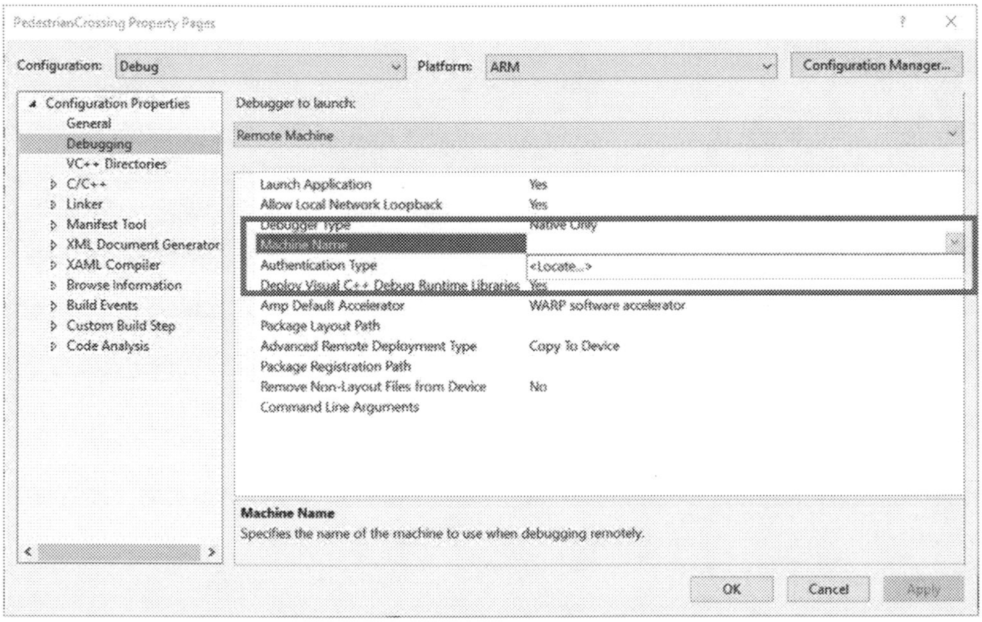

Figure 9-9. *Using the Locate option in the debug properties*

This is what opens the Remote Connections dialog that you are used to seeing with C# and Visual Basic. Recall, the dialog permits you to select the device from a list as shown in Figure 9-10. Simply select the device and click *Select* to select it.

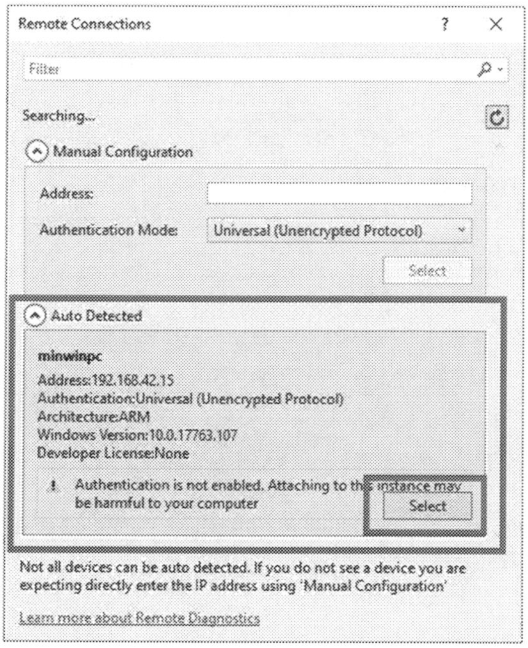

Figure 9-10. *Remote Connections dialog excerpt*

OK, we're almost ready to deploy. There is just one more thing to change. I find it helpful to rename the package name using the package manifest. This is the name that will be displayed in the Device Portal. By default, it uses a generated name, but I find renaming it to the same as the project name helpful for easier tracking. Thus, this is an optional step.

If you look at the right side of Visual Studio, you will see one file in the tree view named Package.appxmanifest. Double-click that file to open it. As you will see, it is a set of options you can set for your package. The one we want is on the last tab named Packaging. Click that and change the Package name, as shown in Figure 9-11.

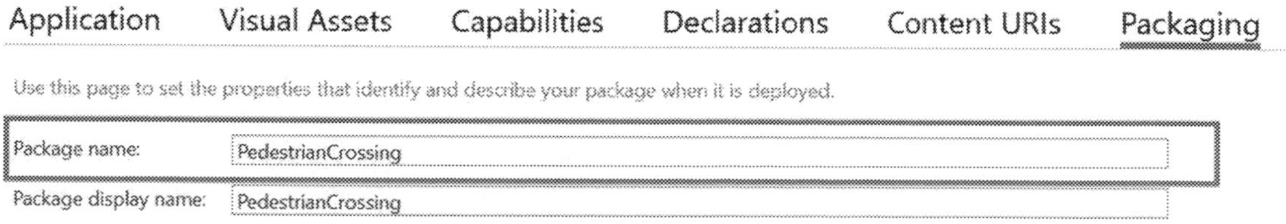

Figure 9-11. *Set the package name (package manifest)*

Now all that is left is to deploy the application. Simply click *Build* ➤ *Deploy Solution*, and the deploy will begin. If this is the first time you're deploying the application (or any application) to your device, the deployment could take a while.

Once complete, Visual Studio will indicate the build and deploy was successful similar to the output shown as follows:

```
Creating a new clean layout...
Copying files: Total 7 mb to layout...
Checking whether required frameworks are installed...
Registering the application to run from layout...
Deployment complete (0:00:42.425). Full package name:
"PedestrianCrossing_1.0.0.0_arm__mvq1akfegk7g6"
========== Build: 1 succeeded, 0 failed, 0 up-to-date, 0 skipped ==========
========== Deploy: 1 succeeded, 0 failed, 0 skipped ==========
```

Tip   If you have difficulty deploying your application, see the troubleshooting section in Chapter 6 for details on some of the things that may go wrong and their solutions.

Now, you can go to the Device Portal and observe your application in the Apps manager as shown in Figure 9-12. Recall, you can start it by selecting *Start* from the *Actions* drop-down box. Go ahead and do that now. You should see your application start, and you can experiment with it.

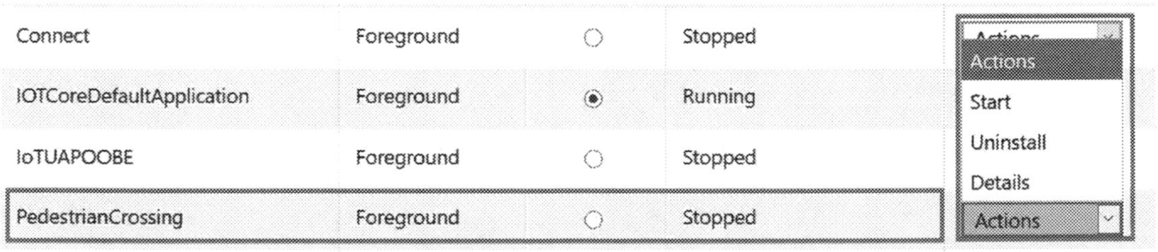

Figure 9-12. *Controlling the application in the Apps manager (Device Portal)*

If everything worked correctly, you should see the lights change when the button is pressed, and the user interface should mimic the behavior. If the lights don't change, be sure you have the pushbutton oriented correctly. Once it is working, try it out a few times to ensure that you see the light sequence several times. If you like this project, I encourage you to experiment with it further by either adding another button or adding a second traffic light for an intersection rather than a one-way street.

Tip   For instructions on how to debug your C++ application, see Chapter 5.

Summary

Learning how to build IoT projects requires a number of electronic components. Fortunately, vendors such as Adafruit sell electronic component kits that have a wide array of common electronic components, such as LEDs, resistors, jumper wires, breadboards, and more. If you are planning to explore Windows IoT projects and especially if you want to work on the more advanced projects in the next few chapters, you should consider buying one (or both) of the kits described in this chapter or a similar kit from another vendor. Just be sure you get the data sheets for all the components in the kits.

You also explored a project that introduces how to read sensors. If you have been implementing all the projects in this book so far or have been reading through them in some detail, you now have the skills (or at least examples of how) to read and write values on GPIO pins. Indeed, you have now seen the basics for writing any form of IoT project.

The next chapter begins a series of example IoT projects that you can use to learn more about building IoT solutions. I recommend working through as many of them as you can. Some require additional hardware that you may or may not want to acquire. Should you decide to not implement some of the projects, you can read through the projects to learn how to use the components and get ideas for your own designs.

## CHAPTER 10

Project 1: Building an LED Power Meter

You've seen a lot of examples of powering LEDs—turning them on and off from our IoT device.[1] There are a lot of interesting things you can do with LEDs. For example, have you wondered how a power meter works? If you have used high-end audio equipment, such as a studio sound board, you may have seen a power meter that has several segments ranging from green to yellow to red where green means the level is low, yellow is medium, and red is high. You can duplicate this behavior with a set of LEDs and your IoT device. Yes, you're going to build a fancy LED power meter!

While this seems very simple and in concept it is, this project helps you learn quite a lot. You will see how to use a potentiometer as a variable input device read from an analog-to-digital converter (ADC), learn how to set up and use a Serial Peripheral Interface (SPI), discover a powerful debugging technique, and learn how to create a class to encapsulate functionality. Clearly, there is a lot to discover, so let's get started.

Note Although this project is written in C#, other than the syntax and mechanics of building, the concepts of using a class are the same in C++.

Overview

You will design and implement an LED power meter that allows you to simulate controlling power and displaying the result as a percentage of a set of LEDs where no LEDs on means minimum and all LEDs on means maximum. More specifically, you use

---

[1]And here's another one!

a component called a *potentiometer* to read and interpret its value as a percentage of its maximum range. You will use the potentiometer through the ADC to control the LEDs.

The potentiometer is a special rotary component that can vary resistance as you turn it. Thus, potentiometers are rated at their maximum resistance. There are a variety of potentiometers packaged in a variety of ways, but you will use a simple 10K ohm potentiometer that you can plug into a breadboard.

Although you will use a series of LEDs for the power meter, you can buy components where LEDs are arranged in a bar. In fact, they're often called *bar graph LEDs*. If you want to take this project a bit further, you can find a number of different bar graph LEDs like those from Adafruit (`www.adafruit.com/categories/279`).

You need to use the ADC because you want to read analog values. The Raspberry Pi (and many other boards) does not have analog-to-digital logic. That is, the GPIO pins are digital only. The ADC acts like a "bridge" between digital and analog devices. The ADC you will use is the MCP3008 from the Adafruit Microsoft IoT Pack for Raspberry Pi (see Chapter 9). The ADC is an integrated circuit (or chip) that you will plug into a breadboard and wire it to your Raspberry Pi. As a side benefit, the MCP3008 has eight analog input pins, but you only need four pins on the GPIO to access it, so you're gaining four more pins. This may not be important for this simple project, but for a project with many sensors or devices, saving four pins may enable you to fully implement your ideas.

The MCP3008 uses the Serial Peripheral Interface (SPI) bus to communicate with the Raspberry Pi. The SPI bus is an interface developed by Motorola as a synchronous (clocked) serial, full-duplex master-slave protocol.[2] In other words, data is transmitted in a synchronized manner timed to a clock signal. The protocol supports full duplex (which means both transmit and receive at the same time). The four wires therefore are one for the clock, one for transmit, one for receive, and one additional wire used to select the chip (the Raspberry Pi can support two SPI buses but only in master mode—the device that sets the clock signal is called the master).

One interesting thing about the SPI is that in order to receive data (say a byte), you must first send data (a byte), which sounds really weird, but it turns out the first transmission can be thought of as a command to read data and the response is the return of the command. You'll see how this works in the code.

Let's look at the components that you need and then look at how to wire everything together.

---

[2]For more information about SPI, see `https://en.wikipedia.org/wiki/Serial_Peripheral_ Interface_Bus`.

Required Components

The following lists the components that you need. You can find these components in either of the kits mentioned in Chapter 9, or you can purchase the components separately from Adafruit (`www.adafruit.com`), SparkFun (`www.sparkfun.com`), or any electronics store that carries electronic components.

- (1) 10K ohm potentiometer (breadboard pin spacing)

- (2) red LEDs

- (2) yellow LEDs (or blue is OK)

- (1) green LED

- (5) 150 ohm resistors (or appropriate for your LEDs)

- MCP3008 ADC chip

- Jumper wires: (8) male-to-male, (11) male-to-female

- Breadboard (full size recommended but half size is OK)

- Raspberry Pi 2 or 3

- Power supply

Set Up the Hardware

This project like the project in Chapter 9 has a lot of connections. Thus, you will plan for how things should connect. To connect the components to the Raspberry Pi, you need four pins for the ADC, five for the LEDs, and one each for power and ground. You will also need to make a number of connections on the breadboard to configure the ADC chip and connect the potentiometer to the ADC. Table 10-1 shows the map I designed for this project. I list the physical pin numbers in parentheses for the named pins. You will use male-to-female jumper wires to make these connections.

Table 10-1. *Connection Map for the Power Meter Project*

GPIO	Connection	Function	Notes
3.3V (1)	Breadboard power rail	Power	
GND (6)	Breadboard ground rail	GND	
MOSI (19)	SPI	MCP3008 pin 11	
MISO (21)	SPI	MCP3008 pin 12	
SCLK (23)	SPI	MCP3008 pin 13	
CC0 (24)	SPI	MCP3008 pin 10	
17	Red LED #1	Meter 81–100%	
18	Red LED #2	Meter 61–80%	
19	Yellow LED #1	Meter 41–60%	
20	Yellow LED #2	Meter 21–40%	
21	Green LED	Meter 0–21%	

Tip   Refer to Chapter 9 for the size of resistors needed for your LEDs.

Next, you need to make a number of connections on the breadboard. For these, you use male-to-male jumpers. Table 10-2 shows the connections needed on the breadboard.

Table 10-2. *Connections on the Breadboard*

From	To	Notes
Breadboard power	Potentiometer pin #1	
Potentiometer pin #2	ADC Channel 0 (MCP3008 pin 1)	
Breadboard GND	Potentiometer pin #3	
Breadboard power	ADC VDD (MCP3008 pin 16)	
Breadboard power	ADC VREF (MCP3008 pin 15)	
Breadboard GND	ADC AGND (MCP3008 pin 14)	
Breadboard GND	ADC GND (MCP3008 pin 9)	
Breadboard GND rail	Breadboard GND rail jump	

Clearly, that's a lot of wires! Don't worry too much about neatness when you build this project. Rather, concentrate on making sure everything is connected correctly. I recommend spending some time to carefully check your connections. There's so many that it is easy to get some plugged in the wrong place. Figure 10-1 shows what my project looked like.

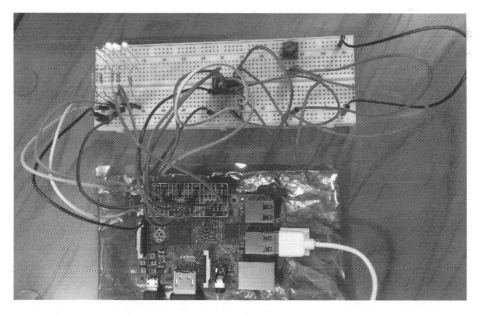

Figure 10-1. *Example power meter connections*

As you can see in the photo, I'm cheating a bit by using the Adafruit GPIO Reference Card for Raspberry Pi 2 or 3 (`www.adafruit.com/products/2263`), which makes locating the SPI pins much easier than counting pin numbers.

The MCP3008 chip is in the center of the breadboard. You cannot see it in this photo, but chips have a small semicircular notch on one end. This indicates the side that contains pin 1 so that you can orient the chip correctly. In this case, you orient the chip on the breadboard with pins 1–8 on the far side of the breadboard (away from the Raspberry Pi as shown in the photo). This is because the SPI interface pins are located on the nearer side (pins 9–16). Orienting the pins closest to the Raspberry Pi makes the connections a bit easier. Figure 10-2 shows the pin layout of the MCP3008. Notice how the pins are numbered. Most are self-explanatory like the channel pins, but all are documented on the data sheet from Adafruit (`https://cdn-shop.adafruit.com/datasheets/MCP3008.pdf`).

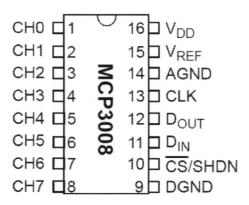

Figure 10-2. *MCP3008 pinout*

Figure 10-3 shows all the connections needed.

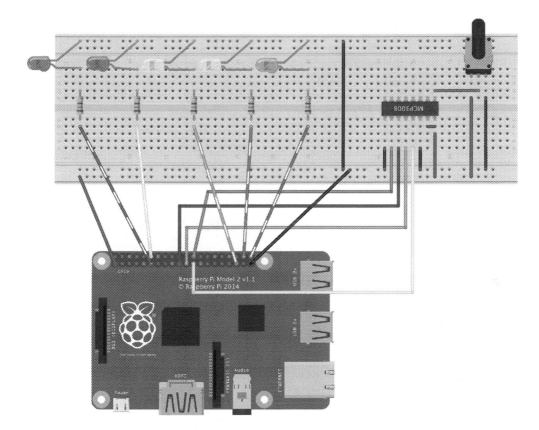

Figure 10-3. *Connections for the power meter project*

Notice how I arranged the LEDs. I placed the red LEDs to the left, the yellow in the middle, and the green to the right. More specifically, I plugged the negative leg of each LED into the GND rail on the breadboard. This allows me to make one connection from one of the ground pins on the GPIO to the breadboard rail, which I can then use to plug in the GND leg of the components. For example, each positive leg of each LED is plugged into the breadboard so that you can plug the corresponding resistor across the DIP trough and connect those to the appropriate GPIO pin. Also, I placed the MCP3008 in the center spanning the DIP trough. Make sure that you plug it in so that pins 1–8 are on one side of the breadboard and that pins 9–16 are on the other side.

If you are following along with this chapter working on the project, make the hardware connections now. Don't power on the board yet; but do double- and triple-check the connections.

Write the Code

Now it's time to write the code for our example. Since you are working with several new components, I introduce the code for each in turn. The code isn't overly complicated but may not be as clear as some of the code from previous projects. I've decided to use C#, but you could implement this project in C++ or Visual Basic. If you are a big fan of one of those languages, I encourage you to do just that using the following as a pattern!

Note   Since you have learned all the basics of creating projects in Visual Studio, including how to build and deploy applications, I omit the details of the common operations for brevity.

The project uses a potentiometer. You read its value (via the ADC) and convert that to a scale that you can use to decide how many LEDs to turn on (and consequently those that need to be turned off). You also use a new technique for debugging the code. Thus, you need to start a new project, set up the SPI interface, write code to communicate with the ADC, and code to turn the LEDs on and off. You will use the `DispatcherTimer` class to periodically check the potentiometer and control the light sequence like you did in the last project.[3] Let's talk about the debugging feature first, and then we will walk through building the project.

Debug Output

Visual Studio supports an interesting and very powerful feature that permits you to insert print statements (and more!) in the code to help you debug your code. It is similar in some ways to the old-school print statement trace (a log of statements written to a file), which is written as the code runs. However, in this case, the statements appear in the output window of Visual Studio. To use the feature, add the following namespace to your code:

```
using System.Diagnostics;        // add this for debugging
```

---

[3]There are other ways to do this, but this uses a technique we've seen previously.

You use the Debug class from this namespace to write out values so you can see them in the output window as you run the code in debug. The following code shows some examples of the methods you use to report data. Some are informational (the proverbial, "I'm here!"), while others show how you can print out the values of variables.

```
Debug.WriteLine("Sorry, the GPIO cannot be initialized. Drat.");
Debug.Write("Val read = ");
Debug.WriteLine(valRead);
```

Here, I use two methods: Write(), which writes a string without a carriage return/line feed (CRLF) symbol (sometimes called a *newline*), and Writeline(), which writes the CRLF.

The output of this code is shown in the output window. The following shows an excerpt of the output from this project. It is to see the progress of the code. There is much more to this class, but this gives you a taste for what is possible with this alternative debugging technique.

```
...
GPIO ready.
Setting up GPIO pin 21.
Setting up GPIO pin 20.
Setting up GPIO pin 19.
Setting up GPIO pin 18.
Setting up GPIO pin 17.
Setting up the MCP3008.
...
```

Tip   For more information about the system diagnostics debug class, see https:// docs.microsoft.com/en-us/dotnet/api/system.diagnostics.debug? view=netcore-3.1.

You must compile your code in debug mode and run the debugger to see the output. If you compile in release mode, the debug code is ignored automatically and not included in the executable code (binary file).

New Project

You will use the same project template as the last project—the *C# Blank App (Universal Windows)* template. Use the name *PowerMeter* for the project name. You can save the project wherever you like or use the default location. Once the project opens, double-click the MainPage.xaml.cs file. This is where you put most of the code for the project. There are three namespaces you need to include. Go ahead and add those now, as shown next:

```
using Windows.UI.Xaml;
using Windows.UI.Xaml.Controls;
using Windows.Devices.Gpio;        // add this for GPIO pins
using System.Diagnostics;          // add this for debugging
```

Tip   You may have noticed some of the namespaces are grayed out. This indicates the namespace may not be needed. If you click one, then press *Alt+Enter*, you can remove any or all of the unnecessary namespaces. I recommend you do this once your project is working correctly.

Next, you need to add some variables and constants. You add constants for the number of LEDs and maximum potentiometer value (1023). You also add variables for the meter pin numbers and the GPIO pin variables. Finally, you add a variable for the timer. Recall, these go after the public sealed partial class MainPage : Page statement and before the public MainPage() statement.

```
// Power meter pins, variables
private const int numLEDs = 5;
private const float maxPotVal = 1021;  // Adjust this value as needed
private int[] METER_PINS = { 21, 20, 19, 18, 17 };
private GpioPin[] Meter = new GpioPin[numLEDs];

// Add a Dispatch Timer
private DispatcherTimer meterTimer;
```

Notice the comment on maxPotVal. If you find your project does not illuminate the fifth LED (last red one), you can adjust this value to match the maximum value read from the potentiometer during execution. Check the debug messages and you should see a message like Val read = 1021 appear. This is the maximum value. Not all ADC chips or potentiometers are perfect, so some small adjustment may be needed.

Next, you'll add a variable for the MCP3008 and instantiate an instance for the new class. In this case, you create a new class to encapsulate the code for the ADC. You'll see this in a later section.

```
// Instantiate the new ADC Chip class
ADC_MCP3008 adc = new ADC_MCP3008();
```

Last, you'll add a constant for the channel on the ADC that you will use. In this case, you use channel zero.

```
// Channel to read the potentiometer
private const int POT_CHANNEL = 0;
```

The code in the MainPage() function initializes the components, the GPIO, and the new ADC class. You also set up the timer. In this case, you use a value of 500, which is one-half a second. If you adjust this lower, the solution is a bit more responsive. You can try this once you've got it to work and tested. Listing 10-1 shows the complete MainPage() method.

Listing 10-1. MainPage Method

```
public MainPage()
{
    this.InitializeComponent();
    InitGPIO();                    // Initialize GPIO
    adc.Initialize();              // Call initialize() method for ADC

    // Setup the timer
    this.meterTimer = new DispatcherTimer();
    this.meterTimer.Interval = TimeSpan.FromMilliseconds(500);
    this.meterTimer.Tick += meterTimer_Tick;
    this.meterTimer.Start();
}
```

Finally, you must add the reference to the Windows 10 IoT Extensions from the project property page. You do this by right-clicking the **References** item on the project item in the Solution Explorer.

Initialize GPIO

As usual, you must provide the InitGPIO() method you called in the MainPage() method.[4] Recall, you want to set up the GPIO controller and initialize the pins you are going to use. In this case, you must initialize the five LED pins, and since you have them in an array, you can use a for loop to accomplish the task.

While you are at it, you'll also use a few diagnostic statements to record whether the GPIO controller succeeds or not and the state of the pins. You can see these calls are made using the Debug.Write(), which does not write an end of line, and Debug.Writeline(), which does write the end of line character. Listing 10-2 shows the completed InitGPIO() method.

Listing 10-2. Initialize GPIO Method

```
// Setup the GPIO initial states
private void InitGPIO()
{
    var gpio = GpioController.GetDefault();

    // Show an error if there is no GPIO controller
    if (gpio == null)
    {
        Debug.WriteLine("Sorry, the GPIO cannot be initialized. Drat.");
        return;
    }
    Debug.WriteLine("GPIO ready.");
    // Initialize the GPIO pins
    for (int i = 0; i < numLEDs; i++)
    {
        this.Meter[i] = gpio.OpenPin(METER_PINS[i]);
```

---

[4]You don't have to name it InitGPIO(), but that is the common theme. Actually, any reasonable name is fine; but if you are going to share your code, InitGPIO() is a good name.

```
        this.Meter[i].SetDriveMode(GpioPinDriveMode.Output);
        this.Meter[i].Write(GpioPinValue.Low);
        Debug.Write("Setting up GPIO pin ");
        Debug.Write(METER_PINS[i]);
        Debug.WriteLine(".");
    }
}
```

Controlling the LEDs

Next, you need code to control the LEDs. You're using the timer class to fire an event every 500 milliseconds (one-half of a second). In the MainPage() method, you assigned the name meterTimer_Tick() to the event. In this method, you must read the value from the potentiometer via the new ADC class (explanation is in the next section) and then calculate how many LEDs to turn on.

Since you are using five LEDs, you convert the value read from the potentiometer as a percentage and then convert it to a scale from 0 to 5. You do this so you can use two loops: one to turn on those LEDs that should be on and another to turn the remainder off. Check the code to ensure that you understand how this works. You also see a copious amount of Debug.Write* methods to display the state of this method. Listing 10-3 shows the complete code for the meterTimer_Tick() method.

Listing 10-3. Controlling the LEDs

```
private void meterTimer_Tick(object sender, object e)
{
    float valRead = 0;
    int numLEDs_On = 0;

    Debug.WriteLine("Timer has fired the meterTimer_Tick() method.");

    // Read value from the ADC
    valRead = adc.getValue(POT_CHANNEL);
    Debug.Write("Val read = ");
    Debug.WriteLine(valRead);
```

```
    float percentCalc = (valRead / maxPotVal) * (float)10.0;
    numLEDs_On = (int)percentCalc / 2;
    Debug.Write("Number of LEDs to turn on = ");
    Debug.WriteLine(numLEDs_On);

    // Adjust power meter LEDs On or Off based on value read
    for (int i = 0; i < numLEDs_On; i++)
    {
        this.Meter[i].Write(GpioPinValue.High);
        Debug.Write("Setting pin ");
        Debug.Write(METER_PINS[i]);
        Debug.WriteLine(" HIGH.");
    }
    for (int i = numLEDs_On; i < numLEDs; i++)
    {
        this.Meter[i].Write(GpioPinValue.Low);
        Debug.Write("Setting pin ");
        Debug.Write(METER_PINS[i]);
        Debug.WriteLine(" LOW.");
    }
}
```

Now that you've seen all the code for the main code file, let's look at how the methods are placed in the class. Listing 10-4 shows the skeleton of the MainPage class. Use this as a guide to place the code in the right places. I omit the details of the methods for brevity.

Listing 10-4. MainPage Code Layout

```
...

namespace PowerMeter
{
    public sealed partial class MainPage : Page
    {
...
```

```
        public MainPage()
        {
...
        }

        // Setup the GPIO initial states
        private void InitGPIO()
        {
...
        }

        private void meterTimer_Tick(object sender, object e)
        {
...
        }
    }
}
```

In the next section, you learn how to write the code to communicate with the ADC chip.

Code for the MCP3008

While you can write the code in this section in the project class MainPage, you'll discover an example of making your code easier to maintain using a class module to group the code for the ADC. You've already seen a couple of references to this in the previous code, including the constructor (ADC_MCP3008 adc = new ADC_MCP3008();), initialization (adc.Initialize();), and reading a value on a channel (valRead = adc. getValue(POT_CHANNEL);). Thus, you will need these functions, at the minimum.

This class is pretty simple in scope, but for larger classes and the objects they model, you want to create a design for the class. That is, you should consider everything the class needs to do to model the object or concept.

To add a new class, right-click the project and choose *Add* ➤ *New Item*.... Name the class ADC_MCP3008, as shown in Figure 10-4.

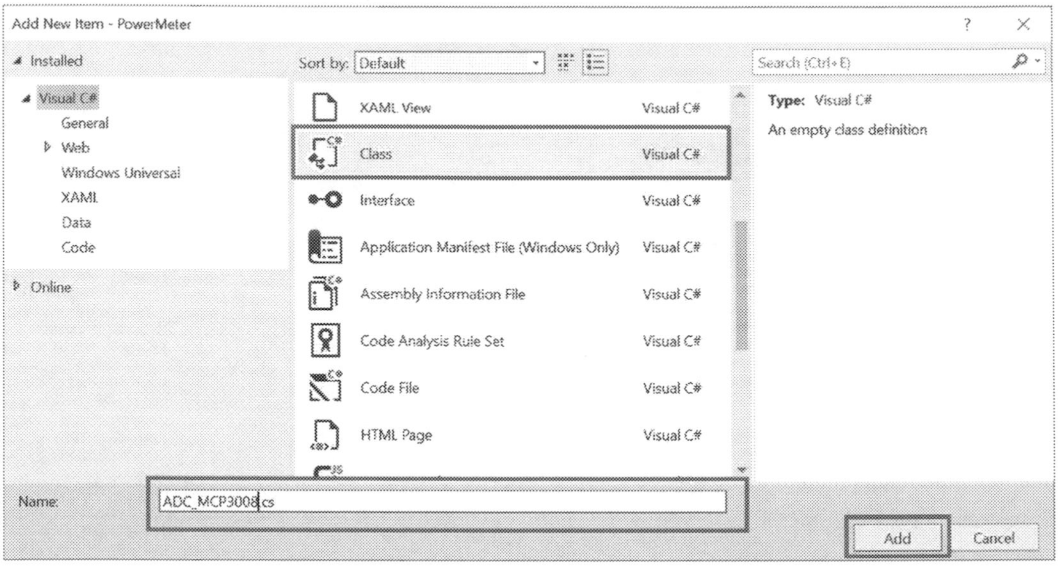

Figure 10-4. *Adding a new class*

When you create a class, you create a new file, as shown in Figure 10-4. Inside this new code file, you see a familiar skeleton. Of note is the same namespace as the main code and a list of namespaces that you will use. Open the file now and add the following namespaces. The need for each is shown in the comments.

```
using Windows.Devices.Spi;          // add this for SPI communication
using System.Diagnostics;           // add this for debugging
using Windows.Devices.Enumeration;  // add this for DeviceInformation
```

Next, you need some variables and constants. You need a variable for the SPI device and constants for the MCP3008. In this case, you set constants for helping you read data as you shall see later in this section.

```
// SPI controller interface
private SpiDevice mcp_var;
const int SPI_CHIP_SELECT_LINE = 0;
const byte MCP3008_SingleEnded = 0x08;
const byte MCP3008_Differential = 0x00;
```

Next, you create the constructor method. You don't need to do anything here, so you'll use a debug statement to indicate the class has been instantiated as follows:

```
public ADC_MCP3008()
{
    Debug.WriteLine("New class instance of ADC_MCP3008 created.");
}
```

Now that the preliminaries are set up, you can concentrate on the Initialize() and getValue() methods. The Initialize() method is used to set up the SPI interface for communicating with the ADC. I provide all the code that you need, but it is good to know what is going on. Listing 10-5 shows the Initialize() method.

Listing 10-5. Initializing the SPI Interface

```
// Setup the MCP3008 chip
public async void Initialize()
{
    Debug.WriteLine("Setting up the MCP3008.");
    try
    {
        // Settings for the SPI bus
        var SPI_settings = new SpiConnectionSettings(SPI_CHIP_SELECT_LINE);
        SPI_settings.ClockFrequency = 3600000;
        SPI_settings.Mode = SpiMode.Mode0;

        // Get the list of devices on the SPI bus and get a device instance
        string strDev = SpiDevice.GetDeviceSelector();
        var spidev = await DeviceInformation.FindAllAsync(strDev);

        // Create an SpiDevice with our bus controller and SPI settings
        mcp_var = await SpiDevice.FromIdAsync(spidev[0].Id, SPI_settings);

        if (mcp_var == null)
        {
            Debug.WriteLine("ERROR! SPI device {0} may be in used.",
            spidev[0].Id);
            return;
        }
```

```
    }
    catch (Exception e)
    {
        Debug.WriteLine("EXEPTION CAUGHT: " + e.Message + "\n" +
        e.StackTrace);
        throw;
    }
}
```

First, you set some variables of the SPI device class, including the line you will use (chip select = 0), the clock frequency (3.6 MHz or 3.6 million times a second), and the mode (0), which is the default. The mode defines how the values are sampled.[5] Next, you get the device name (string) and then call a system method named `await()` that spawns a thread to wait for the chip's SPI interface to wake and set up. You do this because the SPI device is actually running on a separate thread.

You place all of this code in a try block in case the SPI bus fails (say if you misconnect one or more wires) using debug statements to track the progress and report success or failure. Don't worry too much if this code seems very strange. I should note that this may not be the only way to set up an SPI communication, but it is compact and captures a failed initialization.

Now you're ready to learn how to get data from the ADC. While the last method was a bit complex, reading a value is a bit more complex. Recall that you must write before you can read. In this case, you send an array of three bytes and you get back an array of three bytes, where the first byte sets the start bit, the second byte contains the command (shifted 4 bits to the left), the channel select, and the third byte doesn't matter.

How did I discover this information? I read the data sheet! It takes a bit of patience because many data sheets are written in a style known as "by engineers for engineers," which means that they can be a bit terse and not very easy to read. However, in this case, you see a very good explanation of the data needed for reading and writing data. Figure 10-5 shows an excerpt from the data sheet.

---

[5]For mode explanations, see www.byteparadigm.com/applications/introduction-to-i2c-and-spi-protocols/.

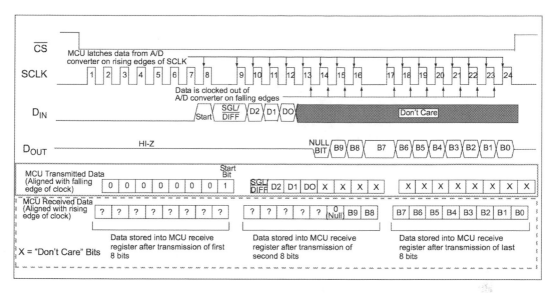

Figure 10-5. *Communicating with the MCP_3008*

I highlight the relevant portion for sending data to the chip in a solid box and the portion for receiving data in the dashed box. Here, you see that the second byte for sending data holds the information for the single-ended command and the command (0x8) shifted to the upper 4 bits (the order of bytes in memory is called *endianness*[6]).

Notice the data you get back from the chip. Here, you see you get 10 bits followed by a 0 (null) that are returned in the third and part of the second byte. Thus, you need to write code to capture the 10 bits in the second and third byte and then mask out the bits starting with the null bit. Since the bits in the second byte are the higher bits, you need to mask out the bits that you don't need and then shift them so that you can interpret the two bytes as an integer (a common trick when working with byte streams). Listing 10-6 shows the completed getValue() method.

Tip   Although this may seem confusing, the number and order of the bits are defined by the manufacturer and not a general case. You should always read the data sheet for each device to learn how to implement its protocol. Fortunately, when you are able to use custom libraries (as you will see in later chapters), the protocol is implemented for you.

---

[6]https://en.wikipedia.org/wiki/Endianness

Listing 10-6. Reading Data from the ADC

```
// Communicate with the chip via the SPI bus. You must encode the command
// as follows so that you can read the value returned (on byte boundaries).
public int getValue(byte whichChannel)
{
    byte command = whichChannel;
    command |= MCP3008_SingleEnded;
    command <<= 4;

    byte[] commandBuf = new byte[] { 0x01, command, 0x00 };
    byte[] readBuf = new byte[] { 0x00, 0x00, 0x00 };
    mcp_var.TransferFullDuplex(commandBuf, readBuf);

    int sample = readBuf[2] + ((readBuf[1] & 0x03) << 8);
    int s2 = sample & 0x3FF;
    return sample;
}
```

The first thing you may notice is the method takes the channel number as a parameter. This allows you to use the method to read other channels, and indeed you do so in the next project.

Notice also how you perform the byte manipulations to set up the command and transmit it with the TransferFullDuplex() method, which returns three bytes to the read buffer. You then shift bytes and mask out the parts that you don't need and then shift the remaining bytes so that you can read the value as an integer.

Once again, don't worry too much if this code seems overly complex. You can find out more about the chip from the data sheet at https://cdn-shop.adafruit.com/datasheets/MCP3008.pdf.

Now that you've seen the individual methods of this class, let's see how they are placed in context in the class. Listing 10-7 shows the skeleton for the ADC_MCP3008 class. I omit the details of the methods for brevity.

Listing 10-7. ADC_MCP3008 Class Layout

```
...

namespace PowerMeter
{
    class ADC_MCP3008
    {
...

        public ADC_MCP3008()
        {
            Debug.WriteLine("New class instance of ADC_MCP3008 created.");
        }

        public async void Initialize()
        {
...
        }

        public int getValue(byte whichChannel)
        {
...
        }
    }
}
```

Tip   If you'd like to explore all the capabilities of the MCP3008, see the data sheet at https://cdn-shop.adafruit.com/datasheets/MCP3008.pdf.

OK, the code is ready for compilation! Be sure to check the earlier listings to ensure that you have all the code in the right place. Once you have entered all the code, you should now attempt to compile the code. Correct any errors that you find until the code compiles without errors or warnings.

Deploy and Execute

Now it is time to deploy the application! Be sure to fix any compilation errors first. You may want to compile the application in debug first (but you can compile in release mode if you'd prefer). In fact, in order to see the debug messages, you will need to be using the debug build.

Recall from Chapter 6 (see the "Deploy and Debug" section and Figure 6-11), to deploy our application in debug mode, we must set the debug settings for our application, setting the following in the debug dialog:

- *Target device*: Remote Machine

- *Remote Machine*: The name of your device

- (Optional) *Uninstall and then re-install*: Checked

- (Optional) *Deploy optional packages*: Checked

- *Application process*: Managed Only

- *Background task process*: Managed Only

Once you make any changes in this dialog, be sure to save your solution. Building or deploying does not automatically save these settings.

We should also change the package name in the package manifest. Recall from Chapter 6 (see the "Deploying Your Application (No Debugging)" section and Figure 6-9), we can change the package name. Do that now so you can find your application easier in the Device Portal.

Now you can deploy your application. Go ahead and do that now. You can run the deployment from the Debug menu, and when you do, you see the debug statements in the output window. Listing 10-8 shows an excerpt of the data you should see. When you start the debugger, the output window automatically opens and displays the debug output. If it does not, you can open the window manually and select debug from the drop-down menu.

Listing 10-8. Example Debug (Output Window)

```
...
New class instance of ADC_MCP3008 created.
...
GPIO ready.
Setting up GPIO pin 21.
Setting up GPIO pin 20.
Setting up GPIO pin 19.
Setting up GPIO pin 18.
Setting up GPIO pin 17.
Setting up the MCP3008.
...
Timer has fired the meterTimer_Tick() method.
Val read = 327
Number of LEDs to turn on = 1
Setting pin 21 HIGH.
Setting pin 20 LOW.
Setting pin 19 LOW.
Setting pin 18 LOW.
Setting pin 17 LOW.
...
Timer has fired the meterTimer_Tick() method.
Val read = 570
Number of LEDs to turn on = 2
Setting pin 21 HIGH.
Setting pin 20 HIGH.
Setting pin 19 LOW.
Setting pin 18 LOW.
Setting pin 17 LOW.
Timer has fired the meterTimer_Tick() method.
Val read = 571
```

Of course, you can run the application by starting it on your device using the Apps pane. Use the small triangle or arrow next to the application name to start the application, the square icon to stop it, and the trash can icon to delete it.

If everything worked correctly, you should see the lights change when the potentiometer is turned. If the lights don't change, be sure you have the potentiometer oriented correctly. Once it is working, try it out a few times to ensure that you see the lights sequence several times.

Tip   If the deployment did not work or you get errors, refer to the "C# Application Deployment Troubleshooting" section in Chapter 6 for troubleshooting help.

Summary

Learning how to build complex IoT projects requires learning how to use some components that may require a bit of effort to set up and use. You saw this with a very simple component—a potentiometer. In this case, since the Raspberry Pi does not have an ADC, you had to use the ADC to "bridge" the devices. Fortunately, you were able to use the potentiometer with the ADC using the SPI bus to control several LEDs as a power meter simulation.

Tip   You may encounter some resistance when plugging components into your breadboard. This can happen when the breadboard is new. If this occurs, try using a smaller-diameter wire to open the pin socket and then try inserting your component.

You saw a number of new things in this chapter, including the ADC, how to connect and set up an SPI device, how to read a potentiometer, and finally how to use the debug feature to write out statements to the output window. While the project itself is rather simplistic, the code clearly was not.

The next chapter continues our set of example IoT projects that you can use to learn more about building IoT solutions. You learn how to use another sensor to measure ambient light. Indeed, you will be making a cool, automatic nightlight.

CHAPTER 11

Project 2: Measuring Light

The projects that you've encountered thus far have used GPIO pins to turn LEDs on or off. But what if you wanted to control the brightness of an LED? Given that you can only turn pins on (high: 3.3V) or off (low: 0V), you have no way to send less power to the LED; the lower the power, the dimmer the LED. If you consider solutions (or features) of devices that are sensitive to ambient light—such as a backlit keyboard—you may have noticed it changes brightness as the room becomes darker. In this project, you're going to explore how such a feature works. That is, you're going to build a fancy LED nightlight.

Unlike the last project, this project is a bit more complex in concept. You must measure the ambient light in the room and then calculate how much power to send to the LED using a technique called *pulse-width modulation* (PWM). You will also use the same analog-to-digital converter (ADC) and SPI interface from the last project, but you will implement it in a slightly different way. Let's get started.

Note   Although this project is written in C#, other than the syntax and mechanics of building, the concepts of using a class are similar in C++ and Visual Basic.

Overview

You will design and implement an LED nightlight using a light-dependent resistor (LDR, also called a *photocell* or *photoresistor*)[1] read from an ADC (the LDR is an analog component like a potentiometer). The LDR is a special resistor that changes resistance depending on the intensity of ambient light.

---

[1] See https://en.wikipedia.org/wiki/Photoresistor.

© Charles Bell 2021

C. Bell, *Windows 10 for the Internet of Things*, https://doi.org/10.1007/978-1-4842-6609-0_11

You need to use the ADC because you want to read analog values, and the Raspberry Pi (and many other boards) does not have analog-to-digital logic. That is, the GPIO pins are digital only. The ADC acts like a "bridge" between digital and analog devices. You will use the MCP3008 that you used in the last chapter.

The real trick to this solution is you must use a special library (namespace) in order to access the pulse-width modulation features of your Raspberry Pi. The library you will use is a contributed library and not part of the standard Visual Studio built-in libraries for Windows 10 IoT Core. However, the library isn't difficult to install, and you will reuse the class you built in the last chapter as the basis for implementing a class for controlling an LED (fading), as well as the ADC. The library is named the Microsoft IoT Lightning Provider (also called the Lightning software development kit or SDK).

Note   In order to use the Lightning SDK, you need Windows 10 IoT Core Version 10.0.10586.0 or later.

Since the LDR responds to light and the ambient light can vary, you will also use a simple user interface to allow you to adjust the sensitivity of the light sensor by restricting its range. That is, you will be able to set the minimum and maximum values for the range. Thus, you will be able to tune the solution to fit the ambient light levels.

Let's look at the components that you need and then see how to wire everything together.

Required Components

The following lists the components that you need. You can find these components in either of the kits mentioned in Chapter 9 or you can purchase the components separately from Adafruit (`www.adafruit.com`), SparkFun (`www.sparkfun.com`), or any electronics store that carries electronic components. Since this solution is a headed application, you will also need a monitor, a keyboard, and a mouse. If you want to run this application headless, you can but I recommend experimenting with the user interface to help tune the values that you need to make the fade effect work properly.

- (1) LED (any color)
- (1) 10K ohm resistor
- (1) 150 ohm resistor (or appropriate for your LEDs)

- (1) light-dependent resistor (photocell)—LDR

- MCP3008 ADC chip

- Jumper wires: (5) male-to-male, (8) male-to-female

- Breadboard (full size recommended but half size is OK)

- Raspberry Pi 2 or 3

- Power supply

- Monitor

- Keyboard and mouse

Set Up the Hardware

Once again, in order to help get everything connected correctly, you will plan for how things should connect. To connect the components to the Raspberry Pi, you need four pins for the ADC, one for the LED, one for the LDR, and one each for power and ground. You will also need to make a number of connections on the breadboard to configure the ADC chip and connect the LDR to the ADC. Table 11-1 shows the map I designed for this project. I list the physical pin numbers in parentheses for the named pins. You will use male/female jumper wires to make these connections.

Table 11-1. *Connection Map for Power Meter Project*

GPIO	Connection	Function	Notes
3.3V (1)	Breadboard power rail	Power	
5V (2)	10K resistor	Power to LDR	
GND (6)	Breadboard ground rail	GND	
MOSI (19)	SPI	MCP3008 pin 11	
MISO (21)	SPI	MCP3008 pin 12	
SCLK (23)	SPI	MCP3008 pin 13	
CC0 (24)	SPI	MCP3008 pin 10	
27	LED	Resistor for LED	

Next, you need to make a number of connections on the breadboard. For these, you will use male-to-male jumpers. Table 11-2 shows the connections needed on the breadboard.

Table 11-2. *Connections on the Breadboard*

From	To	Notes
Breadboard GND	LDR	
Breadboard power	ADC VDD (MCP3008 pin 16)	
Breadboard power	ADC VREF (MCP3008 pin 15)	
Breadboard GND	ADC AGND (MCP3008 pin 14)	
Breadboard GND	ADC GND (MCP3008 pin 9)	
LDR	ADC Channel 0 (MCP3008 pin 1)	

Tip   See Chapter 9 for details on how to determine the correct resistor for your LED.

Once again, I am cheating a bit by using the Adafruit GPIO Reference Card for Raspberry Pi 2 or 3 (`www.adafruit.com/products/2263`), which makes locating GPIO pins much easier than counting pin numbers.

Figure 11-1 shows all the connections needed. Take a close look at this drawing because some connections may not be obvious (e.g., sending 5V power to the LDR).

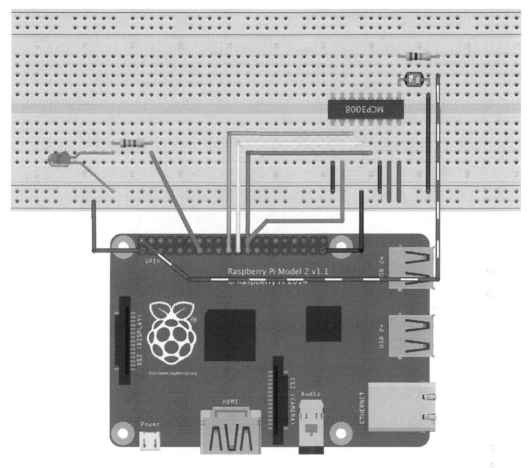

Figure 11-1. *Connections for the nightlight project*

Notice how the LDR and 10K resistor are configured. This is not a mistake. You want the resistor to connect to the second leg of the LDR and the LDR connecting to the channel 0 pin of the ADC as a pull-up resistor. Also note how power to the LDR comes from the 5V pin (2) on the GPIO. The SPI connections are the same as the last project.

WHAT IS PULSE-WIDTH MODULATION?

IoT devices such as the Raspberry Pi can switch between exactly two voltages on their GPIO pins (0.0V or 3.3V). The problem is that many components can operate on a range of voltages. For example, a fan can rotate at different speeds based on the current fed to it. The higher the current, the faster the fan spins. The same is true for other components, such as LEDs, servos, and motors.

It is possible to simulate sending different current levels to a component using a technique called *pulse-width modulation* (PWM). The process works by rapidly switching a pin (line) on and off at different frequencies (hence the pulse) for a given very short period of time (hence the width). The end effect is the device appears to be operating at lower voltages. For example, if you want only 3.0V from 3.3V, you must set the pin to operate at 90% or the pin will be on (through pulsing) at most 90% of the time.

This may sound a little strange, and you may expect to see an LED fed by PWM flicker. The fact is it is flickering, but it is doing so very quickly—too fast for humans to detect. In fact, many of the lights in your home or cffice cycle in a similar manner. The following diagram (see https://en.wikipedia.org/wiki/Pulse-width_modulation)[2] is an illustration of how PWM generates current.

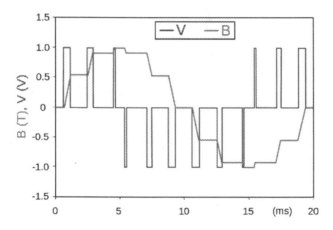

This shows a voltage source (V) modulated as a series of pulses creating a sine-like wave. The result is a sine-like current in the inductor (B). The current waveform is the integral of the voltage waveform. While this sounds all scientific/mathematical, just remember PWM is a technique for simulating different current values by turning the pin on and off rapidly in a specific pattern.

If you are following along with this chapter working on the project, go ahead and make the hardware connections now. Don't power on the board yet but do double- and triple-check the connections.

---

[2]Creative Commons Attribution-Share Alike 3.0 Unported License.

Write the Code

Now it's time to write the code for our example. Since you are working with several new concepts as well as a new component, I will introduce the code for each in turn. I've decided to use C#, but you could implement this project in C++.

Note   Since you have learned all the basics of creating projects in Visual Studio, including how to build and deploy applications, I omit the details of the common operations for brevity.

The project uses an LDR to measure ambient light read via an ADC (MCP3008); the value read is used to control the percentage of current sent to an LED via PWM. Unfortunately, while the IoT devices you have for Windows 10 IoT Core have PWM support, Windows 10 IoT Core has only rudimentary support for PWM. Fortunately, there is an add-on library called the Microsoft IoT Lightning Providers that get you what you need to create a PWM output on a GPIO pin.

I like to use modularization to help make the code easier to write and maintain. It also helps to focus your efforts by concentrating on a specific concept to model. You will see this as you walk through the code. You will create a class for the ADC and a class for the PWM control of the LED. Finally, you use the `DispatcherTimer` class to periodically check the LDR and set the PWM percentage for the LED.[3]

Let's talk about starting a new project and adding the resources that you need first (e.g., the Microsoft IoT Lightning Providers), and then we will walk through the entire code.

New Project

You will use the same project template as the last project—the *C# Blank App (Universal Windows)* template. Use the name `NightLight` for the project name. You can save the project wherever you like or use the default location. Once the project opens, double-click the `MainPage.xaml.cs` file. There are two namespaces you need to include. Go ahead and add those now, as shown next:

---

[3]There are other ways to do this, but this uses a technique you've seen previously.

```
using System.Diagnostics;                  // add this for debugging
using Microsoft.IoT.Lightning.Providers;   // add for Lightning interface
using Windows.Devices;
```

Next, you need to add some variables and constants. First, you create a constant for the GPIO pin for the LED (27) and an instance of one of the new classes that you create named LED_Fade. You see this class in a later section.

```
// Constants and variables for pin
private const int LED_PIN = 27;
private LED_Fade fader = new LED_Fade(LED_PIN);
```

Next, you create an instance for the timer.

```
// Timer to refresh the LED brightness
private DispatcherTimer refreshTimer;
```

Next, you add an instance of the ADC class, which you name MCP3008 to distinguish it from the class you created in the last project. You will see this described in a later section.

```
// Add the new ADC Chip class
private MCP3008 adc = new MCP3008();
```

Finally, you need some variables and a constant for the LDR. You use these variables in conjunction with the user interface to adjust the lower and upper bounds for the light sensitivity. This is really nice because it allows you to experiment with the project while it runs rather than using constants or variables that you must modify, compile, and then rerun to test.

```
// Channel to read the photocell with initial min/max settings
private const int LDR_CHANNEL = 0;
private int max_LDR = 300;    // Tune this to match lighting
private int min_LDR = 100;    // Tune this to match lighting
```

The code in the MainPage() function initializes the components, the new LED fade class (fader), and the new MCP3008 class (adc). Each of these classes has an Initialize() method. There is also a method to initialize the lightning providers that must be called first. You will see this method in the next section.

You also set up the timer. In this case, you use a value of 500, which is one-half a second. If you adjust this lower, the solution will be a bit more responsive. You can try this once you've got it to work and tested. Listing 11-1 shows the complete `MainPage()` method.

Listing 11-1. MainPage Method

```
public MainPage()
{
    InitializeComponent();      // init UI
    InitLightningProvider();    // setup lightning provider
    fader.Initialize();         // setup PWM
    adc.Initialize();           // setup ADC

    // Add code to setup timer
    refreshTimer = new DispatcherTimer();
    refreshTimer.Interval = TimeSpan.FromMilliseconds(500);
    refreshTimer.Tick += refreshTimer_Tick;
    refreshTimer.Start();
}
```

You need two references added to the solution. You need the Windows 10 IoT Extensions from the project property page. You do this by right-clicking the *References* item under the project in the Solution Explorer. You also need to install the Microsoft IoT Lightning Providers library. But first, we must turn on the lightning provider on our device.

Lighting Providers

The Microsoft IoT Lightning Provider is a set of providers to interface with GPIO, SPI, PWM, and I2C support via a direct memory access driver on the device. The lightning providers are implemented as a set of classes you can use to interface with components. In order to use the library, you must first turn on the Lightning driver on your device via the Device Portal. Figure 11-2 shows the Devices tab and the drop-down menu that you use to change the driver.

Figure 11-2. *Setting the Direct Memory Mapped Driver on the Device Portal*

When you select the direct memory mapped driver from the drop-down menu and click *Update Driver*, the portal warns you that the driver is a development-level prerelease and may not perform at peak efficiency. For our purposes, you can ignore the warning, as it has no bearing on the project. The device needs to reboot to complete the change.

You will receive a dialog to restart your device. If the device doesn't restart, you can power off the device and power it back on. Once restarted, the device will be using the updated driver.

Remember, if you want to return your device to normal, you will have to go back to the Device Portal and change the drive back to *Inbox Driver*. Be sure to do that after you are finished with this project if you are not continuing on to Chapter 12.

Caution   You want to reverse this step once you have finished the project and before starting a new project that does not use the lightning library.

To add the lightning library, use the *NuGet Package Manager* from the *Tools* ➤ *NuGet Package Manager* ➤ *Manage NuGet Packages for Solution...* menu. Click the *Browse* tab and then type `iot.lightning` in the search box. After a moment, the list updates. You should see a list with names that include lightning, as shown in Figure 11-3.

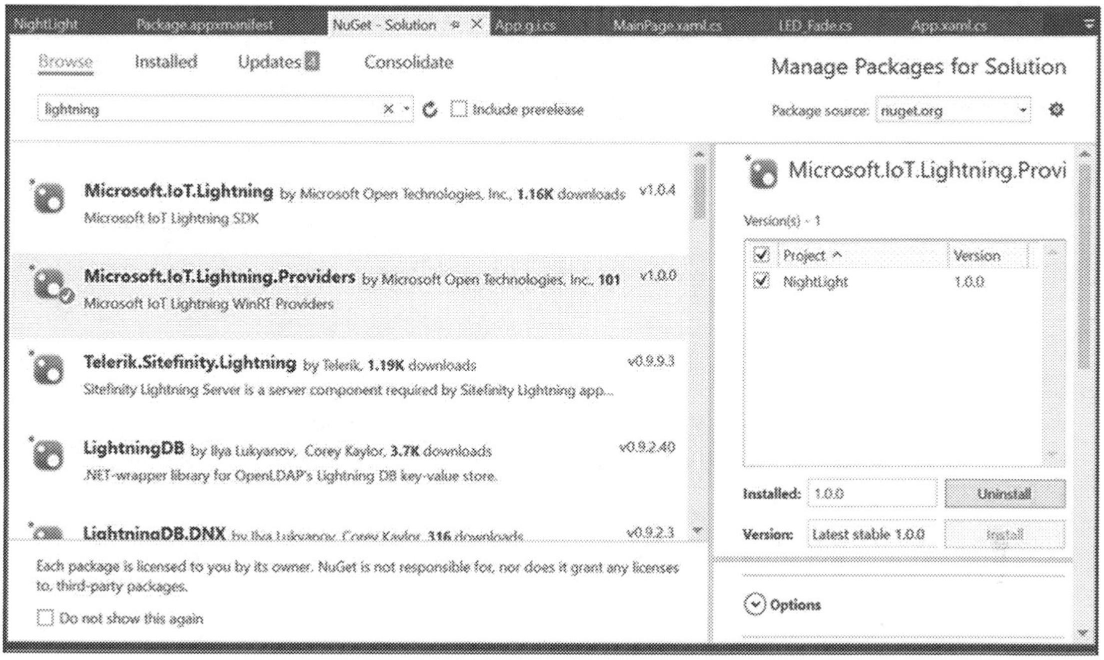

Figure 11-3. *NuGet Package Manager*

Select the entry named *Microsoft.IoT.Lightning* in the list, tick the project name (solution) in the list on the right, and finally click *Install*. The installation starts and you may get a confirmation dialog where you click *OK* to continue. Visual Studio downloads a number of packages and then asks your permission to install them. Go ahead and let the installation complete. A dialog box tells you when the installation is complete.

Tip   You can also download this library and install it manually if the NuGet manager fails. See `https://docs.microsoft.com/en-us/windows/iot-core/develop-your-app/lightningproviders` for more details.

Since this library isn't integrated with Visual Studio or the Windows UWP libraries, you must make a small alteration to the project files to grant permission to use the library (reference) in the project. Failure to do this step leads to some strange execution issues (but it compiles without errors or warnings). More specifically, the `InitLightningProvider()` method fails because the `LightningProvider.IsLightningEnabled` property is False.

In order to use the library, you need to manually update the `Package.appxmanifest` file to reference the Lightning device interface. To edit the file, right-click it in the Solution Explorer and then choose *View Code*. In the code editor, change the code at the top of the file as follows. I have noted the required changes in bold. The first line is a capability setting that enables the application to access custom devices, and the second line is the globally unique identifier (GUID) for the lightning interface. Finally, you add iot to the namespaces list.

```
<Package
  xmlns="http://schemas.microsoft.com/appx/manifest/foundation/windows10"
  xmlns:mp="http://schemas.microsoft.com/appx/2014/phone/manifest"
  xmlns:uap="http://schemas.microsoft.com/appx/manifest/uap/windows10"
  xmlns:iot="http://schemas.microsoft.com/appx/manifest/iot/windows10"
  IgnorableNamespaces="uap mp iot">
...
```

Next, scroll to the bottom of the file and make the following changes. This adds the capabilities to match the changes at the top of the file.

```
...
  <Capabilities>
    <Capability Name="internetClient"  />
    <iot:Capability Name="lowLevelDevices" />
    <DeviceCapability Name="109b86ad-f53d-4b76-aa5f-821e2ddf2141"/>
  </Capabilities>
</Package>
```

You can also make this last change using the package manifest. Simply open the package manifest, click the *Capabilities* tab, and tick the *Low Level Devices* in the list as shown in Figure 11-4.

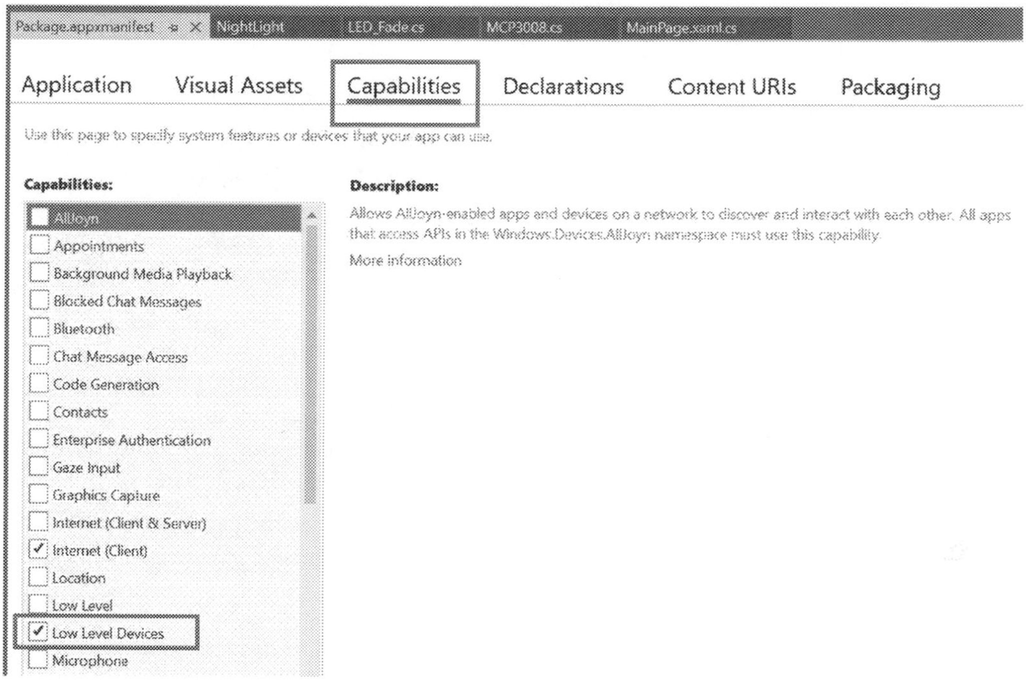

Figure 11-4. *Enabling capabilities in package manifest*

Now you're ready to add code to initialize the lightning providers. Since you are using the lightning providers, you must initialize the provider library in the `MainPage` class. The following method shows how to initialize the lightning provider:

```
private void InitLightningProvider()
{
    // Set the Lightning Provider as the default if Lightning driver is
    enabled on the target device
    if (LightningProvider.IsLightningEnabled)
    {
        LowLevelDevicesController.DefaultProvider = LightningProvider.
        GetAggregateProvider();
    }
}
```

The bulk of the code for the SPI and PWM are contained in the new classes you will add shortly. But first, let's implement the user interface and complete the `MainPage` class.

User Interface

The user interface is deliberately simple consisting of only a few labels and two sliders. The sliders control the minimum and maximum values for the range of sensitivity for the LDR. One of the labels displays the value read from the LDR. You can use this value to help tune the fade effect.

That is, you can set the lower value to the value read during ambient light levels. This effectively sets the percent power to 0 where the LED is off (or very dim). You can use the higher value slider to set the maximum value for the scale. You can find out what that value is by placing your hand over the LDR and reading the value. Thus, you can tune the nightlight effect to match ambient light readings.

Listing 11-2 shows the code for the user interface. Open the `MainPage.xaml` file and make the changes as shown. This code uses a stacked panel to contain the controls. Note the lower value slider. You add a reference to a method to be called when the slider is updated. This helps you control the slider to avoid a nasty surprise should the lower value exceed the higher value. Speaking of which, I have added code to restrict the range to 0.0–1.0.

Listing 11-2. User Interface (XAML) Code

```
<Page
    x:Class="NightLight.MainPage"
    xmlns="http://schemas.microsoft.com/winfx/2006/xaml/presentation"
    xmlns:x="http://schemas.microsoft.com/winfx/2006/xaml"
    xmlns:local="using:NightLight"
    xmlns:d="http://schemas.microsoft.com/expression/blend/2008"

    mc:Ignorable="d">

    <Grid Background="{ThemeResource ApplicationPageBackgroundThemeBrush}">
        <StackPanel Width="400" Height="400">
            <TextBlock x:Name="title" Height="60" TextWrapping="NoWrap"
                    Text="Night Light Experiment" FontSize="28"
                    Foreground="Blue"
                    Margin="10" HorizontalAlignment="Center"/>
            <TextBlock Height="60" TextWrapping="NoWrap"
                    Text="High Value" FontSize="28" Foreground="Blue"
                    Margin="10" HorizontalAlignment="Center"/>
```

```
<Slider x:Name="ldrHigh" Width="400" Value="350"
        Orientation="Horizontal" HorizontalAlignment="Left"
        Maximum="1023"/>
<TextBlock Height="60" TextWrapping="NoWrap"
        Text="Low Value" FontSize="28" Foreground="Blue"
        Margin="10" HorizontalAlignment="Center"/>
<Slider x:Name="ldrLow" Width="400" Value="125"
        Orientation="Horizontal" HorizontalAlignment="Left"
        Maximum="1023"
        ValueChanged="ldrLow_ValueChanged"/>
<TextBlock x:Name="status" Height="60" TextWrapping="NoWrap"
        Text="LDR Value = 0" FontSize="28" Foreground="Blue"
        Margin="10" HorizontalAlignment="Center"/>
    </StackPanel>
  </Grid>
</Page>
```

Figure 11-5 shows an example of the user interface.

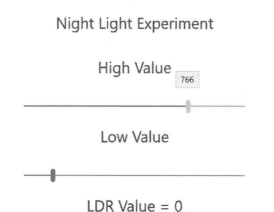

Figure 11-5. *Sample user interface*

The sliders can be moved independently. This presents a problem if the user slides the lower value to a setting greater than the higher value. Thus, you want to control the sliders so that the lower value never exceeds the higher value. To do this, you complete the code for the ldrLow_ValueChanged() method as follows. This illustrates a common technique to manage the range of user input.

```
private void ldrLow_ValueChanged(object sender, Windows.UI.Xaml.Controls.
Primitives.RangeBaseValueChangedEventArgs e)
{
    // Make sure the low value doesn't exceed the high value
    if (ldrLow.Value > ldrHigh.Value)
    {
        ldrHigh.Value = ldrLow.Value + 1;
    }
}
```

Tip   This code needs one extra step to ensure that you stay within the limits of the slider. See if you can spot it yourself. Hint: What happens when the high value is already at the highest setting? Although not essential for our experiment, you should strive to improve error handling to include the fringe values and conditions.

Now let's see the code to read from the value of the LDR and fade the LED.

Controlling the LED

You are using a timer to refresh the project. This method, named `refreshTimer_Tick()`, first reads the value of the LDR from the MCP3008 class and then calculates a percentage based on the minimum and maximum value scale as set via the user interface. You may have noticed that I used some constant values to set the initial range. Listing 11-3 shows the completed `refreshTimer_Tick()` method.

Listing 11-3. Timer Code

```
private void refreshTimer_Tick(object sender, object e)
{
    float valRead = 0;
    float brightness = 0;

    Debug.WriteLine("Timer has fired the refreshTimer_Tick() method.");
```

```
// Read value from the ADC
valRead = adc.getValue(LDR_CHANNEL);
Debug.Write("Val read = ");
Debug.WriteLine(valRead);
status.Text = "LDR Value = " + valRead;

// Get min, max from sliders
min_LDR = (int)ldrLow.Value;
max_LDR = (int)ldrHigh.Value;
Debug.WriteLine("Min LDR = " + min_LDR);
Debug.WriteLine("Max LDR = " + max_LDR);

// Calculate the brightness
brightness = ((valRead - min_LDR) / (max_LDR - min_LDR));
// Make sure the range stays 0.0 - 1.0
if (brightness > 1)
{
    brightness = (float)1.0;
}
else if (brightness < 0)
{
    brightness = (float)0.0;
}
Debug.Write("Brightness percent = ");
Debug.WriteLine(brightness);

// Set the brightness
fader.set_fade(brightness);

// For extra credit, show the voltage returned from the LDR
// convert the ADC readings to voltages to make them more friendly.
float ldrVolts = adc.ADCToVoltage((int)valRead);

// Let us know what was read in.
Debug.WriteLine(String.Format("Voltage for value read: {0}, {1}v",
                              valRead, ldrVolts));
}
```

Wow, that's a lot of code. For the most part, it is similar to the timer event from the last project. However, the calculation for the percentage to use for the PWM is new. Here, the percentage represents the brightness of the LED. The higher the percentage calculated means the faster the LED pulses and therefore appears brighter.

I have added an interesting bit of data at the end of the method. This demonstrates how to calculate the voltage that is flowing through the LDR. You may need this technique for any projects where you are reading analog voltage.

Take some time to read through the code. You will see a number of additional debug statements that you can use when testing the project in debug mode. There is code to update the label in the user interface with the value read and the voltage calculated. If you are curious, you can use a multimeter to measure the voltage on the ADC side of the LDR (positive lead on the LDR and negative lead on the ground). You should see the same voltage on the multimeter as shown in the user interface.

Now let's see how to put all of these pieces together.

Completing the Main Class

Now that you've seen all the code for the main code file, let's look at how the methods are placed in the class. Listing 11-4 shows the skeleton of the MainPage class. Use this as a guide to place the code in the right places. I omit the details of the methods for brevity.

Listing 11-4. MainPage Code Layout

```
...

namespace NightLight
{
    public sealed partial class MainPage : Page
    {
...
        public MainPage()
        {
...
        }
```

```
        private void InitLightningProvider()
        {
...
        }

        private void refreshTimer_Tick(object sender, object e)
        {
...
        }

        private void ldrLow_ValueChanged(object sender,
                Windows.UI.Xaml.Controls.Primitives.
                RangeBaseValueChangedEventArgs e)
        {
...
        }
    }
}
```

Now let's see how to create the code for the two new classes.

Code for the MCP3008

The code for the ADC class is very similar to the class you used in the last project. The difference is you will use the lightning providers to communicate with the ADC rather than the Windows 10 IoT extension. This is because the Windows 10 IoT extension and the lightning provider are incompatible. That is, you cannot use the same code from the last project together with the PWM code from the lightning provider. Fortunately, the lightning provider has SPI support.

The methods that you use in this class are the same as you had in the last project. The following summarizes the methods that you need. The last method is a bonus method that you may need in the future if you reuse this code for more advanced projects:

- Constructor—MCP3008(): Log instantiation with a debug statement

- Initialize(): Initialize the SPI interface

- getValue(): Read a value from the ADC on a specified channel

- ADCToVoltage(): Return the voltage read for a specific value

Recall from Chapter 10 that to add a new class, right-click the project and choose *Add* ➤ *New Item....* Name the class MCP3008 to distinguish it from the similar class in the last project. When you create a class, you create a new file, as shown earlier. Inside this new code file, you see a familiar skeleton. Of note is the same namespace as the main code and a list of namespaces that you will use. Open the file now and add the following namespaces. The need for each is shown in the comments.

```
using Windows.Devices.Spi;          // add this for SPI communication
using System.Diagnostics;           // add this for debugging
```

Next, you need some variables and constants. You need a variable for the SPI device and constants for the MCP3008. In this case, you set constants for helping you read data.

```
// SPI controller interface
private SpiDevice mcp_var;
const int SPI_CHIP_SELECT_LINE = 0;
const byte MCP3008_SingleEnded = 0x08;
const byte MCP3008_Differential = 0x00;
```

You also add a variable to store the reference voltage for the value read. You also store the maximum value of the component read from the ADC (1023). Of note is how I converted the constant 5.0 to a float by adding F to the end. This is a common shortcut that you can use instead of casting with (float). Also, be sure to change the reference voltage if you use a different voltage to power the components connected to the ADC channels. Recall from the wiring layout in Figure 11-1, you're using 5V on the LDR.

```
// These are used when you calculate the voltage from the ADC units
float ReferenceVoltage = 5.0F;
private const int MAX = 1023;
```

The constructor is simple; you just announce that you've instantiated the class via a debug statement. You will see this in the class layout later in the section. The really interesting changes appear in the Initialize() method. Listing 11-5 shows the updated method that uses the lightning provider class methods.

Listing 11-5. The Initialize() Method

```
// Setup the MCP3008 chip
public async void Initialize()
{
    Debug.WriteLine("Setting up the MCP3008.");
    try
    {
        // Settings for the SPI bus
        var SPI_settings = new SpiConnectionSettings(SPI_CHIP_SELECT_LINE);
        SPI_settings.ClockFrequency = 3600000;
        SPI_settings.Mode = SpiMode.Mode0;

        SpiController controller = await SpiController.GetDefaultAsync();
        mcp_var = controller.GetDevice(SPI_settings);
        if (mcp_var == null)
        {
            Debug.WriteLine("ERROR! SPI device may be in use.");
            return;
        }
    }
    catch (Exception e)
    {
        Debug.WriteLine("EXEPTION CAUGHT: " + e.Message + "\n" +
        e.StackTrace);
        throw;
    }
}
```

This method is very similar to the class that you used in the last project. The biggest difference is you can get the SPI device with a single class to the provider. This code is a pattern for the other providers in the Lightning Providers library. In fact, you see very similar code in the PWM class.

The getValue() method is the same as the class used in the project from Chapter 10 (the ADC_MCP3008.cs file). You can simply copy it from the last project and paste it in the new class. I omit the details from the list for brevity.

Next, we will need a method to read the voltage, which is simply a function where we take the value and multiply it by the reference voltage and divide it by the maximum voltage as shown in the following. Go ahead and add this method to the class.

```
// Utility function to get voltage from ADC
public float ADCToVoltage(int value)
{
    return (float)value * ReferenceVoltage / (float)MAX;
}
```

Now, let's see how all of these pieces fit together. Listing 11-6 shows the class layout for the MCP3008 class.

Listing 11-6. The MCP3008 Class Layout

```
...
namespace NightLight
{
    class MCP3008
    {
...

        public MCP3008()
        {
            Debug.WriteLine("New class instance of MCP3008 created.");
        }

        // Setup the MCP3008 chip
        public async void Initialize()
        {
...
        }
        public int getValue(byte whichChannel)
        {
...
        }
```

```
        // Utility function to get voltage from ADC
        public float ADCToVoltage(int value)
        {
            return (float)value * ReferenceVoltage / (float)MAX;
        }
    }
}
```

Notice the last method. This is a helper method that you can use to calculate the voltage read from the channel. You used this in the MainPage class to display the voltage when the value is read from the LDR.

Tip   If you want to copy a class from another project, you can copy the file into the project folder and optionally rename the file. To add the class to the solution, right-click the project and choose *Add ➤ Existing Item…* and select the class file. If you renamed the class file, you may also have to rename the class itself as well as change the namespace to match the current project/solution.

Now let's look at the code for the PWM class.

Code for the PWM

The code to implement PWM on a GPIO pin is very straightforward and models the code in the MCP3008 class. The following summarizes the methods that you need. The last method is a bonus method that you may need in the future if you reuse this code for more advanced projects:

- Constructor—LED_Fade(): Log instantiation with a debug statement

- Initialize(): Initialize the PWM interface

- set_fade(): Set the duty cycle (percentage of time the LED is on) for the GPIO (LED) pin

Recall from Chapter 10, to add a new class, right-click the project and choose *Add ➤ New Item….* Name the class LED_Fade. When you create a class, you create a new file, as shown. Inside this new code file, you will see a familiar skeleton. Of note is the same namespace as the main code and a list of namespaces that you will use. Open the file now and add the following namespaces. The need for each is shown in the comments.

```
using System.Diagnostics;      // add for Debug.Write()
using Windows.Devices.Pwm;     // add for PWM control (10586 and newer)
using Microsoft.IoT.Lightning.Providers;  // add for Lightning driver for Pwm
```

There are only two variables required: one for the GPIO (LED) pin and another for the PWM class instance from the lightning provider. You use this variable to communicate with the PWM features.

```
// Variables for controlling the PWM class
private int LED_pin;
private PwmPin Pwm;
```

The constructor sets the pin number as a parameter and prints a debug statement indicating the instance has been created. The code is shown next:

```
public LED_Fade(int pin_num = 27)
{
    LED_pin = pin_num;   // GPIO pin
    Debug.WriteLine("New class instance of LED_Fade created.");
}
```

The Initialize() method is very similar to the same method in the MCP3008 class. In this case, you are setting up the PWM provider. Listing 11-7 shows the complete Initialize() method.

Listing 11-7. The Initialize() Method

```
// Initialize the PwmController class instance
public async void Initialize()
{
    try
    {
        var pwmControllers = await
            PwmController.GetControllersAsync(LightningPwmProvider.
            GetPwmProvider());
```

```
        var pwmController = pwmControllers[1];    // the device controller
        pwmController.SetDesiredFrequency(50);

        Pwm = pwmController.OpenPin(LED_pin);
        Pwm.SetActiveDutyCyclePercentage(0);  // start at 0%
        Pwm.Start();
        if (Pwm == null)
        {
            Debug.WriteLine("ERROR! Pwm device {0} may be in use.");
            return;
        }
        Debug.WriteLine("GPIO pin setup for Pwm.");
    }
    catch (Exception e)
    {
        Debug.WriteLine("EXCEPTION CAUGHT: " + e.Message + "\n" +
        e.StackTrace);
        throw;
    }
}
```

This code is a bit different than the SPI code. In this case, you need to open a pin for controlling the PWM and set the starting duty cycle (percent). In this case, you set it to 0, which effectively turns off the LED. Finally, you start the PWM class.

To control the fading effect, you provide a method named set_fade() that takes a floating-point value (percent) that you use to change the duty cycle of the PWM. The following shows the complete code:

```
// Set percentage of brightness (or how many cycles are pulsed) where
public void set_fade(float percent)
{
    if (Pwm != null)
    {
        Pwm.SetActiveDutyCyclePercentage(percent);
        Debug.WriteLine("Pwm set.");
    }
```

```
    else
    {
        Debug.WriteLine("Cannot trigger Pwm.");
    }
}
```

Now, let's see how all of these pieces fit together. Listing 11-8 shows the class layout for the LED_Fade class.

Listing 11-8. The LED_Fade Class Layout

```
...
namespace NightLight
{
    class LED_Fade
    {
...
        public LED_Fade(int pin_num = 27)
        {
            LED_pin = pin_num;    // GPIO pin
        }

        // Initialize the PwmController class instance
        public async void Initialize()
        {
...
        }

        // Set percentage of brightness (or how many cycles are pulsed) where
        // 100 is fast or "bright" and 0 is slow or "dim"
        public void set_fade(float percent)
        {
...
        }
    }
}
```

That's it! The code is complete and ready for compilation. Be sure to check the prior listings to ensure that you have all the code in the right place. Once you have entered all the code, you should then attempt to compile the code. Correct any errors that you find until the code compiles without errors or warnings.

Deploy and Execute

Now it is time to deploy the application! Be sure to fix any compilation errors first. You may want to compile the application in debug first (but you can compile in release mode if you'd prefer). In fact, in order to see the debug messages, you will need to be using the debug build.

Recall from Chapter 6 (see the "Deploy and Debug" section and Figure 6-11), to deploy our application in debug mode, we must set the debug settings for our application, setting the following in the debug dialog:

- *Target device*: Remote Machine

- *Remote Machine*: The name of your device

- (Optional) *Uninstall and then re-install*: Checked

- (Optional) *Deploy optional packages*: Checked

- *Application process*: Managed Only

- *Background task process*: Managed Only

Once you make any changes in this dialog, be sure to save your solution. Building or deploying does not automatically save these settings.

We should also change the package name in the package manifest. Recall from Chapter 6 (see the "Deploying Your Application (No Debugging)" section and Figure 6-9), we can change the package name. Do that now so you can find your application easier in the Device Portal.

Now you can deploy your application. Go ahead and do that now. You can run the deployment from the *Debug* menu, and when you do, you see the debug statements in the output window. Listing 11-9 shows an excerpt of the data you should see. When you start the debugger, the output window automatically opens and displays the debug output. If it does not, you can open the window manually and select debug from the drop-down menu.

Listing 11-9. Example Debug (Output Window)

```
...
New class instance of MCP3008 created.
New class instance of LED_Fade created.
Setting up the MCP3008.
GPIO pin setup for Pwm.
...
Timer has fired the refreshTimer_Tick() method.
commandBuf = 11280
readBuf = 00239
Val read = 239
Min LDR = 235
Max LDR = 384
Brightness percent = 0.02684564
Pwm set.
Voltage for value read: 239, 1.168133v
Timer has fired the refreshTimer_Tick() method.
commandBuf = 11280
readBuf = 00239
Val read = 239
Min LDR = 235
Max LDR = 384
Brightness percent = 0.02684564
Pwm set.
Voltage for value read: 239, 1.168133v
Timer has fired the refreshTimer_Tick() method.
commandBuf = 11280
readBuf = 00238
Val read = 238
Min LDR = 235
Max LDR = 384
Brightness percent = 0.02013423
Pwm set.
Voltage for value read: 238, 1.163245v
```

Of course, you can run the application by starting it on your device using the Apps pane. You use the small triangle or arrow next to the application name to start it and the square icon to stop it or the trashcan icon to delete the application.

If everything worked correctly, you should be able to place your hand over the LDR and see the LED brighten. If it doesn't (and there is a good chance it won't), don't worry. Conduct an experiment by clearing away anything that can cast a shadow on the LDR and take note of the value in the user interface. Let it run for a few minutes and note the values. You should see it bounce around a bit but should settle to within about 20–30 of normal. For example, in my office, the ambient light level value was about 220. Next, place your hand over the LDR and note the change in value. Let the values settle and note the value range. In my office, the value was about 450.

Now, adjust the high value slider to the value when your hand was over the LDR and the low value to the value for ambient light. Once set, you can experiment with the LED brightness by slowly placing your hand over the LDR. The LED should go from a very dim setting to full (or nearly) brightness. It may take some additional runs to fine-tune the high and low values, but once set, the project acts as a light-sensing nightlight. How cool is that?

Summary

IoT solutions often require using analog components. Moreover, you may need to control analog components by adjusting the voltage to the component. This often requires using additional libraries and features that may not be in the standard libraries. In this case, you had to use a special library named the Microsoft IoT Lightning Providers to get access to PWM and SPI interfaces.

You saw a number of new things in this project, including another way to use an ADC with an SPI interface, how to read values from an LDR, and how to use PWM to control the brightness of an LED. While the project itself is rather simplistic, the code clearly was not, and it was a bit more code than you've seen so far.[4]

The next chapter continues the set of example IoT projects that you can use to learn more about building IoT solutions. You learn how to write code to read more sensors. In Chapter 12, you create a project to report weather data from several sensors, and then in Chapter 13, you learn how to collect and store the data generated.

---

[4]This pattern of ever-increasing complexity continues in the next chapters—so strap in and hang on!

## CHAPTER 12

Project 3: Using Weather Sensors

IoT solutions often employ a number of sensors to observe the world around us, and while you've explored a project with a simple sensor, you have yet to see how to work with more sophisticated sensors, such as those available as breakout boards. One of the more popular choices of sensors includes those you use to observe weather. In this case, you'll start out with a sensor that measures temperature and barometric pressure and calculates the altitude based on sea level pressure. There's a lot you learn with only those two measurements.

The sensor that you will use is the BMP280 I2C or the SPI barometric pressure and altitude sensor breakout board from Adafruit (`www.adafruit.com/products/2651`). This sensor comes with Microsoft IoT Pack for Raspberry Pi (but you can buy it separately). Although the board can be used with I2C or SPI, you will use the sensor with the I2C interface.

Note Some newer releases of the Microsoft IoT Pack for Raspberry Pi come with a BME280 sensor instead of the BMP280. In which case, you cannot use it with this project because the library that we use only works with the BMP280 (but it can be modified to work; see the data sheet for the BME280). However, the project in the next chapter permits either the BMP280 or the BME280.

This project also represents another escalation of complexity from the last project. While you will learn how to use the I2C interface, you will also discover how to work with sensors and breakout boards that have a complex communication protocol. Recall from our work with the ADC chip, you had to interpret the data read from the chip. The same is true for most breakout boards and sensors. In the case of the BMP280, the protocol is

© Charles Bell 2021
C. Bell, *Windows 10 for the Internet of Things*, https://doi.org/10.1007/978-1-4842-6609-0_12

quite a bit more complex. Fortunately, you are able to make use of code made by others and thus reduce the burden of having to figure it out for yourself.

Finally, you take a different tactic in this project and implement the project in Visual Basic. However, since the third-party code was written in C#, you also discover how to build a Windows 10 IoT solution that uses mixed languages. While this sounds like unnecessary complexity, it allows you to master leveraging one of the most impressive features of Visual Studio—mixing C# and VB in the same solution. With so much to work on and learn about, let's take a moment to examine the goals of the project and what you need to get started.

Overview

Weather projects can be a lot of fun since they are not only easy to wire up because the sensors are packaged in breakout boards, but also because they are more practical than projects that teach techniques alone. While the projects thus far have been mainly in that category,[1] this project is something you can build and use. You may even want to take it a bit further and add additional sensors and features.

To make it more practical, this project is written with a basic user interface, so you can adapt it to a desktop solution or even a wall-mounted weather station. For example, you could use the Raspberry Pi 7" Touch LCD Panel or similar LCD to display the weather conditions.

The hardware for this project is very simplistic. All you need are the BMP280 sensor and four wires to connect it to your device. What isn't simplistic is the code needed to read data from the BMP280. Fortunately, there are two solutions available for you to use. More specifically, you do not have to write any code to use the BMP280 with these solutions (but you have to modify it slightly). One is a C# class that you can get from a tutorial from Adafruit, and the other is a library that you can download and install from NuGet. Both of which are written in/for C#.

You use VB to build this project to not only learn how to build Windows 10 IoT solutions in VB but also to learn how to consume code and libraries built for other languages in the process. But don't be alarmed. The process you need to use to enable this cross-language solution is not difficult (but requires a precise set of steps and settings to get correct).

---

[1]But still fun, I think.

This project is part one of two for writing a weather solution. In this chapter, you use one of the solutions for the BMP280, and in the next chapter, you use the other solution. In fact, you will see a pure C# implementation in the next chapter. For this chapter, you use the code class solution so that you can see how to modify it slightly for use in a VB (or C++) solution. Fortunately, the changes are minor and help understand what you may need to do to use a C# code library in other applications.

I demonstrate all the code necessary and more in the following sections. But first, let's talk briefly about the hardware that you need for this project.

OTHER WEATHER IOT SOLUTIONS

You can find a number of weather-themed example projects on the Internet. In fact, there are three that you may want to look at to see how to use other sensors or libraries to build a weather project. The following are three of the better sample projects that offer the most interesting and complete solutions:

- `www.hackster.io/windows-iot/weather-station-67e40d`

- `https://github.com/ms-iot/samples/tree/develop/WeatherStation`

- `https://github.com/ms-iot/BusProviders/tree/develop/` `Microsoft.IoT.Lightning.Providers/WeatherStation`

If you decide to explore these solutions, be sure to download the sample code rather than building it from scratch. In fact, the projects on GitHub do not have very clear documentation. Those projects use a different sensor too (see `www.adafruit.com/products/1899`) but are written in Visual Basic, C#, C++, and Python, so they show a wide array of possible implementations. The last one uses the Lightning driver you saw in the last chapter.

Required Components

The following lists the components that you need. You can get the BMP280 sensor from Adafruit (`www.adafruit.com`) either in the Microsoft IoT Pack for Raspberry Pi or purchased separately, SparkFun (`www.sparkfun.com`), or any electronics store that carries electronic components. However, if you use a sensor made by someone other than Adafruit, you may need to alter the code to change the I2C address. Since this solution is a headed application, you also need a monitor, a keyboard, and a mouse.

- Adafruit BMP280 I2C or SPI barometric pressure and altitude sensor
- (4) jumper wires: male-to-female
- Breadboard (full size recommended but half size is OK)
- Raspberry Pi 2 or 3
- Power supply
- Monitor
- Keyboard and mouse

Note   Some BMP280 modules come unassembled, and you will have to solder the header pins to the module.

Set Up the Hardware

Although there are only four connections needed for this project, you will plan for how things should connect, which is good practice to hone. To connect the components to the Raspberry Pi, you need four pins for the BMP280 sensor, which requires only power, ground, and two pins for the I2C interface. Table 12-1 shows the map I designed for this project. I list the physical pin numbers in parentheses for the named pins. You use male-to-female jumper wires to make these connections.

Table 12-1. *Connection Map for Weather Display Project*

GPIO	Connection	Function	Notes
5V (2)	BMP280 VIN	Power to breakout board	
GND (6)	BMP280 GND	GND on breakout board	
SDA1 (3)	SDI on breakout board	I2C	
SCL1 (5)	SCK on breakout board	I2C	

Figure 12-1 shows all the connections needed.

Figure 12-1. *Connections for the WeatherDisplay project*

So, what is this I2C interface? I2C[2] (pronounced "eye-two-see" or the less popular "eye-squared-see") is a fast digital protocol that uses two wires (plus power and ground) to read data from circuits, sensors, or devices. One pin is used to read the data, and the other is used as a clock to control how data is sent. There are several forms of how to read data from I2C, but fortunately the Windows 10 IoT Core has libraries that you can use to communicate with I2C devices. The IoT boards in this book all support the interface.

If you are following along with this chapter working on the project, go ahead and make the hardware connections now. Don't power on the board yet but do double- and triple-check the connections.

---

[2]See https://en.wikipedia.org/wiki/I%C2%B2C.

Write the Code

Now it's time to write the code for our project. Since you are working with several new concepts as well as a new component (BMP280), I introduce the code for each in turn. I've decided to use a VB project that consumes a C# project, which encapsulates a class library for the BMP280. You'll use a timer to read the data and update the user interface.

Note   You may read that it is impossible to use a C# library in a VB project. Sadly, that is a bit of misinformation. You can use a C# library in a VB (or C++) project provided they are compatible (same platform, etc.). However, you should implement the project in this chapter strictly as described. For example, if you use the wrong project template for the C# project, the solution will build and deploy, but may throw an exception or it will simply not work.

Recall that our project uses code to read the BMP280 from a tutorial written by someone else (Adafruit). In fact, the tutorial is one of the sample projects that Adafruit includes as an introduction to their Microsoft IoT Pack for the Raspberry Pi. Rather than simply present their solution, I present a VB implementation that reuses their code for the BMP280. The code that you need can be downloaded from `https://github.com/ms-iot/adafruitsample/tree/master/Lesson_203/FullSolution`. In this case, you only need the file named `BMP280.cs`. If you are following along with this chapter writing the code for the project, go ahead and download the sample code now and then extract it. You will need the `BMP280.cs` shortly.

Once built, reading the temperature and pressure from the sensor is really easy. Rather than display the data using a debug interface, you will use a simple user interface to present the data. Yes, that means you will write some XAML code in our VB project.

You are going to write the code in a slightly different order than previous projects. You first create the new projects in the same solution file and then work on the C# code to get it correct, and then you'll return to the VB project to fill in the user interface and code to read the sensor. As you will see, there isn't much code to write, but there are a few things you must do in order to get everything working together. Let's start with the first project.

Caution   If you have just finished the project from Chapter 11, be sure to switch the driver back to the default controller driver (*Inbox Driver*); otherwise, your project may not execute properly.

New Project

Since you are using VB for the active project (called the startup project), you use the *VB Blank App (Universal Windows)* template. Use the name WeatherDisplay for the project name. You can save the project wherever you like or use the default location.

Caution   Do not use the name "Weather" for the application because there is already an application named "Weather" that comes with the Windows 10 IoT Core. If you name your projects the same as an existing application like this, you may not be able to run your application, or you may corrupt your existing device image.[3]

Add a C# Runtime Component Project

With the *WeatherDisplay* VB project open, right-click the solution and choose *Add ➤ New Project....* For this project, you want a C# Runtime Component project. Use the *C# Windows Runtime Component (Universal Windows)* template. Name the project WeatherSensor. You can save the project wherever you like or use the default location.

What you should see at this point is one solution entitled Solution 'WeatherDisplay' with two projects: *WeatherDisplay* and *WeatherSensor*. Figure 12-2 shows an example of what you should see in the Solution Explorer.

---

[3]Guess how I know this.

Figure 12-2. Solution Explorer with WeatherDisplay and WeatherSensor projects

The *WeatherDisplay* project is the startup project. If the *WeatherDisplay* project is not the startup project, you can right-click it and set it as the startup project.

At this point, you have the projects created, but you still must add the code for the BMP280 class and add references for the projects, the user interface, and code to read the sensor and display it in the user interface.

BMP280 Class

You are going to use the Adafruit code in our project. If you haven't downloaded the sample project, do that now. You need the file named BMP280.cs that you downloaded earlier. To add it to the project, right-click the *WeatherSensor* project and choose *Add* ➤ *Existing Item...* and then browse for the file and select it. Visual Studio copies the file to your project folder.

You will have to edit this file because it is not in the correct state for use in a runtime component project. But first, you can delete the Class1.cs file that the template provided. You do not need it. Simply right-click the file and choose *Delete*.

The modifications you need to make include changes to the class decorations and some of the methods. You want to allow the code to read from the sensor to run as a task; however, the code as written won't work. You must start the tasks from inside the project rather than from other code as it was originally written. You will change the following things. I go through each of them in turn. As you will see, the changes aren't complex, but they are required for the C# project to be used with the VB project.

- Class declarations must be sealed.

- The Initialize() method must run the setup code as a task.

- The Read* methods must be normal methods that run the Begin() code as a task.

Note   The following listings show unified difference output generated from diff.[4] Essentially, each line marked with a minus sign is replaced with the line immediately below marked with a plus sign (but you delete the plus sign).

Let's begin with changes to the classes. Open the BMP280.cs file. You see two classes: BMP280_CalibrationData and BMP280. Change the class declaration to a sealed class, as shown in Listing 12-1. This is an excerpt of the affected code for brevity and adjacent lines to show context.

Listing 12-1. Changes to the Class Declarations

```
...
 using Windows.Devices.Gpio;
 using Windows.Devices.I2c;
-namespace Lesson_203
+namespace WeatherSensor
 {
-    public class BMP280_CalibrationData
+    public sealed class BMP280_CalibrationData
     {
         //BMP280 Registers
         public UInt16 dig_T1 { get; set; }

...

         public Int16 dig_P9 { get; set; }
     }
```

---

[4]A utility that is very familiar to Linux and Mac developers. See http://gnuwin32.sourceforge.net/packages/diffutils.htm for a Windows port of this tool.

```
-    public class BMP280
+    public sealed class BMP280
     {
         //The BMP280 register addresses according the datasheet: http://
         www.adafruit.com/datasheets/BST-BMP280-DS001-11.pdf
         const byte BMP280_Address = 0x77;
...
```

Caution   Notice that you changed the namespace. If you skip this or forget to change it, your code will compile but fail to link or deploy, presenting a strange and confusing set of errors regarding names missing from the template. If you see such, check your namespace declaration to make sure that it is correct.

Next, you must change the Initialize() method to call a new Setup() method as a task. Listing 12-2 shows the changes needed for that method.

Listing 12-2.   Changes to the Initialize() Method

```
     //Method to initialize the BMP280 sensor
-    public async Task Initialize()
+    public void Initialize()
     {
         Debug.WriteLine("BMP280::Initialize");
+        Task t_setup = Task.Run(() => Setup());
+        t_setup.Wait();
+    }
+
+    private async Task Setup()
+    {
+        Debug.WriteLine("BMP280::Setup");
...
```

You are splitting the setup code out as a separate method decorated to run as an asynchronous task. This is necessary to ensure that the code does not run in the same thread as the user interface (it can't).

Next, you must change the ReadTemperature() method to remove the async decorator and to call the Begin() method as a task. This protects the thread and ensures that the code executes outside of the user interface thread. Listing 12-3 shows the changes needed.

Listing 12-3. Changes to the ReadTemperature() Method

```
...
-        public async Task<float> ReadTemperature()
+        public float ReadTemperature()
         {
             //Make sure the I2C device is initialized
-            if (!init) await Begin();
+            if (!init)
+            {
+                Task t_begin = Task.Run(() => Begin());
+                t_begin.Wait();
+            }
...
```

You must change the ReadPressure() method in the same manner as ReadTemperature(). However, since this method must call ReadTemperature() and it is no longer a task, you can call the method directly. Listing 12-4 shows the changes needed.

Listing 12-4. Changes to the ReadPressure() Method

```
...
-        public async Task<float> ReadPreasure()
+        public float ReadPreasure()
         {
             //Make sure the I2C device is initialized
-            if (!init) await Begin();
+            if (!init)
+            {
+                Task t_begin = Task.Run(() => Begin());
+                t_begin.Wait();
+            }
             //Read the temperature first to load the t_fine value for
             compensation
```

```
          if (t_fine == Int32.MinValue)
          {
-             await ReadTemperature();
+             ReadTemperature();
          }
...
```

Finally, you must change the ReadAltitude() method in the same manner.
Listing 12-5 shows the changes needed.

Listing 12-5. Changes to the ReadAltitude() Method

```
...
-       public async Task<float> ReadAltitude(float seaLevel)
+       public float ReadAltitude(float seaLevel)
        {
            //Make sure the I2C device is initialized
-           if (!init) await Begin();
+           if (!init)
+           {
+               Task t_begin = Task.Run(() => Begin());
+               t_begin.Wait();
+           }
            //Read the pressure first
-           float pressure = await ReadPreasure();
+           float pressure = ReadPreasure();
            //Convert the pressure to Hectopascals(hPa)
            pressure /= 100;
...
```

That's it! Not too bad, eh? Remember, the complete source code is available
for download from the Apress website, so if you don't want to make these changes
yourself, you can download the sample code and get the completed file. However, I do
recommend you build the solution yourself otherwise so you can learn how to mix VB
and C# projects in the same solution.

You use a couple of tasks in the code. You do this because the initialization of
the libraries (e.g., the Wait() method) should not run in the UI thread. This is a very
common technique when working with libraries and other assemblies.

But wait, how does this class work to read the data from the sensor? I mentioned the code needs to read multiple bytes from the sensor and decode them. The authors of the donor class have done their homework well and have implemented the code based on the recommendations from the manufacturer's data sheet. Take a moment while you are in the code to look around. There is a whole lot more code in there than what you've seen in the last few code listings!

Tip The data sheet for the BMP280 can be found at `https://cdn-shop.adafruit.com/datasheets/BST-BMP280-DS001-11.pdf`.

So, how did the authors know what to do? Figure 12-3 shows an excerpt from the data sheet that outlines how to read data from the sensor.

5.2.1 I²C write

Writing is done by sending the slave address in write mode (RW = '0'), resulting in slave address 111011X0 ('X' is determined by state of SDO pin. Then the master sends pairs of register addresses and register data. The transaction is ended by a stop condition. This is depicted in Figure 7.

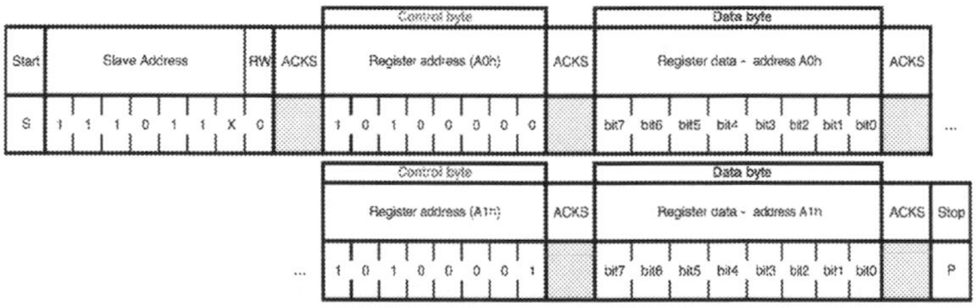

Figure 7: I²C multiple byte write (not auto-incremented)

5.2.2 I²C read

To be able to read registers, first the register address must be sent in write mode (slave address 111011X0). Then either a stop or a repeated start condition must be generated. After this the slave is addressed in read mode (RW = '1') at address 111011X1, after which the slave sends out data from auto-incremented register addresses until a NOACKM and stop condition occurs. This is depicted in Figure 8, where two bytes are read from register 0xF6 and 0xF7.

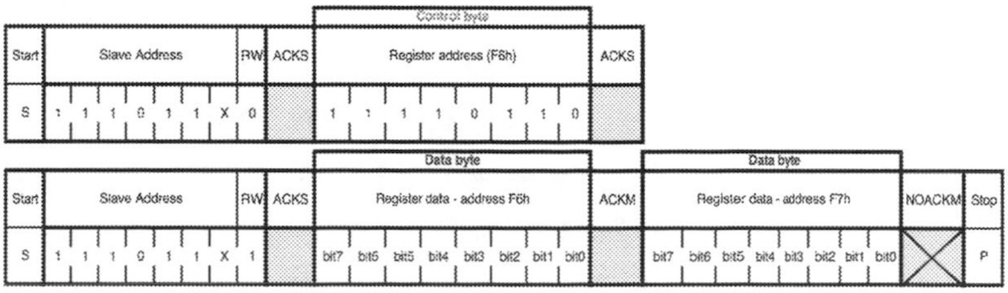

Figure 8: I²C multiple byte read

Figure 12-3. *Excerpt from BMP280 data sheet*

Clearly, there is a lot going on here! What complicates the code are features in the BMP280 to provide calibration data, which is used to correct the values read. Given you must read multiple bytes and interpret them, the code is quite complex. Fortunately, all the hard work has been done for us, and you can simply use the code provided. This is often the case for complex breakout boards like the BMP280 from Adafruit. If you ever have to implement your own code for such a component, be sure to read and study the data sheet. You should also consider searching the Internet for solutions. Oftentimes, you can find a solution that may not be written in your language of choice, but you can either rewrite or, as you've seen in this chapter, use the power of Visual Studio to combine C# and VB in the same solution.

Now, let's add the user interface code.

User Interface

The user interface code is very simple—just four labels: one for the title and one for each of the values read (temperature, pressure, and altitude). Listing 12-6 shows the XAML code for the interface. Open the `MainPage.xaml` file in the weather project and edit the file as shown. The new lines you need to add are shown in bold.

Listing 12-6. User Interface Code

```
...
    <Grid Background="{ThemeResource ApplicationPageBackgroundThemeBrush}">
        <StackPanel Width="400" Height="400">
            <TextBlock x:Name="Title" Height="60" TextWrapping="NoWrap"
                    Text="Weather Station" FontSize="28" Foreground="Blue"
                    Margin="10" HorizontalAlignment="Center"/>
            <TextBlock x:Name="Temp" Height="60" TextWrapping="NoWrap"
                    Text="Initializing..." FontSize="28" Foreground="Blue"
                    Margin="10" HorizontalAlignment="Center"/>
            <TextBlock x:Name="Press" Height="60" TextWrapping="NoWrap"
                    Text="Initializing..." FontSize="28" Foreground="Blue"
                    Margin="10" HorizontalAlignment="Center"/>
            <TextBlock x:Name="Alt" Height="60" TextWrapping="NoWrap"
                    Text="Initializing..." FontSize="28" Foreground="Blue"
                    Margin="10" HorizontalAlignment="Center"/>
```

```
    </StackPanel>
  </Grid>
...
```

Figure 12-4 shows an example of what the user interface will look like when the application runs on the Raspberry Pi.

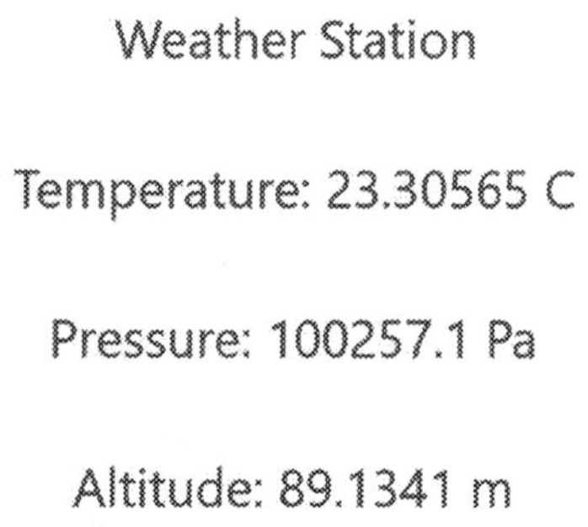

Weather Station

Temperature: 23.30565 C

Pressure: 100257.1 Pa

Altitude: 89.1341 m

Figure 12-4. *Example user interface*

Add References

There is one reference needed. You must add a reference in the *WeatherDisplay* project that references the *WeatherSensor* project. Right-click *References* in the project tree in the Solution Explorer and add the *WeatherSensor* project, as shown in Figure 12-5. Be sure to tick the box before clicking *OK*.

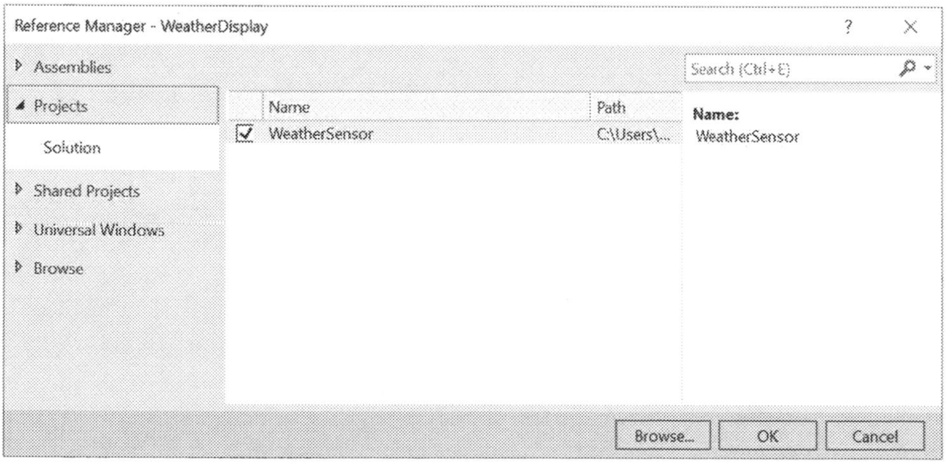

Figure 12-5. *Adding the WeatherSensor project reference*

You're almost there! You just need to add the code to use the BMP280 class and update the user interface. Let's do that now.

Reading the Weather Data

Now let's add the code to read the temperature, pressure, and altitude from the BMP280 class and display it in the user interface you implemented. Let's begin by adding the references we need in the form of local variables. Open the `MainPage.xaml.vb` in the *WeatherDisplay* project and add the references shown as follows:

```
Private bmpTimer As DispatcherTimer    ' Timer
Private bmp280 As WeatherSensor.BMP280      ' Instance Of BMP280 Class
```

Here, we see one local variable for the `DispatchTimer` and another for the BMP280. You add these to the private section by convention. You also add a variable for a timer that you will use to read and update the user interface and the method, which is called on each "tick" of the timer event.

Next, you edit the constructor code by creating a function named `New()` to call the `Initialize()` method on the BMP280 instance and then set up the timer code. You use a small update (delay), but you can increase this once you are comfortable the solution is working correctly. If you have been following along with the projects in the previous chapters, this code should look familiar in a not-C#-but-similar manner.

Next, you add the code for the `OnTick()` method. You use this method to read the data and update the user interface. You first read each of the values and then update the corresponding label in the user interface. Listing 12-7 shows an excerpt of the file with all the code needed for the constructor and the new method.

Listing 12-7. The MainPage.xaml.vb Source File

```
Public NotInheritable Class MainPage
    Inherits Page
    Private bmpTimer As DispatcherTimer    ' Timer
    Private bmp280 As WeatherSensor.BMP280      ' Instance Of BMP280 Class

    Public Sub New()
        Me.InitializeComponent()
```

```vbnet
        ' Instantiate a New BMP280 class instance
        bmp280 = New WeatherSensor.BMP280()
        bmp280.Initialize()

        bmpTimer = New DispatcherTimer()
        bmpTimer.Interval = TimeSpan.FromMilliseconds(5000)
        AddHandler bmpTimer.Tick, AddressOf OnTick
        bmpTimer.Start()
    End Sub

    Private Sub OnTick(sender As Object, e As Object)
        Dim tempRead As Double = 0
        Dim pressRead As Double = 0
        Dim altRead As Double = 0

        ' Create a constant for pressure at sea level.
        ' This Is based on your local sea level pressure (Unit: Hectopascal)
        ' To find the sea level pressure for your area, go to:
        ' weather.gov And enter your city then read the pressure from the
        ' history.
        Const seaLevelPressure As Double = 1013.25F

        ' Read samples of the data every 10 seconds
        tempRead = bmp280.ReadTemperature()
        pressRead = bmp280.ReadPreasure()
        altRead = bmp280.ReadAltitude(seaLevelPressure)

        Temp.Text = "Temperature: " + tempRead.ToString("F2") + " C"
        Press.Text = "Pressure: " + pressRead.ToString("F2") + " Pa"
        Alt.Text = "Altitude: " + altRead.ToString("F2") + " m"
    End Sub

End Class
```

You supply a constant for the sea level pressure. This value may be different for your location. Check the weather websites for your location to get the latest calculated sea level pressure (e.g., www.wunderground.com/, www.weatherwx.com, or https://weather. com/). Substitute your value in the code to get more accurate altitude calculations. That is, the altitude is the result of a calculation based on the barometric pressure and sea level pressure values. Thus, it may not be accurate but should be pretty close if you use the correct value for your area.

Also, notice the formatting for the temperature, pressure, and altitude. Here, we use the format specifier F2, which formats the string to two decimal places. For example, 12.87653531 would display as 12.88. Once you get the project working, you can experiment with more decimal places or even other formats.

That's it! The code is complete and ready for compilation. Be sure to check the preceding listings to ensure that you have all the code in the right place. Once you have entered all the code, you should now attempt to compile the solution. You can compile the projects individually, but if you do so, start with the *WeatherSensor* project. Once that is compiling correctly, you can compile the *WeatherDisplay* project. Or you can just build the entire solution. Correct any errors you find until the code compiles without errors or warnings.

Deploy and Execute

Now it is time to deploy the application! Be sure to fix any compilation errors first. You may want to compile the application in debug first (but you can compile in release mode if you'd prefer). In fact, in order to see the debug messages, you will need to be using the debug build.

Recall from Chapter 7 (see the "Deploy and Debug" section and Figure 7-10), to deploy our application in debug mode, we must set the debug settings for our application, setting the following in the debug dialog:

- *Target device*: Remote Machine

- *Remote Machine*: The name of your device

- (Optional) *Uninstall and then re-install*: Checked

- (Optional) *Deploy optional packages*: Checked

- *Application process*: Managed Only

- *Background task process*: Managed Only

Once you make any changes in this dialog, be sure to save your solution. Building or deploying does not automatically save these settings.

We should also change the package name in the package manifest. Recall from Chapter 7 (see the "Deploying Your Application (No Debugging)" section and Figure 7-8), we can change the package name. Do that now so you can find your application easier in the Device Portal.

Now you can deploy your application. Go ahead and do that now. You should see the user interface appear on the device after a few moments and then the screen update as the timer tick event fires.

If you do not see the user interface or you see the box with the X in it, then the default application reloads, chances are there is something wrong in your code. In this case, you should run the code in debug mode and observe the exception thrown. There are a number of things that can go wrong. The following summarizes some of the things (mistakes) I've encountered in designing this project:

- Ensure that you are compiling the entire solution for *ARM*.

- Ensure that the namespace of the BMP280 class is set correctly if you copied it from the Adafruit sample.

- Double-check your references to ensure they are correct.

- Ensure that the BMP280 sensor is wired correctly to your device.

- If you connected the sensor with 3.3V by mistake, it should still work, but double-check the GND pin is connected correctly.

- Use the Device Portal to start and stop the application after deployment and do not set the application to run as a startup until you have successfully run and debugged the code.

Once the project is running correctly, enjoy it for a while and then put the hardware aside. You will need it again for the next chapter.

Tip   If you encounter problems deploying your application, see the "Visual Basic Application Deployment Troubleshooting" section in Chapter 7 for help.

Summary

Weather IoT solutions can be a lot of fun to develop, and unlike an experiment that teaches techniques, it can be a very practical solution. Indeed, you can show it to your friends and family and possibly get more than a noncommittal "that's nice" accolade.[5]

In this chapter, you've seen a depth of complexity that presented a number of advanced tools and techniques from using a code library written by someone else to building a VB headed application to incorporating a C# and VB project in the same solution. Combining all of these together makes this project the most complex in the book.[6]

In the next chapter, you'll step back a bit from escalating the complexity of this project by adding more functionality to this project by recording the data read in a MySQL database. Adding a database storage feature makes the project much more interesting and useful for post-event analysis. That is, you cannot perform any form of analysis without having stored the data collected over time. You could do this with a log file (and many solutions do), but having the ability to use a powerful query language like SQL can give your solution a great deal of sophistication and capability for free.

---

[5]You do show off your projects, don't you? Well, this time you can show them something they can actually use!

[6]But you see less complex yet more diverse projects in the following chapters.

CHAPTER 13

Project 4: Using MySQL to Store Data

IoT solutions, by definition if not implementation, can generate a lot of data. Indeed, most IoT solutions observe the world in one or often several ways. Those observations generate data at whatever rate the solution specifies (called a *sample rate*). To make the data most useful for historical or similar analytics, you have to save the data for later processing. Database systems provide a perfect solution for storing IoT data and making it available for later use.

With the exception of the last project, you wouldn't consider storing the data from the example projects you've seen thus far for any length of time. That changed when we started using sensors in our projects. The project from Chapter 12 is a perfect example; it observes weather data that could be useful for later analysis.

The project in this chapter takes the same project goals from the last chapter and implements them in a different way. There are two methods for connecting to an I2C device. You used one method in the last project; you will use the other in this project. You will also add a database element storing the data generated in a MySQL database. I demonstrate how to retrieve that data.

Tip   I cover incorporating database systems and specifically using MySQL for the IoT in my book *MySQL for the Internet of Things* (Apress, 2015). If you want to explore the theory and application of database systems in the IoT, this book will get you started, even if you know very little about IoT or MySQL.

Before we get into the project design and how to implement the code, let's take a moment to discover MySQL. If you already have a lot of experience with MySQL, you

© Charles Bell 2021
C. Bell, *Windows 10 for the Internet of Things*, https://doi.org/10.1007/978-1-4842-6609-0_13

may want to skim the section as a refresher. If you have never used MySQL, the following brief primer gets you started and covers everything you need to know in order to complete this project.

What Is MySQL?

MySQL is the world's most popular open source database system for many excellent reasons. First and foremost, it is open source, which means anyone can use it for a wide variety of tasks for free.[1] Best of all, MySQL is included in many platform repositories, making it easy to get and install. If your platform doesn't include MySQL in the repository (such as aptitude), you can download it from the MySQL website (`http://dev.mysql.com`).

Oracle Corporation owns MySQL. Oracle obtained MySQL through an acquisition of Sun Microsystems, which acquired MySQL from its original owners, MySQL AB. Despite fears to the contrary, Oracle has shown excellent stewardship of MySQL by continuing to invest in the evolution and development of new features as well as faithfully maintaining its open source heritage. Although Oracle also offers commercial licenses of MySQL—just as its prior owners did in the past—MySQL is still open source and available to everyone.

WHAT IS OPEN SOURCE? IS IT REALLY FREE?

Open source software grew from a conscious resistance to the corporate-property mindset. While working for MIT, Richard Stallman, the father of the free software movement, resisted the trend of making software private (closed) and left MIT to start the GNU (GNU Not Unix) project and the Free Software Foundation (FSF).

Stallman's goal was to reestablish a cooperating community of developers. He had the foresight, however, to realize that the system needed a copyright license that guaranteed certain freedoms. (Some have called Stallman's take on copyright "copyleft," because it guarantees freedom rather than restricts it.) To solve this, Stallman created the GNU General Public License (GPL). The GPL, a clever work of legal permissions that permits the code to be copied and modified without restriction, states that derivative works (the modified copies) must be distributed under the same license as the original version without any additional restrictions.

---

[1]According to GNU (`www.gnu.org/philosophy/free-sw.html`), "free software is a matter of liberty, not price. To understand the concept, you should think of 'free' as in 'free speech,' not as in 'free beer.'"

There was one problem with the free software movement. The term *free* was intended to guarantee freedom to use, modify, and distribute; it was not intended to mean "no cost" or "free to a good home." To counter this misconception, the Open Source Initiative (OSI) formed and later adopted and promoted the phrase *open source* to describe the freedoms guaranteed by the GPL license. For more information about open source software, visit `www.opensource.org`.

MySQL runs as a background process (or as a foreground process if you launch it from the command line)[2] on your system. Like most database systems, MySQL supports Structured Query Language (SQL). You can use SQL to create databases and objects (using data definition language [DDL]), write or change data (using data manipulation language [DML]), and execute various commands for managing the server.

To issue these commands, you must first connect to the database server. MySQL provides an excellent, very powerful environment called the MySQL Shell (executed as *mysqlsh*). You can install this on your server or your PC. The MySQL Shell (or simply the shell) allows you to use one of three modes indicated with a command-line option: SQL (`--sql`), JavaScript (`--js`), or Python (`--py`). As you can surmise, the JavaScript and Python modes allow for the use of scripting languages to interact with data, and the SQL mode is the more familiar mode that allows you to issue SQL commands. For our purposes, we will use the SQL mode.

Tip   For more information about the MySQL Shell, see my book *Introducing MySQL Shell* (Apress 2019) or visit the Oracle MySQL documentation website (`https://dev.mysql.com/doc/mysql-shell/8.0/en/`).

Oracle also provides the older client application named `mysql` and is known as the `mysql` *client* (previously the `mysql` monitor). Either application enables you to connect to and run commands on the server.

Listing 13-1 shows examples of each type of command in action using the `mysqlsh` client.

---

[2]And use the `--console` command-line option on Windows systems.

Listing 13-1. Commands Using the mysqlsh

```
C:\Users\cbell> mysqlsh --uri root@localhost --sql
Please provide the password for 'root@localhost': ****
Save password for 'root@localhost'? [Y]es/[N]o/Ne[v]er (default No):
MySQL Shell 8.0.19

Copyright (c) 2016, 2019, Oracle and/or its affiliates. All rights reserved.
Oracle is a registered trademark of Oracle Corporation and/or its affiliates.
Other names may be trademarks of their respective owners.

Type '\help' or '\?' for help; '\quit' to exit.
Creating a session to 'root@localhost'
Fetching schema names for autocompletion... Press ^C to stop.
Your MySQL connection id is 13 (X protocol)
Server version: 8.0.19 MySQL Community Server - GPL
No default schema selected; type \use <schema> to set one.

>  CREATE DATABASE testme;
Query OK, 1 row affected (0.0312 sec)

>  CREATE TABLE testme.table1 (sensor_node char(30), sensor_value int,
sensor_event timestamp);
Query OK, 0 rows affected (0.1058 sec)

>  INSERT INTO testme.table1 VALUES ('living room', 23, NULL);
Query OK, 1 row affected (0.0212 sec)

>  SELECT * FROM testme.table1;
+-------------+--------------+--------------+
| sensor_node | sensor_value | sensor_event |
+-------------+--------------+--------------+
| living room |           23 | NULL         |
+-------------+--------------+--------------+
1 row in set (0.0007 sec)

>  SET @@global.server_id = 111;
Query OK, 0 rows affected (0.0004 sec)

>  \q
Bye!
```

In this example, you see DDL in the form of the CREATE DATABASE and CREATE TABLE statements, DML in the form of the INSERT and SELECT statements, and a simple administrative command to set a global server variable. Next, you see the creation of a database and a table to store the data, the addition of a row in the table, and finally retrieval of the data in the table.

A great many commands are available in MySQL. Fortunately, you need master only a few of the more common ones. The following are the commands that you use most often. The portions enclosed in <> indicate user-supplied components of the command, and [...] indicates that additional options are needed:

- CREATE DATABASE <database_name>: Creates a database

- USE <database>: Sets the default database

- CREATE TABLE <table_name> [...]: Creates a table or structure to store data

- INSERT INTO <table_name> [...]: Adds data to a table

- UPDATE [...]: Changes one or more values for a specific row

- DELETE FROM <table_name> [...]: Removes data from a table

- SELECT [...]: Retrieves data (rows) from the table

Tip   If you use the mysql client, you must terminate each command with a semicolon (;) or \G.

Although this list is only a short introduction and nothing like a complete syntax guide, there is an excellent online reference manual that explains each and every command (and much more) in great detail. You should refer to the online reference manual whenever you have a question about anything in MySQL. You can find it at http://dev.mysql.com/doc/.

If you are thinking that there is a lot more to MySQL than a few simple commands, you are absolutely correct. Despite its ease of use and fast startup time, MySQL is a full-fledged relational database management system (RDBMS). There is much more to it than you've seen here. For more information about MySQL, including all the advanced features, see the reference manual.

MYSQL: WHAT DOES IT MEAN?

The name MySQL is a combination of a proper name and an acronym. SQL is Structured Query Language. The *My* part isn't the possessive form—it is a name. In this case, My is the name of the founder's daughter. As for pronunciation, MySQL experts pronounce it "My-S-Q-L" and not "my sequel." Indeed, the mark of a savvy MySQL user is in their pronunciation of the product. There is a corollary with Mac OS X: Is it "Mac O-S Ex" or "Mac O-S Ten"? Check to see.

Getting Started with MySQL

Now that you know what MySQL is and how it is used, you need to know a bit more about RDBMSs and MySQL in particular before you start building your first database server. This section discusses how MySQL stores data (and where it is stored), how it communicates with other systems, and some basic administration tasks required in order to manage your new MySQL server.

WHAT'S A RELATIONAL DATABASE MANAGEMENT SYSTEM?

An RDBMS is a data storage-and-retrieval service based on the Relational Model of Data as proposed by E. F. Codd in 1970. These systems are the standard storage mechanism for structured data. A great deal of research is devoted to refining the essential model proposed by Codd, as discussed by C. J. Date in *The Database Relational Model: A Retrospective Review and Analysis*.[3] This evolution of theory and practice is best documented in *The Third Manifesto*.[4]

The relational model is an intuitive concept of a storage repository (database) that can be easily queried by using a mechanism called a *query language* to retrieve, update, and insert data. The relational model has been implemented by many vendors because it has a sound systematic theory, a firm mathematical foundation, and a simple structure. The most

---

[3]C. J. Date, *The Database Relational Model: A Retrospective Review and Analysis* (Reading, MA: Addison-Wesley, 2001).

[4]C. J. Date and H. Darwen, *Foundation for Future Database Systems: The Third Manifesto* (Reading, MA: Addison-Wesley, 2000).

commonly used query mechanism is SQL, which resembles natural language. Although SQL is not included in the relational model, it provides an integral part of the practical application of the relational model in RDBMSs.

The data are represented as related pieces of information (attributes or *columns*) about a certain event or entity. The set of values for the attributes is formed as a *tuple* (sometimes called a *record* or a *row*). Tuples are stored in tables that have the same set of attributes. Tables can then be related to other tables through constraints on keys, attributes, and tuples.

Tables can have special mappings of columns called *indexes* that permit you to read the data in a specific order. Indexes are also very useful for fast retrieval of rows that match the value(s) of the indexed columns.

How and Where MySQL Stores Data

The MySQL database system stores data via an interesting mechanism of programmatic isolation; it is called a *storage engine*, which is governed by the handler interface. The handler interface permits the use of interchangeable storage components in the MySQL server so that the parser, the optimizer, and all manner of components can interact in storing data on disk using a common mechanism. This is also referred to as a *pluggable storage engine.*[5] However, the default storage engine, InnoDB, is the default and you should not need to change it (but you can).

```
CREATE TABLE `books` (
  `ISBN` varchar(15) DEFAULT NULL,
  `Title` varchar(125) DEFAULT NULL,
  `Authors` varchar(100) DEFAULT NULL,
  `Quantity` int(11) DEFAULT NULL,
  `Slot` int(11) DEFAULT NULL,
  `Thumbnail` varchar(100) DEFAULT NULL,
  `Description` text
) ENGINE=INNODB;
```

---

[5]If you would like to know more about storage engines and what makes them tick, see my book *Expert MySQL* (Apress, 2012).

Great! Now, what storage engines exist on MySQL? You can discover which storage engines are supported by issuing the following command. As you can see, there are a lot to choose from. I cover a few that may be pertinent to planning sensor networks:

```
> SHOW STORAGE ENGINES \G
*************************** 1. row ***************************
      Engine: MEMORY
     Support: YES
     Comment: Hash based, stored in memory, useful for temporary tables
Transactions: NO
          XA: NO
  Savepoints: NO
*************************** 2. row ***************************
      Engine: MRG_MYISAM
     Support: YES
     Comment: Collection of identical MyISAM tables
Transactions: NO
          XA: NO
  Savepoints: NO
*************************** 3. row ***************************
      Engine: CSV
     Support: YES
     Comment: CSV storage engine
Transactions: NO
          XA: NO
  Savepoints: NO
*************************** 4. row ***************************
      Engine: FEDERATED
     Support: NO
     Comment: Federated MySQL storage engine
Transactions: NULL
          XA: NULL
  Savepoints: NULL
```

```
*************************** 5. row ***************************
      Engine: PERFORMANCE_SCHEMA
     Support: YES
     Comment: Performance Schema
Transactions: NO
          XA: NO
  Savepoints: NO
*************************** 6. row ***************************
      Engine: MyISAM
     Support: YES
     Comment: MyISAM storage engine
Transactions: NO
          XA: NO
  Savepoints: NO
*************************** 7. row ***************************
      Engine: InnoDB
     Support: DEFAULT
     Comment: Supports transactions, row-level locking, and foreign keys
Transactions: YES
          XA: YES
  Savepoints: YES
*************************** 8. row ***************************
      Engine: BLACKHOLE
     Support: YES
     Comment: /dev/null storage engine (anything you write to it disappears)
Transactions: NO
          XA: NO
  Savepoints: NO
*************************** 9. row ***************************
      Engine: ARCHIVE
     Support: YES
     Comment: Archive storage engine
Transactions: NO
          XA: NO
  Savepoints: NO
9 rows in set (0.0008 sec)
```

Wrapping the statements in a transaction ensures that no data is written to disk until and unless all statements are completed without errors. Transactions in this case are designated with a BEGIN statement and concluded with either a COMMIT to save the changes or a ROLLBACK to undo the changes. InnoDB stores its data in a single file (with some additional files for managing indexes and transactions).

The MyISAM storage engine is optimized for reads. MyISAM has been the default for some time and was one of the first storage engines available. In fact, a large portion of the server is dedicated to supporting MyISAM. It differs from InnoDB in that it does not support transactions and stores its data in an indexed sequential access method format. This means it supports fast indexing. You would choose MyISAM over InnoDB if you did not need transactions and you wanted to be able to move or back up individual tables.

Another storage engine that you may want to consider, especially for sensor networks, is Archive. This engine does not support deletes (but you can drop entire tables) and is optimized for minimal storage on disk. Clearly, if you are running MySQL on a small system like a Raspberry Pi, small is almost always better! The inability to delete data may limit more advanced applications, but most sensor networks merely store data and rarely delete it. In this case, you can consider using the Archive storage engine.

There is also the CSV storage engine (where CSV stands for *comma-separated values*). This storage engine creates text files to store the data in plain text that can be read by other applications, such as a spreadsheet application. If you use your sensor data for statistical analysis, the CSV storage engine may make the process of ingesting the data easier.

So where is all this data? If you query the MySQL server and issue the SHOW VARIABLES LIKE "datadir"; command, you see the path to the location on disk that all storage engines use to store data. In the case of InnoDB, this is a single file on disk located in the data directory. InnoDB also creates a few administrative files, but the data is stored in the single file. For most other storage engines except NDB and MEMORY, the data for the tables is stored in a folder with the name of the database under the data directory. Listing 13-2 shows how to determine the location of the data directory (it is typically in a protected folder).

Listing 13-2. Finding Where Your Data Is Located

```
> SHOW VARIABLES LIKE 'datadir';
+---------------+-----------------------------------------+
| Variable_name | Value                                   |
+---------------+-----------------------------------------+
| datadir       | C:\ProgramData\MySQL\MySQL Server 8.0\Data\ |
+---------------+-----------------------------------------+
1 row in set (0.0069 sec)
```

If you navigate to the location (path) shown and (with elevated privileges) issue a directory listing command, you can see the InnoDB files identified by the `ib` and `ibd` prefixes. You may also see a number of directories, all of which are the databases on this server.

For more information about storage engines and the choices and features of each, please see the online MySQL Reference Manual section "Storage Engines" (`http://dev.mysql.com/doc/`).

The MySQL Configuration File

The MySQL server can be configured using a configuration file, similar to the way you configure other Windows applications. On Windows, the MySQL configuration file is named `my.ini` and located in the program data folder under the MySQL server version (e.g., `C:\ProgramData\MySQL\MySQL Server 8.0`). This file contains several sections, one of which is labeled `[mysqld]`. The items in this list are key-value pairs: the name on the left of the equal sign is the option, and its value on the right. The following is a typical configuration file (with many lines suppressed for brevity):

```
[mysqld]
port=3306
basedir="C:/Program Files/MySQL/MySQL Server 8.0/"
datadir=C:/ProgramData/MySQL/MySQL Server 8.0/Data
default_authentication_plugin=caching_sha2_password
default-storage-engine=INNODB
sql-mode="STRICT_TRANS_TABLES,NO_ENGINE_SUBSTITUTION"
log-output=FILE
general-log=0
```

```
general_log_file="DESKTOP-VJ2PPOK.log"
slow-query-log=1
slow_query_log_file="DESKTOP-VJ2PPOK-slow.log"
long_query_time=10
log-error="DESKTOP-VJ2PPOK.err"
log-bin="DESKTOP-VJ2PPOK-bin"
server-id=1
...
```

As you can see, this is a simple way to configure a system. This example sets the TCP port, base directory (the root of the MySQL installation, including the data as well as binary and auxiliary files), data directory, and server ID (used for replication, as discussed shortly) and turns on the general log (when the Boolean switch is included, it turns on the log). There are many such variables you can set for MySQL. See the online MySQL Reference Manual for details concerning using the configuration file.

How to Get and Install MySQL

The MySQL server is available for a variety of platforms, including most Linux and Unix platforms, Mac OS X, and Windows. To download MySQL server, visit `http://dev.mysql.com/downloads/` and click the *MySQL Community (GPL) Downloads* link (scroll down) and then *MySQL Community Server*. This is the GPLv2 license of MySQL. The page automatically detects your operating system. If you want to download for another platform, you can select it from the drop-down menu.

Oracle has provided a special installation packaging for Windows named the Windows Installer. This package includes all the MySQL products available under the community license, including MySQL Server, Workbench, Utilities, Fabric, and all the available connectors (program libraries for connecting to MySQL). This makes installing on Windows a one-stop, one-installation affair. How cool is that? Figure 13-1 shows the download page for the Windows installer.

Figure 13-1. *Download page for Windows installer*

To install MySQL, begin by choosing either the Windows installer 32- or 64-bit installation package that matches your Windows version. Once the file is downloaded, click the file to begin installation. Note that some browsers, such as the new Edge browser, may ask you if you want to launch the installation. You may need to reply to a dialog permitting the installation.

The installation is fully automated with a series of dialogs to help you configure your installation. Like most Windows installation packages, you can choose what you want to install as well as where to install it. I recommend accepting the defaults since these are optimized for typical use cases, but be sure to install the MySQL Shell. You can even choose to automatically start MySQL. You may also be given a temporary password for the root user account. Be sure to write that down and change it later (with the ALTER USER command; see the following).

How to Start, Stop, and Restart MySQL

While working with your databases and configuring MySQL on your Windows system, you may need to control the startup and shutdown of the MySQL server. The default mode for installing MySQL is to automatically start on boot and stop on shutdown, but you may want to change that, or you may need to stop and start the server after changing a parameter. In addition, when you change the configuration file, you need to restart the server to see the effect of your changes.

You can start, pause (stop), and restart the MySQL server with the *Services* application in Windows. Simply type **Services** in the search box and choose the application. You can then scroll down, choose *MySQL80* from the list, and start, stop, or restart the server using the links provided. Figure 13-2 shows an example of the application with the commands highlighted.

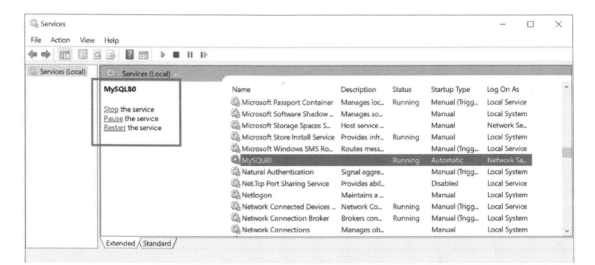

Figure 13-2. *Services application*

Creating Users and Granting Access

You need to know about two additional administrative operations before working with MySQL: creating user accounts and granting access to databases. More specifically, you first have to issue a CREATE USER command followed by one or more GRANT commands. For example, the following shows the creation of a user named sensor1 and grants the user access to the database room_temp:

```
CREATE USER 'sensor1'@'%' IDENTIFIED BY 'secret';
GRANT SELECT, INSERT, UPDATE ON room_temp.* TO 'sensor1'@'%';
```

The first command creates the user named sensor1, but the name also has an @ followed by another string. This second string is the hostname of the machine with which the user is associated. That is, each user in MySQL has both a username and a hostname, in the form user@host, to uniquely identify them. That means the user and host sensor1@10.0.1.16 and the user and host sensor1@10.0.1.17 are not the same. However, the % symbol can be used as a wildcard to associate the user with any host. The IDENTIFIED BY clause sets the password for the user.

A NOTE ABOUT SECURITY

It is always a good idea to create a user for your application that does not have full access to the MySQL system. This is so you can minimize any accidental changes and also to prevent exploitation. For sensor networks, it is recommended that you create a user with access only to those databases where you store (or retrieve) data. You can change MySQL user passwords with the following command:

```
ALTER USER sensor1@"%" IDENTIFIED BY "secret123";
```

Also, be careful about using the wildcard % for the host. Although it makes it easier to create a single user and let the user access the database server from any host, it also makes it much easier for someone bent on malice to access your server (once they discover the password).

Another consideration is connectivity. If you connect a database to your network and the network is in turn connected to the Internet, it may be possible for other users on your network or the Internet to gain access to the database. Don't make it easy for them—change your root user password and create users for your applications.

The second command allows access to databases. There are many privileges that you can give a user. The example shows the most likely set that you would want to give a user of a sensor network database: read (SELECT), add data (INSERT), and change data (UPDATE). See the online reference manual for more about security and account access privileges.

The command also specifies a database and objects to which to grant the privilege. Thus, it is possible to give a user read (SELECT) privileges to some tables and write (INSERT, UPDATE) privileges to other tables. This example gives the user access to all objects (tables, views, etc.) in the room_temp database.

Now that you know more about MySQL, let's talk about the project and how you will develop the database portion.

Overview

This project is a rewrite of the weather project from the last chapter. The project uses a BMP280 sensor breakout board with an I2C interface. Unlike the last project, you will not be writing a user interface. Similarly, you will implement the project in C# rather than C++.

You will write this project purely in C# so that you can take advantage of a NuGet package to read data from the sensor using an I2C interface. This is because the library you want is written in C# and only works with C# applications. Interestingly, the library you will use provides a bit more data than the previous code library producing some different units for barometric precision. The extra data are extrapolated from the BMP280 data and may be interesting if you're a real weather buff.

Otherwise, the project goals are the same as the previous project. You want to read data from the sensor, but instead of displaying it for an instant and replacing it with new values, you will store the data in a MySQL database.

You will use the same hardware, as you did in the previous project except you won't need a monitor, a keyboard, or a mouse since this project will run headless.

I demonstrate all the code necessary and more in the following sections. But first, let's discover how to set up the MySQL database for the project.

Set Up the Database

If you have not installed MySQL, you should consider doing that before you complete the code. You do not have to run MySQL on your laptop. You could run it on another machine, across the Internet, or wherever. You just need to have your IoT device configured so that it can connect to the MySQL server via TCP/IP.

I normally set up the database I want to use as a first step. That is, I design the database and table(s)[6] before I write the code. In this case, I've done the design work for you. To keep things simple, you will store data in a single table one row at a time. That is, as you read data from the sensor, you will store it together as a single entry (via an INSERT statement).

In order to keep the rows unique, you will use an auto increment column, which is a surrogate key technique that generates an increasing integer value. This allows you to refer to a row by the key for faster access (via indexing). You will also use a special column called a *timestamp* that the server will automatically supply the current date and time you inserted the row. This is a great technique and a best practice for generating database tables that store data that may be used for historical analysis.

Finally, you add a column for each of the data elements you collect. This includes temperature (in both Celsius and Fahrenheit—the library converts it for you) and

---

[6]And any additional database objects that I need.

bars, hectopascals, and atmospheres for barometric pressure. Listing 13-3 shows the commands you issue to create a database named weather with a single table named history.

Listing 13-3. SQL Commands for Weather Database

```
> CREATE DATABASE weather;
Query OK, 1 row affected (0.0326 sec)

> USE weather;
Default schema set to `weather`.
Fetching table and column names from `weather` for auto-completion... Press
^C to stop.

> CREATE TABLE history (
->   id int not null auto_increment primary key,
->   dateRecorded timestamp,
->   degreesCelsius float,
->   degreesFarenheit float,
->   bars float,
->   hectopascals float,
->   atmospheres float
->   );
Query OK, 0 rows affected (0.1255 sec)
```

There is just one more step to accomplish. It is always a good practice to create a new user account that has rights to update the database (and only the one database). This allows you to ensure that casual intrusion via your application limits exposure of other data on your system. The following command creates a user named w_user with a specific password and access only to the weather database. You can use a different password if you'd like; just be sure to remember it when you write the code.

```
> CREATE USER w_user@'%' IDENTIFIED BY 'secret';
Query OK, 0 rows affected (0.04 sec)

> GRANT ALL ON weather.* TO w_user@'%';
Query OK, 0 rows affected (0.03 sec)
```

Now that you have the database table set up, let's review the hardware you need for this project. I repeat the data from the last chapter for clarity.

Required Components

The following lists the components that you need. You can get the BMP280 sensor from Adafruit (`www.adafruit.com`) either in the Microsoft IoT Pack for Raspberry Pi or purchased separately, SparkFun (`www.sparkfun.com`), or any electronics store that carries electronic components. However, if you use a sensor made by someone other than Adafruit, you may need to alter the code to change the I2C address. Since this solution is a headless application, you do not need a monitor, a keyboard, or a mouse.

- Adafruit BMP280 I2C or SPI barometric pressure and altitude sensor

- Jumper wires: (4) male-to-female

- Breadboard (full size recommended but half size is OK)

- Raspberry Pi 2 or 3

- Power supply

Set Up the Hardware

Although there are only four connections needed for this project, you will plan for how things should connect, which is good practice to hone. To connect the components to the Raspberry Pi, you need four pins for the BMP280 sensor, which requires only power, ground, and two pins for the I2C interface. Refer to Chapter 12 for how to wire the sensor. I include Figure 13-3 as a reminder of how things are connected.

Figure 13-3. *Connections for the weather sensor project*

If you are following along with this chapter working on the project, go ahead and make the hardware connections now. Don't power on the board yet but do double- and triple-check the connections.

Write the Code

Now it's time to write the code for the project. You will use the same BMP280 sensor from the last project, but you will use a different library that you can download from NuGet. You will also be using the Connector/Net database connector from Oracle to connect to your MySQL server. You'll use a timer to read the data and update the database. Once you launch the new project, you can monitor its progress by running a SELECT query on the MySQL server.

Note   While this example is written in Visual Basic, I include a C# version written slightly differently but using the same logic with the book source code. I leave the investigation of the C# code for an exercise.

Let's begin with setting up the new project, and then you'll see how to add the new resources and complete the code for both reading from the sensor and writing the data to the database.

New Project

You will use a Visual Basic project template for this project—the *VB Blank App (Universal Windows)* template. Use the name WeatherDatabase for the project name. You can save the project wherever you like or use the default location.

We begin with the user interface, which is similar to the interface in Chapter 12, but this time we will display all of the raw data collected from the sensor. Once the project opens, double-click the MainPage.xaml file and enter the user interface XAML code as shown in Listing 13-4.

Listing 13-4. User Interface Code

```
...
    <Grid Background="{ThemeResource ApplicationPageBackgroundThemeBrush}">
        <StackPanel Width="400" Height="400">
            <TextBlock x:Name="Title" Height="40" TextWrapping="NoWrap"
                    Text="Weather Database" FontSize="28"
                    Foreground="Blue"
                    Margin="10" HorizontalAlignment="Center"/>
            <TextBlock x:Name="Celcius" Height="30" TextWrapping="NoWrap"
                    Text="Initializing..." FontSize="24" Foreground="Blue"
                    Margin="10" HorizontalAlignment="Center"/>
            <TextBlock x:Name="Fahrenheit" Height="30"
            TextWrapping="NoWrap"
                    Text="Initializing..." FontSize="24" Foreground="Blue"
                    Margin="10" HorizontalAlignment="Center"/>
```

```
            <TextBlock x:Name="PressBars" Height="30" TextWrapping="NoWrap"
                    Text="Initializing..." FontSize="24" Foreground="Blue"
                    Margin="10" HorizontalAlignment="Center"/>
            <TextBlock x:Name="PressHecto" Height="30"
            TextWrapping="NoWrap"
                    Text="Initializing..." FontSize="24" Foreground="Blue"
                    Margin="10" HorizontalAlignment="Center"/>
            <TextBlock x:Name="PressAtmos" Height="30"
            TextWrapping="NoWrap"
                    Text="Initializing..." FontSize="24" Foreground="Blue"
                    Margin="10" HorizontalAlignment="Center"/>
            <TextBlock x:Name="DBStatus" Height="30" TextWrapping="NoWrap"
                    Text="Not connected" FontSize="20" Foreground="Blue"
                    Margin="10" HorizontalAlignment="Center"/>
        </StackPanel>
    </Grid>
...
```

Figure 13-4 shows what the user interface will look like once it is reading data.

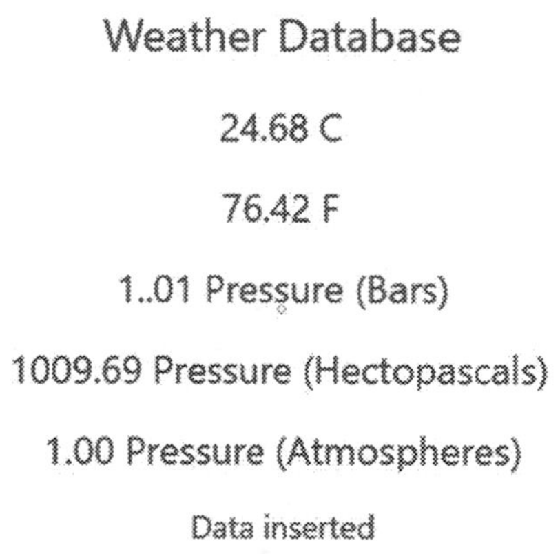

Weather Database

24.68 C

76.42 F

1..01 Pressure (Bars)

1009.69 Pressure (Hectopascals)

1.00 Pressure (Atmospheres)

Data inserted

Figure 13-4. *User interface example*

Next, double-click the `MainPage.xaml.vb` file. There are two namespaces you need to include that relate to the NuGet packages we'll install next. Go ahead and add those now, as shown next:

```
Imports MySql.Data.MySqlClient
Imports Glovebox.IoT.Devices.Sensors
```

Here, you added namespaces for the Glovebox library (for the BMP280 sensor) and the MySQL database client. You'll see how to add the MySQL and Glovebox references in the next sections.

Next, you need to add some variables. You add a variable for the `DispatcherTimer` class, a variable for the `MySQLConnection` class, and a variable for the BMP280 class. The following shows the correct code for defining these variables. These are placed in the `MainPage` class.

```
Private bmpTimer As DispatcherTimer    ' Timer
Private mysql_conn As MySqlConnection ' Connection To MySQL
Private tempAndPressure As BMP280      ' Instance Of BMP280 Class
```

Next, you need to add some constants for the database. First, you add a connection string. This string contains several parts that you need to modify to match your systems. These are the IP or hostname of the server, the user account and password that you want to use to connect to the MySQL server (you can use what is shown if you issued the preceding commands to create the user), the port (3306 is the default), the default database (can be omitted), and the SSL mode (which must be set using `AllowPublicK eyRetrieval=true`). Next, you add a formatted string so that you can issue the `INSERT` statement using parameters for the values.

```
' String constants for database
Private connStr As String = "server=192.168.42.14;user=w_
user;password=secret;" +
                    "port=3306;database=weather;sslMode=None;
                    AllowPublicKeyRetrieval=true"
Private INSERT As String  = "INSERT INTO weather.history VALUES " +
                    "(null, NOW(), " +
                    "{0}, {1}, {2}, {3}, {4})"
```

You may notice that the first two columns for the INSERT statement are null and Now(). You do this to tell the database server to use the default values for the first column and the current date and time for the second column—in this case, the auto increment column named Id and the timestamp column named dateRecorded. I should note that you could have used a different form of the INSERT statement by specifying the columns by name, but passing null is a bit easier if not a bit lazy.

Finally, we add a structure to store the data read from the sensor. Rather than pass the structure around via parameters and return types, we'll allow the methods to access the same structure. This is safe for this example because the code is designed to allow only one part at a time to write or read the data. The following shows the structure and a variable created to store the data. This concludes the preliminary variables and structures portion of the code:

```
Structure WeatherData
    Dim degreesCelsius As Double
    Dim degreesFahrenheit As Double
    Dim bars As Double
    Dim hectopascals As Double
    Dim atmospheres As Double
End Structure

Private dataRead As WeatherData
```

The code in the MainPage() function initializes the components, the BMP280 class, and calls a new imaginative method named Connect() to connect to the database server. You make this a new method to keep things easier to write (and the code easier to understand). You'll see this new method shortly. You also set up the timer. In this case, you use a value of 5000, which is 5 seconds. You may want to consider making value greater if you plan to use the project for practical use cases. Listing 13-5 shows the complete MainPage() method.

Listing 13-5. MainPage Method

```
Public Sub New()
    Me.InitializeComponent()

    ' Instantiate a New BMP280 class instance
    tempAndPressure = New BMP280()
```

```
' Connect to MySQL. If successful, setup timer
If Connect() Then
    bmpTimer = New DispatcherTimer()
    bmpTimer.Interval = TimeSpan.FromMilliseconds(5000)
    AddHandler bmpTimer.Tick, AddressOf OnTick
    bmpTimer.Start()
Else
    Debug.WriteLine("ERROR: Cannot proceed without database connection.")
End If
End Sub
```

Notice that if the `Connect()` method returns false, you issue a debug statement stating you cannot connect to the MySQL server. Thus, you should run this project with debug to ensure that you are connecting to the server properly (see the "Why Can't I Connect?" sidebar).

Before we add the code to read the data, let's add the libraries we need from NuGet. We do this now so that our code completion feature in Visual Studio will detect the new libraries.

We need to add three references from NuGet to the solution. You need the *Glovebox. IoT.Devices* library, which in turn requires the *Units.NET* library. You also need to add the reference to the *MySQL.Data* library. Let's discuss each of these in turn.

Glovebox.IoT.Devices

Glovebox.IoT.Devices is a library that supports a host of sensors—each presented as a separate class to make it easy to use for specific sensors, like the BMP280. It also takes care of the I2C communication for you. All you need to do is instantiate the class and call the methods to read the data.

To add the *Glovebox.IoT.Devices* library, use the NuGet Package Manager from the *Tools ➤ NuGet Package Manager ➤ Manage NuGet Packages for Solution...* menu. Click the *Browse* tab and then type `glovebox` in the search box, as shown in Figure 13-5.

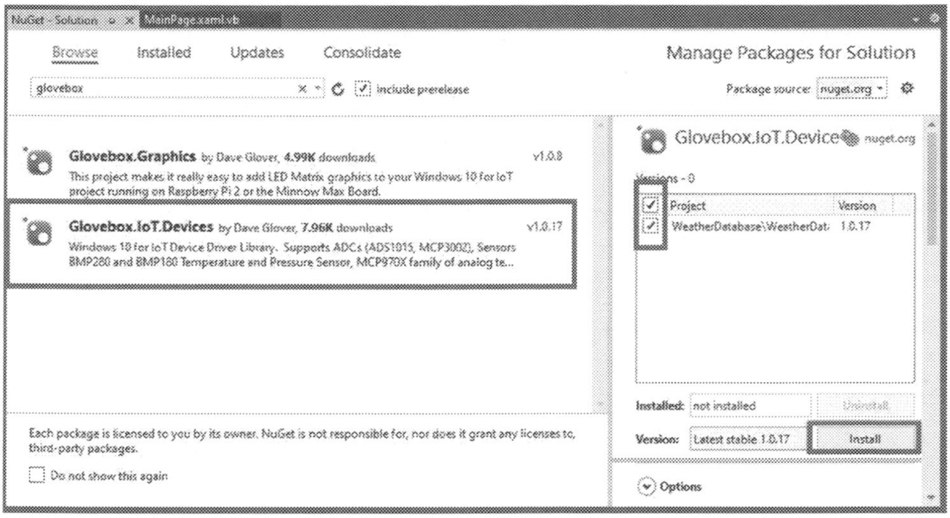

Figure 13-5. *NuGet Package Manager, results for Glovebox*

Select the entry named *Glovebox.IoT.Devices* in the list, tick the project name (solution) in the list on the right, and finally click *Install*. The installation starts and Visual Studio downloads a number of packages and then asks your permission to install them. Go ahead and let the installation complete. A dialog box will open to tell you that the installation is complete.

Unlike the lightning library you used in Chapter 11, there are no additional steps needed to use the library; however, the library requires another library—the Units.NET library, also available via NuGet.

Units.NET

Glovebox.IoT.Devices requires another library that provides helper methods for converting units of measure. The notes state you must have the *Units.NET* library version 3.34 installed.

Caution   Be sure to select version 3.34 when installing the library. Newer versions are not compatible.

To add the *Units.NET* library, use the NuGet Package Manager from the *Tools ➤ NuGet Package Manager ➤ Manage NuGet Packages for Solution...* menu. Click the *Browse* tab and then type units.net in the search box, as shown in Figure 13-6.

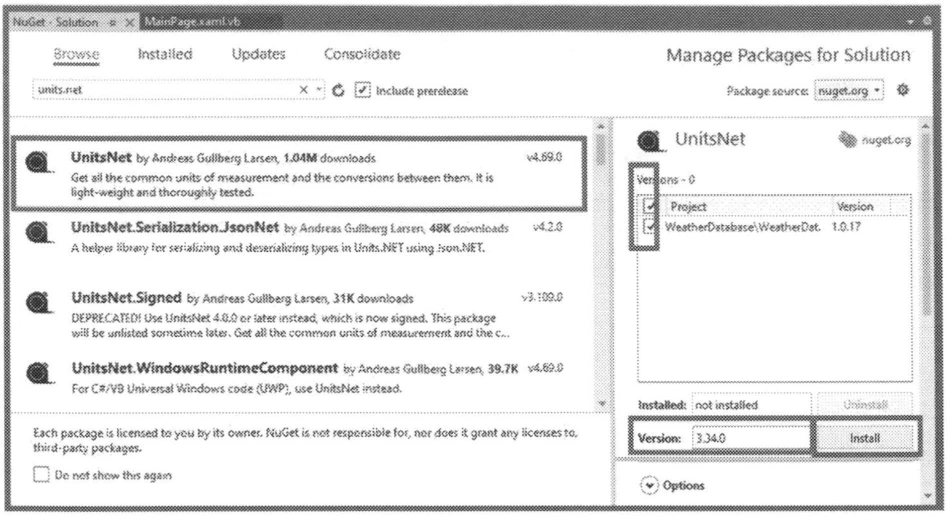

Figure 13-6. *NuGet Package Manager, results for Units.NET*

Select the entry named Units.NET in the list, tick the project name (solution) in the list on the right, then choose the 3.34 version, and finally click *Install*. The installation starts and Visual Studio downloads a number of packages and then asks your permission to install them. Go ahead and let the installation complete. A dialog box opens to tell you when the installation is complete.

MySQL.Data

MySQL.Data (also called Connector/Net or the .Net Connector) is a Microsoft .NET database library for connecting to a MySQL database server. To add the *MySQL.Data* library, use the NuGet Package Manager from the *Tools* ➤ *NuGet Package Manager* ➤ *Manage NuGet Packages for Solution...* menu. Click the *Browse* tab and then type `mysql.data` in the search box, as shown in Figure 13-7.

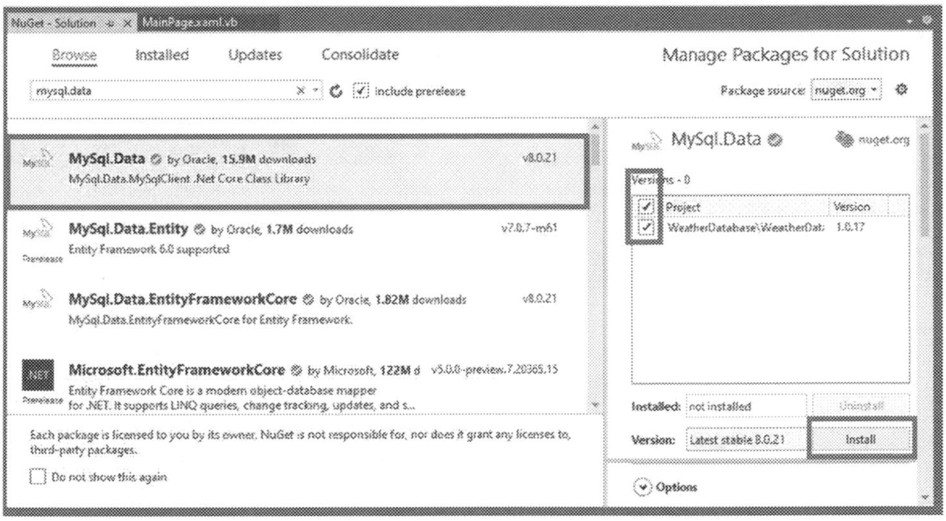

Figure 13-7. *NuGet Package Manager, results for MySQL.Net*

Select the entry named *MySQL.Data* in the list, tick the project name (solution) in the list on the right, and finally click *Install*. The installation starts and Visual Studio downloads a number of packages and then asks your permission to install them. Go ahead and let the installation complete. A dialog box opens to tell you when the installation is complete.

Tip   See `https://dev.mysql.com/doc/connector-net/en/` to learn more about the connector.

Now that we have all of the libraries we need, we can complete the code.

Connecting to MySQL

Connecting to MySQL with Connector/Net is pretty easy. You simply need to instantiate a `MySQLConnection` object, passing it the connection string defined earlier, and then you open the connection. You wrap the `Open()` call in a `try` block so that you can detect whether the connection fails (the library throws an exception). You add some debug statements to help diagnose the problem—just remember to run the project in debug to see the messages. Listing 13-6 shows the `Connect()` method.

Listing 13-6. Connecting to MySQL

```
Private Function Connect() As Boolean
    mysql_conn = New MySqlConnection(connStr)
    Try
        Debug.WriteLine("Connecting to MySQL...")
        mysql_conn.Open()
        Dim msg As String = "Connected to " + mysql_conn.ServerVersion + "."
        Debug.WriteLine(msg)
        DBStatus.Text = msg
    Catch ex As Exception
        Dim msg As String = "ERROR: " + ex.ToString()
        Debug.Write(msg)
        DBStatus.Text = msg
        Return False
    End Try
    Return True
End Function
```

Tip   For complete documentation on Connector/Net, see `http://dev.mysql.com/doc/connector-net/en/`.

Now, let's see the code for reading the weather data.

Reading the Weather Data

Now let's add the code to read the weather data from the BMP280. A timer is used to fire an event every 5 seconds to read (and save) the data. Thus, you need the event for the DispatcherTimer object defined earlier, named `BmpTimer_Tick()`.

However, instead of putting the code to read the data in this event, you use another method and run it as a task (in a new thread). This is because the BMP280 object cannot run in the user interface thread.[7] Thus, you create a new method named getData() to read the data. You also write the data to MySQL in the same method, but you look at that in the next section.

The BMP280 class from the Glovebox.IoT.Devices library provides a variety of data from the sensor, including temperature in both Fahrenheit and Celsius and barometric pressure in several units. You read all of these with the Temperature and Pressure attributes, as shown in Listing 13-7. This listing only shows the code for reading from the sensor.

Listing 13-7. Reading the Weather Data

```
Public Async Sub GetData()
    dataRead.degreesCelsius = tempAndPressure.Temperature.DegreesCelsius
    dataRead.degreesFahrenheit = tempAndPressure.Temperature.
    DegreesFahrenheit
    dataRead.bars = tempAndPressure.Pressure.Bars
    dataRead.hectopascals = tempAndPressure.Pressure.Hectopascals
    dataRead.atmospheres = tempAndPressure.Pressure.Atmospheres
End Sub
```

Next, we need to read the data using the GetData() method as a task and then write the data to the database. But before we do that, let's look at the setup portion of the code for the OnTick() method as shown in Listing 13-8. Recall, this is the method that is fired when the timer fires.

Listing 13-8. The OnTick() Method to Read the Data

```
Private Sub OnTick(sender As Object, e As Object)
    Dim msg As String
    Dim getDataTask As Task = New Task(AddressOf GetData)

    DBStatus.Text = "Reading data..."

    getDataTask.Start()  ' Start the task.
    getDataTask.Wait()   ' Wait for the data to be read.
```

---

[7]You could overcome this by changing the project template, but I like to use the blank app template in case I ever decide to add a user interface.

```
Debug.WriteLine(dataRead.degreesCelsius)
Debug.WriteLine(dataRead.degreesFahrenheit)
Debug.WriteLine(dataRead.bars)
Debug.WriteLine(dataRead.hectopascals)
Debug.WriteLine(dataRead.atmospheres)

Celcius.Text = dataRead.degreesCelsius.ToString("F2") + " C"
Fahrenheit.Text = dataRead.degreesFahrenheit.ToString("F2") + " F"
PressBars.Text = dataRead.bars.ToString("F2") + " Pressure (Bars)"
PressHecto.Text = dataRead.hectopascals.ToString("F2") + " Pressure
(Hectopascals)"
PressAtmos.Text = dataRead.atmospheres.ToString("F2") + " Pressure
(Atmospheres)"
```

...

Now, let's look at the rest of the code for this method.

Writing the Data to the Database

Writing data to the database is also easy to do. You create a string constant that you can use to fill in the data as parameters. Thus, you use the INSERT constant with the String.Format() method and pass in the data you read. You must put these values in the same order as the table columns. This means that you supply temperature in Celsius, temperature in Fahrenheit, bars, hectopascals, and atmospheres for the pressure.

Next, you create a new class instance for a MySQLCommand class passing the query you just formatted. Once the class is instantiated, you then call the method ExecuteNonQuery() to run the query. This method is the one you use if there are no results returned. Other methods are provided for reading data from the database server (see the documentation for more information).

You wrap all of this code in a try block so that you can capture any exceptions and display them using debug statements. Listing 13-9 shows the rest of the getData() method with the database code highlighted.

Listing 13-9. Writing Data to MySQL in the getData() Method

```
...
    Try
        ' Format the query string with data read
        Dim insert_str As String = String.Format(INSERT, dataRead.
        degreesCelsius,
                                                    dataRead.
                                                    degreesFahrenheit,
                                                    dataRead.bars,
                                                    dataRead.hectopascals,
                                                    dataRead.atmospheres)
        Debug.WriteLine(insert_str)
        ' Create a New command And setup the query
        Dim cmd As MySqlCommand = New MySqlCommand(insert_str, mysql_conn)
        ' Execute the query
        cmd.ExecuteNonQuery()
        msg = "Data inserted."
    Catch ex As Exception
        msg = "ERROR: " + ex.ToString()
    End Try
    Debug.Write(msg)
    DBStatus.Text = msg
End Sub
```

A debug statement was added to inform you that the data was inserted in the database. This could be helpful if you are debugging the code.

That's it! The code is complete and ready for compilation. Be sure to check the earlier listings to ensure that you have all the code in the right place. Once you have entered all the code, you should now attempt to compile the solution. Correct any errors you find until the code compiles without errors or warnings.

Deploy and Execute

Now we can deploy the application! Be sure to fix any compilation errors first. You may want to compile the application in debug first (but you can compile in release mode if you'd prefer). In fact, in order to see the debug messages, you will need to be using the debug build.

Recall from Chapter 7 (see the "Deploy and Debug" section and Figure 7-10), to deploy our application in debug mode, we must set the debug settings for our application, setting the following in the debug dialog:

- *Target device*: Remote Machine

- *Remote Machine*: The name of your device

- (Optional) *Uninstall and then re-install*: Checked

- (Optional) *Deploy optional packages*: Checked

- *Application process*: Managed Only

- *Background task process*: Managed Only

Once you make any changes in this dialog, be sure to save your solution. Building or deploying does not automatically save these settings.

We should also change the package name in the package manifest. Recall from Chapter 7 (see the "Deploying Your Application (No Debugging)" section and Figure 7-8), we can change the package name. Do that now so you can find your application easier in the Device Portal.

Now you can deploy your application. Go ahead and do that now. You should see the user interface appear on the device after a few moments and then the screen update as the timer tick event fires.

If you do not see the user interface or you see the box with the X in it, then the default application reloads, chances are there is something wrong in your code. In this case, you should run the code in debug mode and observe the exception thrown. There are a number of things that can go wrong. The following summarizes some of the things (mistakes) I've encountered in designing this project:

- Ensure that you are compiling the entire solution for *ARM*.

- Double-check your references to ensure they are correct.

- Ensure that the BMP280 sensor is wired correctly to your device.

- If you connected the sensor with 3.3V by mistake, it should still work, but double-check the GND pin is connected correctly.

- Use the Device Portal to start and stop the application after deployment and do not set the application to run as a startup until you have successfully run and debugged the code.

Once the project is running correctly, enjoy it for a while and then put the hardware aside. You will need it again for the next chapter.

Tip   If you encounter problems deploying your application, see the "Visual Basic Application Deployment Troubleshooting" section in Chapter 7 for help.

Now you can deploy your application. Go ahead and do that now. You can run the deployment from the *Debug* menu, and when you do, you see the debug statements in the output window. When you start the debugger, the output window automatically opens and displays the debug output. If it does not, you can open the window manually and select debug from the drop-down menu.

If you have errors or something doesn't work, you are most likely going to see problems in making the connection. This is more likely if you use a MySQL server set up differently than the default, managed by another entity, or you don't have the database server or user accounts set up properly. If this is the case, take the following advice and be sure to run all the steps in the previous sections on setting up the database for the project. Correct any errors until the code is running properly in debug.

WHY CAN'T I CONNECT?

One of the most frustrating experiences for those new to writing database applications is trying to discover why the code will not connect to the server. If this happens to you, you should resist the temptation to change your code or blame the connector or database server. Rather, you should perform the following diagnostic steps to discover the problem:

- Use the `mysql` client tool with the values from your connection string to ensure that you have the correct username, password, host, port, and so forth.

- If you can't connect, check your database server to ensure that the user account you're using has privileges to connect to the server (remember, the user and host must match).

- If you can connect, but cannot access the database objects, check the user privileges in MySQL to ensure that everything is set correctly.

- In some cases, you may need to modify your MySQL configuration file to remove the `bind_address` setting or modify your firewall to ensure that the port for MySQL is not blocked.

- Finally, ensure that there are no networking issues from your device to your server (e.g., isolated LAN segment).

When you have isolated the problem, change your connection string or database server settings, and try your code again.

If you run the application in debug and have it stating it is writing data to the database, you can issue a query to see the results, as shown in Listing 13-10.

Listing 13-10. Example Weather Data

```
mysql> select * from weather.history \G
*************************** 1. row ***************************
            id: 52
  dateRecorded: 2020-07-29 17:32:59
 degreesCelsius: 24.69
degreesFarenheit: 76.442
          bars: 1.00965
```

```
      hectopascals: 1009.65
       atmospheres: 0.996448
*************************** 2. row ***************************
               id: 53
     dateRecorded: 2020-07-29 17:33:04
   degreesCelsius: 24.69
degreesFarenheit: 76.442
             bars: 1.00963
     hectopascals: 1009.63
      atmospheres: 0.996428
*************************** 3. row ***************************
               id: 54
     dateRecorded: 2020-07-29 17:33:10
   degreesCelsius: 24.68
degreesFarenheit: 76.424
             bars: 1.00962
     hectopascals: 1009.62
      atmospheres: 0.996416
*************************** 4. row ***************************
               id: 55
     dateRecorded: 2020-07-29 17:33:15
   degreesCelsius: 24.68
degreesFarenheit: 76.424
             bars: 1.00963
     hectopascals: 1009.63
      atmospheres: 0.996423
*************************** 5. row ***************************
               id: 56
     dateRecorded: 2020-07-29 17:33:20
   degreesCelsius: 24.68
degreesFarenheit: 76.424
             bars: 1.00961
     hectopascals: 1009.61
      atmospheres: 0.996409
```

```
*************************** 6. row ***************************
            id: 57
  dateRecorded: 2020-07-29 17:33:25
 degreesCelsius: 24.68
degreesFarenheit: 76.424
          bars: 1.00961
   hectopascals: 1009.61
    atmospheres: 0.996411
*************************** 7. row ***************************
            id: 58
  dateRecorded: 2020-07-29 17:33:30
 degreesCelsius: 24.69
degreesFarenheit: 76.442
          bars: 1.00963
   hectopascals: 1009.66
    atmospheres: 0.996456
*************************** 8. row ***************************
            id: 59
  dateRecorded: 2020-07-29 17:33:35
 degreesCelsius: 24.68
degreesFarenheit: 76.424
          bars: 1.00962
   hectopascals: 1009.62
    atmospheres: 0.996413
*************************** 9. row ***************************
            id: 60
  dateRecorded: 2020-07-29 17:33:40
 degreesCelsius: 24.68
degreesFarenheit: 76.424
          bars: 1.00964
   hectopascals: 1009.64
    atmospheres: 0.996441
```

```
*************************** 10. row ***************************
              id: 61
    dateRecorded: 2020-07-29 17:33:46
   degreesCelsius: 24.68
degreesFarenheit: 76.424
            bars: 1.00963
     hectopascals: 1009.63
      atmospheres: 0.996431
10 rows in set (0.09 sec)
```

There are several things in this listing to note. First, the id column is filled in for you (you passed null for this column in the insert statement) as well as the timestamp column. Not only can you see the values change, you can also see when the sample was taken. Thus, you can issue queries later, such as the following (and many more):

- Dates when the temperature exceeded a certain value

- Average temperature for a day, week, month, and so forth

- Plot temperature or pressure using a spreadsheet program

Tip   If you are interested in learning how to issue queries to find data for these or similar questions, see *MySQL Cookbook: Solutions for Database Developers and Administrators* 3rd Edition by Paul DuBois (O'Reilly, 2014).

If you want to reset the data, you can issue the following command to empty the table, thereby purging the data. Use this command with caution as a DELETE without a WHERE clause has ruined many database administrators' day. Once run, you can't undo it!

```
DELETE FROM weather.history;
```

Once you have the application working correctly, you can deploy it to your device and start or stop the application from the *Apps* tab on the Device Portal using the *Actions* drop-down box, as shown in Figure 13-8.

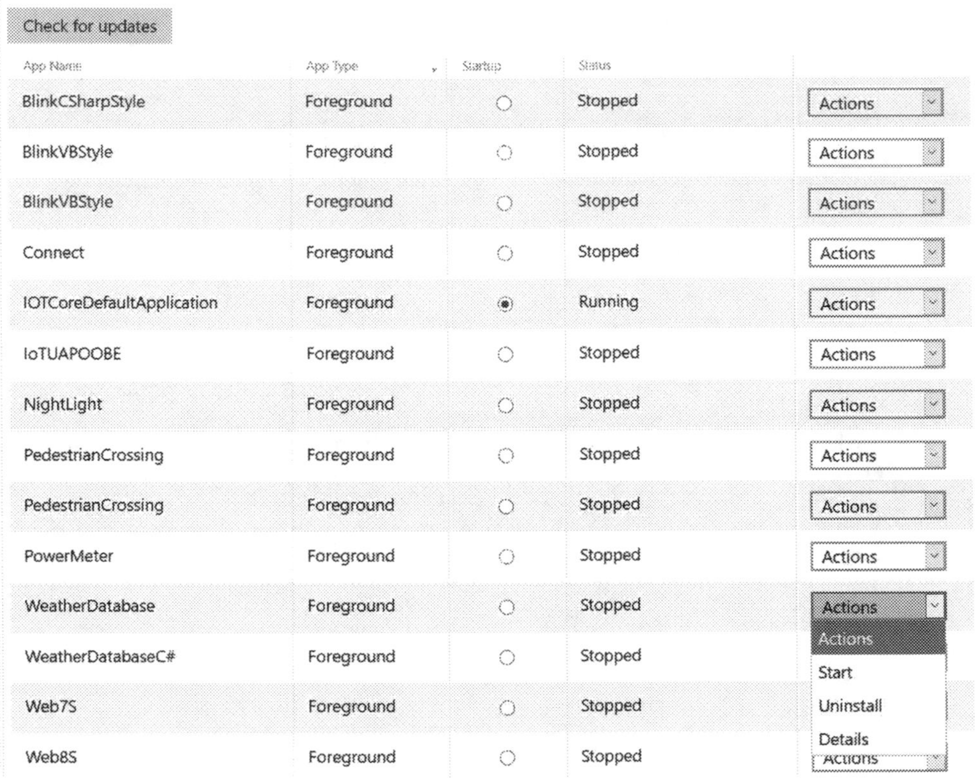

Check for updates				
App Name	App Type	Startup	Status	
BlinkCSharpStyle	Foreground	○	Stopped	Actions ▾
BlinkVBStyle	Foreground	○	Stopped	Actions ▾
BlinkVBStyle	Foreground	○	Stopped	Actions ▾
Connect	Foreground	○	Stopped	Actions ▾
IOTCoreDefaultApplication	Foreground	◉	Running	Actions ▾
IoTUAPOOBE	Foreground	○	Stopped	Actions ▾
NightLight	Foreground	○	Stopped	Actions ▾
PedestrianCrossing	Foreground	○	Stopped	Actions ▾
PedestrianCrossing	Foreground	○	Stopped	Actions ▾
PowerMeter	Foreground	○	Stopped	Actions ▾
WeatherDatabase	Foreground	○	Stopped	Actions ▾
WeatherDatabaseC#	Foreground	○	Stopped	Actions / Start / Uninstall / Details
Web7S	Foreground	○	Stopped	
Web8S	Foreground	○	Stopped	Actions ▾

Figure 13-8. *Starting and stopping apps using the Device Portal*

Now that you have the project working, rejoice! You have just completed the most sophisticated project in this book and have learned how to leverage external storage to keep your data in an organized, reliable storage medium—MySQL!

Summary

Now that you've seen how easy it is to save data to MySQL from your IoT projects, you can leverage the power and convenience of structured storage in a database in your own IoT projects. In fact, there are very few IoT projects that generate meaningful data that I would not consider using a database to store the data. Sure, there are plenty of projects that generate data that is only meaningful in the current context, but those IoT projects that I refer to are those that help you understand the world by drawing conclusions from the data. The best way to do that is to process the data using tools designed for such analysis, and by putting the data in the database, you open the door to a host of tools that can read and process the data (including your own code!).

In this chapter, you saw how to add a database component to a Visual Basic project. In the next chapter, you look at a different technique for IoT solutions—controlling hardware remotely via the Internet through network programming using sockets. The project demonstrates what is possible with only a little bit of work.

CHAPTER 14

Project 5: Remote Control Hardware

Some of the more interesting IoT solutions are those that implement actionable devices or features that allow you to control hardware remotely. This could be controlling a remote control toy (car, truck, drone), moving a camera (pan, tilt, zoom), opening or closing a gate, and so forth or simply controlling lights or locks remotely.

That's the real key and trick to the solution—making the solution available remotely. One of the best ways to do that is to use networking protocols running on the device and control it using an application on another PC or device (via the user interface). More specifically, you implement a server that supports actions (events) that allow you to control the actionable devices, lights, locks, and so forth from a client.

In this chapter, you will see one method for building IoT solutions that control hardware remotely using TCP sockets. You have already discovered the tools and techniques for turning LEDs on/off, reading data from sensors, and implementing PWM-controlled devices.

Unlike previous projects, the code for this project is not overly complicated, but it is different from the techniques you've seen. That is, you will use C# and a library from NuGet to remotely control hardware. There is an optional part that requires some soldering and cutting or drilling to assemble, making the hardware for this project a bit more complicated than previous projects. Regardless, you can take the concepts presented and make all manner of interesting solutions. Let's get started.

© Charles Bell 2021
C. Bell, *Windows 10 for the Internet of Things*, https://doi.org/10.1007/978-1-4842-6609-0_14

Overview

If you work in a cube farm[1] or have a similar office arrangement, you may enjoy this project. You're going to implement an out-of-office sign that you can use to inform your coworkers whether you are in your office or not. To make it interesting, you will use a servo to raise or lower a flag for IN or OUT. You will combine that with a set of LEDs that you can use to indicate why you are out. More specifically, you will use the following states:

- *Available*: You are available.

- *Do not disturb*: You do not want to be interrupted.

- *Out to lunch*: You are away from your desk on lunch break.

- *Be back later*: You are away from your desk but expect to return.

- *Gone for the day*: You are not returning to your desk until the next business day.

A servo operates with pulse-width modulation (PWM).[2] That is, the faster the pulses you send it, the more (further) it will rotate. Typically, you would choose specific patterns to move the servo to one of several positions. Servos are used in all manner of solutions from mechanical movements in toys, robots, remote control planes, cars, and even 3D printers. Basically, if you need a small motor to move a lever, rotate something, steer, or move something in a precision manner, you may want to use a servo.

The servo that you use in this project is a typical miniature servo that you can find at most online electronics stores, including Adafruit (`www.adafruit.com/products/169`), SparkFun (`www.sparkfun.com/products/9065`), and online auction sites. Figure 14-1 shows a typical micro hobby servo.

---

[1]See `http://dilbert.com/search_results?terms=Cube+Farming`.
[2]See `https://en.wikipedia.org/wiki/Pulse-width_modulation`.

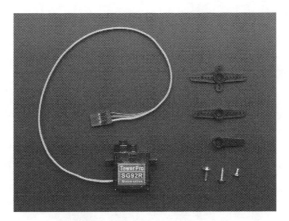

Figure 14-1. *Micro servo (courtesy of Adafruit)*

As you can see in the photo, servos typically come with a number of arms you can use to connect a thin wire or rod from the servo to another component that you want to move. Servos have a range of motion of 90 degrees. A special form of servo—called a *continuous rotation servo*—can rotate 360 degrees. For this project, the normal 90-degree range of motion is all that you need.

When you use a servo, you must discover the positions you want for the feature. For example, in this project, you simply want to raise a flag, so you need to know only two positions: one for when the flag is down and one for when the flag is raised. To do this, you need to know how to use PWM in your projects. Fortunately, you have seen an example of how to do this. Recall that you saw how to use PWM in Chapter 11. However, you're going to cheat a little and use a nifty new product from SparkFun called a *Servo Trigger* (`www.sparkfun.com/products/13118`).

The Servo Trigger is a breakout board that has three small potentiometers that you can use to control the fully counterclockwise position (or "off") and fully clockwise position (or "on"/triggered) as well as the speed at which the servo turns. This is perfect for this project because you only need the two positions. Furthermore, since the board is designed to use a simple trigger event, you can use a single GPIO pin to toggle the board. How cool is that? Figure 14-2 shows the Servo Trigger with the connections highlighted.

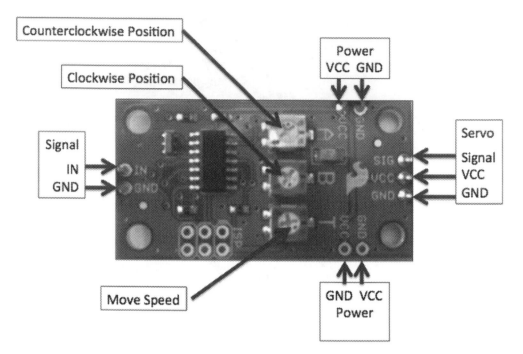

Figure 14-2. *Servo Trigger from SparkFun*

On the left is the signal connector with only two wires—a ground and signal. On the right is the connector for the servo. On the top and bottom are power connectors for the servo. This allows you to power the servo using 5V (most servos can operate on 3–9V). There are two connectors so that you can wire several Servo Triggers in series, so you can use less wire for connecting power.

While the Servo Trigger makes using a servo for this project really easy, it does not come with any headers soldered. Thus, you will have to solder headers to the board. At a minimum, you need male headers on the signal, one of the power connections, and the servo connector. Soldering is not difficult, but if you do not know how to solder, ask a friend to help you. If nothing else, you now have a really good reason to learn how to solder!

Note   The Servo Trigger is customized via the programming interface. See the product site on SparkFun for a complete guide (called the *hookup guide*).

Servos have three wires: one for the signal, one for ground, and another for power. These are normally colored brown, orange, and yellow. However, not all servos have the same colored wires. Table 14-1 shows some of the possible color schemes that vendors use (but there are many such variations). Just make sure that you don't connect it backward! It is always best to check with your vendor on how to wire your servo.

Table 14-1. *Servo Wires*

Ground (Pin 1)	Power (Pin 2)	Signal (Pin 3)
Black	Red	White
Black	Brown	Yellow
Black	Brown	White
Black	Red	Yellow
Brown	Red	White
Brown	Orange	Yellow

Tip   To learn more about servos, see `jameco.com/jameco/workshop/howitworks/how-servo-motors-work.html`.

You will also use a number of LEDs to indicate a message for your coworkers. That is, when the LED is on, it indicates the current state. For example, if the flag is set to IN and the LED for "Do not disturb" is turned on, you are in the office but working on things that require your attention (such as a phone call or similar meeting) and do not want to be interrupted.

Finally, you will use C# to write two solutions: a server that runs on the device and an application that runs on your PC that, together, allow you to control the hardware.

What you should gain from this project is how to write small applications to control hardware via TCP sockets, how to use a servo (with help from SparkFun), and how to create specialized network-based solutions in Windows 10 IoT Core.

The really fun part of this project is building the sign itself. You have two options: (1) build the circuit on the breadboard, which allows you to explore all the basic concepts without the extra work, or (2) build the solution in an enclosure using a simple cardboard box that you can mount the LEDs, servo, and flag. You can hang this box on your cubical wall so that visitors can see your office status at a glance. However, this option does require a bit of soldering.

I present both options; thus, I recommend doing the first (on a breadboard) and then later build the solution in an enclosure. But first, let's look at the hardware needed for this project.

Required Components

The following lists the components that you need. You can get the LEDs and servo from Adafruit (www.adafruit.com), SparkFun (www.sparkfun.com), or any electronics store that carries electronic components. The Servo Trigger board is available from SparkFun (www.sparkfun.com/products/13118). Since this solution is a headless application, you need a monitor, a keyboard, and a mouse.

- SparkFun Servo Trigger (www.sparkfun.com/products/13118)
- Servo (www.adafruit.com/products/169 or www.sparkfun.com/products/9065)
- (4) red LEDs
- (1) green LED
- (5) 150 ohm resistors
- Jumper wires: (10) male-to-female, (1) male-to-male
- Breadboard (full size recommended but half size is OK)
- Raspberry Pi 2 or 3
- Power supply

Tip   The servo mount and flag are available in the source code download. They are 3D printer STL files that you can download and print or have printed for you using a 3D printing service.

Set Up the Hardware

There are a number of connections needed for this project, and as usual, you will plan for how things should connect. To connect the components to the Raspberry Pi, you need five pins for the LEDs, one pin for the Servo Trigger, 5V power, ground, and four jumpers to complete the connections to the Servo Trigger breakout board. Table 14-2 shows the map I designed for this project.

Table 14-2. *Connection Map for Out-of-Office Project*

GPIO	Connection	Function	Notes
5V (2)	Power for Servo Trigger	Power to breakout board	
GND (6)	Ground for Servo Trigger, LEDs	GND on breakout board	
20	Servo Trigger Signal	Engage Servo	
21	Green LED	LED on	
22	Red LED #1	LED on	
23	Red LED #2	LED on	
24	Red LED #3	LED on	
25	Red LED #4	LED on	

Next, you need to make a number of connections on the breadboard for joining the ground rails together and connecting the Servo Trigger. Table 14-3 shows the connections needed.

Table 14-3. *Connections on the Breadboard*

From	To	Notes
Breadboard GND Rail#1	Breadboard GND Rail#2	
Breadboard Power	Servo Trigger VCC	
Breadboard GND	Servo GND (power in)	
Breadboard GND	Servo GND (signal)	
Servo	Servo Trigger servo header	

Figure 14-3 shows all the connections needed.

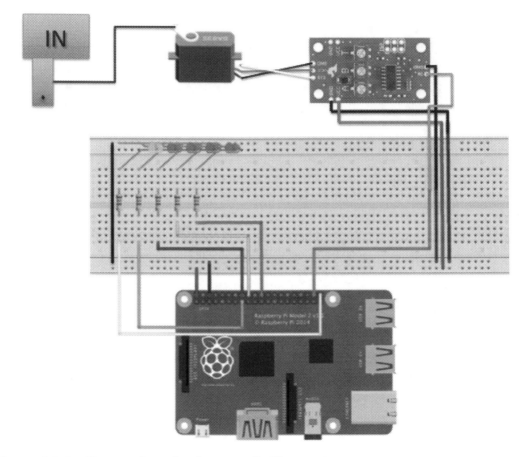

Figure 14-3. *Connections for the out-of-office project*

Here, you see the breadboard implementation of the project. Recall from earlier discussions, this is how most projects start out—as a set of circuits implemented on a breadboard before being transferred to an enclosure, a printed circuit board designed, and so forth.

There are GPIO connections for each of the LEDs with the negative lead plugged into the breadboard for each LED. You also see a GPIO pin for the Servo Trigger as well as power connections for the Servo Trigger and ground for the LEDs. Lastly, you see the servo depicted connected to the Servo Trigger and an arm connected to the flag to raise and lower it on command.

If you are following along with this chapter working on the project, go ahead and make the hardware connections now. Don't power on the board yet but do double- and triple-check the connections.

COOL GADGET

SparkFun has a really neat adapter for working with the GPIO header on a Raspberry Pi with a breadboard. It's called a Raspberry Pi Wedge (`www.sparkfun.com/products/13717`). It comes with a 40-pin ribbon cable that you connect to your Raspberry Pi, thereby allowing you to position the Raspberry Pi farther away from the breadboard or simply reposition the device without disturbing the breadboard circuits.

What I like most about it is it plugs into the breadboard along the DIP channel, allowing you to connect more than one jumper to the pins, which is really helpful for connecting power or ground. The following is a photo of the Pi Wedge.

If you use this device, you can save yourself some time looking for the correct GPIO pin since they are clearly marked on the board (but they are not in the same order as the GPIO header on the board but, in my opinion, grouped together a bit more logically).

Write the Code

Now it is time to write the code for the project. Since it is written in C#, it is very easy to follow; however, the networking portion is not as intuitive as some of the previous projects. Recall, we will create two projects: one for the server that runs on the device (Raspberry Pi) and another that runs on your PC (or another IoT device). Let's begin with creating the server, walking through each of the parts that you need to implement. We'll see the client solution in a later section.

Don't worry if you've never written any code for working with networking protocols like TCP sockets. We'll be using a special library that makes it very easy to use. As you will see, you only need to know the basics of socket programming.

Tip   For more information and a concise overview of network programming, see www.codeproject.com/Articles/10649/An-Introduction-to-Socket-Programming-in-NET-using or www.tutorialspoint.com/Socket-Programming-in-Chash.

Writing the Server Application

You will use a new project template. In this case, we want the application to run in the background, so we will use the *C# Background Application (IoT)* template. Use the name *OfficeSignServer* for the project name. You can save the project wherever you like or use the default location. The majority of the code for this project will be implemented in its of class named Server. But before we write that code, let's add the resources we need for the project.

The resources required include two libraries we can get from NuGet. First, we need the *EasyTcp* library, which encapsulates the code needed to work with TCP sockets (making it very easy). We also need the *System.Text.Json* library, which enables us to work with JavaScript Object Notation (JSON[3]). Let's see how to get and install each of these.

---

[3]https://en.wikipedia.org/wiki/JSON

EasyTcp

EasyTcp is a library that supports working with TCP sockets. It also takes care of the low-level communication for you. All you need to do is instantiate the class and call the methods to read the data to/from the network. We will use version 3.6.3 of this library for this project. Newer versions may not work with the project code.

Tip   See `https://github.com/Job79/EasyTcp` to learn more about EasyTcp.

To add the *EasyTcp* library, use the NuGet Package Manager from the *Tools* ➤ *NuGet Package Manager* ➤ *Manage NuGet Packages for Solution…* menu. Click the *Browse* tab and then type `easytcp` in the search box, as shown in Figure 14-4.

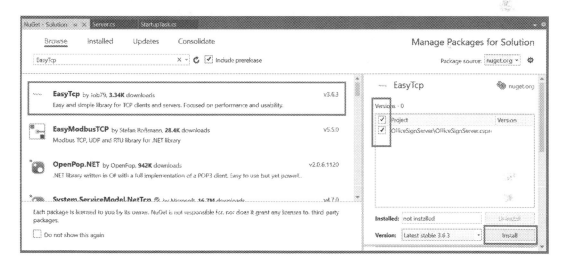

Figure 14-4. *NuGet Package Manager, results for EasyTcp*

Select the entry named *EasyTcp* in the list, choose version 3.6.3 in the *Versions* list, tick the project name (solution) in the list on the right, and finally click *Install*. The installation starts and Visual Studio downloads a number of packages and then asks your permission to install them. A dialog box will open to tell you that the installation is complete.

System.Text.Json

System.Text.Json is a library that supports JSON objects, which we will use for our communication protocol between the server and the client. That is, we will pass JSON data that has a command inside that tells the server how to control the hardware (the sign). Cool, eh?

To add the *System.Text.Json* library, use the NuGet Package Manager from the *Tools ➤ NuGet Package Manager ➤ Manage NuGet Packages for Solution…* menu. Click the *Browse* tab and then type `system.text.json` in the search box, as shown in Figure 14-5.

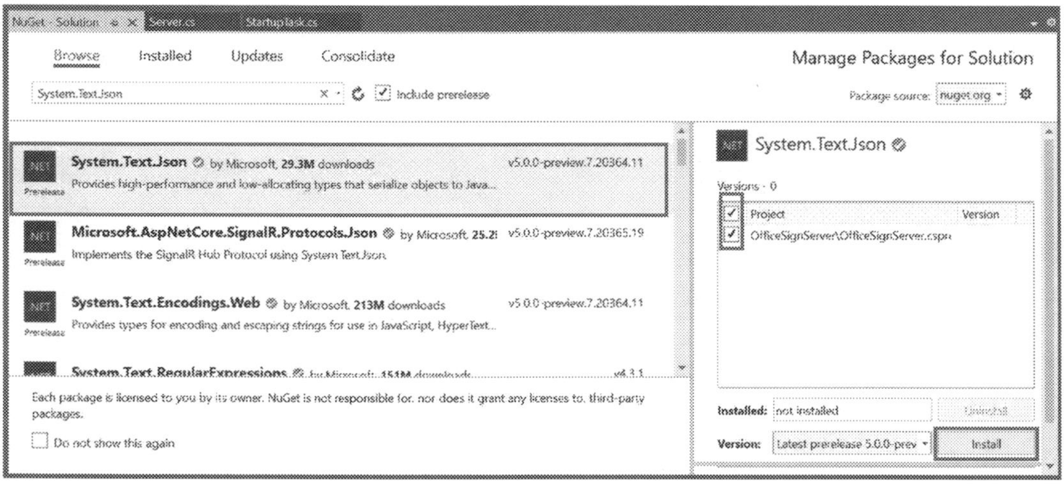

Figure 14-5. *NuGet Package Manager, results for System.Text.Json*

Select the entry named *System.Text.Json* in the list, tick the project name (solution) in the list on the right, and finally click *Install*. The installation starts and Visual Studio downloads a number of packages and then asks your permission to install them. Go ahead and let the installation complete. A dialog box will open to tell you that the installation is complete.

OK, now we have the references we need. Now, let's write the code for the server class.

Server Class

To add the server class, right-click the project and choose *Add ➤ New Item…*. Choose the class entry in the list and name it `Server.cs` and then press the *Add* button as shown in Figure 14-6.

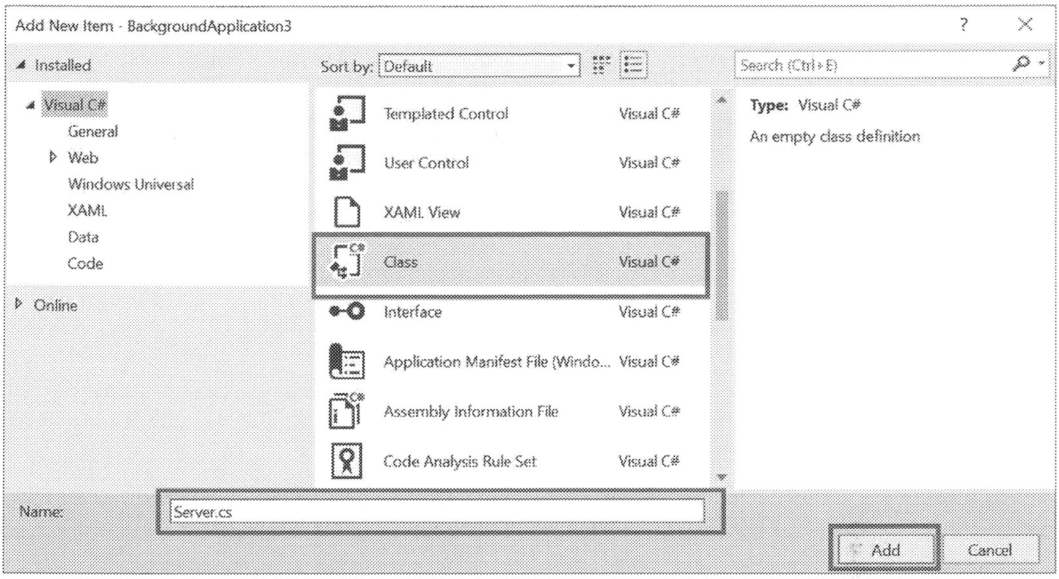

Figure 14-6. Adding the Server class

Once the dialog closes, double-click the Server.cs file. This is where you put most of the code for the project for working with the TCP socket. We will also place the code for the GPIO interactions in this class since it doesn't require many lines of code. If the project were more complex (or for your own projects), you may want to consider placing the GPIO code in its own class. This choice is typical of what a developer would encounter and is often a balance between convenience (all code in one class) and complexity, maintainability, or readability.

We'll begin with the namespaces. Add the following namespaces. Here, we see we need to use namespaces for the GPIO to control the hardware, diagnostics for our debug statements, the JSON library, and several from the EasyTcp library. Go ahead and add those now.

```
using Windows.Devices.Gpio; // Add using clause for GPIO
using System.Diagnostics;    // Add for debug
using System.Text.Json;
using EasyTcp3.Server;
using EasyTcp3.Server.ServerUtils;
using EasyTcp3;
```

Next, we will use a small, embedded (sometimes called nested) class to hold the information about the office sign—specifically, the state of the flag (controlled by the servo) and the location (as indicated by the LEDs). We'll call this class OfficeStatus and place it after the namespace statement as shown in the following:

```
namespace OfficeSignServer
{
    public sealed class OfficeStatus
    {
        public bool inOffice { get; set; }
        public int location { get; set; }
    }
    public sealed class Server
    {
...
```

Notice we use the C# get and set directives to automatically generate the get() and set() methods for the class. This will come in handy later when we read the data from the client.

Next comes the code for the Server class itself. We will need a number of constants and variables. We will use constants for the LED pin numbers, an array to hold the values to make it easier to loop over the LEDs (we'll see that code later), and variables for storing the port and hostname (IP). We also need to define variables for the GPIO controller as well as the servo and LED pins. Listing 14-1 shows the constants and variables for the class.

Listing 14-1. Server Class Constants and Variables

```
    public sealed class Server
    {
        private ushort port;
        private string hostname;

        // Constants for GPIO pins
        private const int SERVO_PIN = 20;
        private const int AVAIL = 21;
        private const int DND = 22;
        private const int BBL = 23;
```

```
    private const int LUNCH = 24;
    private const int GFTD = 25;
    private int[] LED_PIN_VALUES = new int[] { AVAIL, DND, BBL, LUNCH,
    GFTD };

    // Add variables for the GPIO
    private GpioController gpio_ctrl;
    private GpioPin servoPin;
    private GpioPin[] ledPins = new GpioPin[5];
```
...

Next, we will need a few methods including a constructor, methods to set up the GPIO, start the server, and respond to incoming messages. Let's look at each in turn.

The constructor simply accepts the port and hostname and saves them to the embedded class we defined earlier as shown in the following. This permits us to declare a variable (instance) of the class and pass in this data without having to supply it later. Recall, constructors are named the same as the class.

```
public Server(string hostname, ushort port)
{
    this.port = port;
    this.hostname = hostname;
}
```

Next, we will look at the method that starts the server. We will name this method StartServer. This method must create a new instance of the EasyTcpServer class and call its Start() method passing in the hostname and port. Finally, we assign the OnDataReceive method address/callback to a new method we'll create that processes the incoming messages. Listing 14-2 shows the StartServer() method.

Listing 14-2. Server Class Start Server Method

```
public void StartServer()
{
    // Create new easyTcpServer
    var server = new EasyTcpServer();

    // Start server
    server.Start(hostname, port);
```

```
// Process all receiving data
server.OnDataReceive += OnMessageReceived;
}
```

Before going any further, you may be wondering about what sort of message the server class will receive. This message is a data stream that originates in the client application where a string is created in JSON format and sent to the server. The server then accepts the strings, decodes them, and uses the information to manipulate the servo and turn the LEDs on or off. The following is a sample message (in human-readable form) that the client may send:

```
{
    inOffice: true,
    location: 0
}
```

Notice there are two attributes: inOffice, which is a Boolean that we can use to set the flag, and location, which is an integer. The client sets the location based on the radio button chosen for the location. We use an integer starting with 0 so that it can be used to reference the LEDs in the array. That way, what is sent to the server matches the interface on the client. That is, when the radio button at array index 0 is ticked, LED index 0 is turned on. While this message is quite simple, it shows the power you can build into your client/server applications and the level of sophistication you can achieve to pass data.

Notice something familiar here? Yes, that's right. The attributes in the JSON string are the same names as those in the embedded class we created earlier. If you're thinking that is significant, it is! More on that in a moment.

Next, let's look at the method to set up the GPIO on the server. Here, we name the method InitGPIO. Here, we need to do the same sort of operations we've seen in the previous project chapters. The difference is which pins are used for the LEDs and servo. Listing 14-3 shows the code for the method. I leave the explanation and description of the contents as an exercise, but it all should be familiar.

Listing 14-3. Server Class Initialize GPIO Method

```
public void InitGPIO()
{
    // Initialize the office board
    gpio_ctrl = GpioController.GetDefault();

    // Check GPIO state
    if (gpio_ctrl == null)
    {
        Debug.Print("ERROR: No GPIO controller found!");
        return;
    }
    // Setup the Servo pin
    this.servoPin = gpio_ctrl.OpenPin(SERVO_PIN);
    if (servoPin == null)
    {
        Debug.Print("ERROR: Can't get servo pin!");
        return;
    }
    this.servoPin.SetDriveMode(GpioPinDriveMode.Output);
    this.servoPin.Write(GpioPinValue.Low); // turn off
    // Setup the LED pins
    for (int i = 0; i < 5; i++)
    {
        this.ledPins[i] = gpio_ctrl.OpenPin(LED_PIN_VALUES[i]);
        if (this.ledPins[i] == null)
        {
            Debug.Print("ERROR: Can't get led pin " + LED_PIN_VALUES[i] +
            "!");
            return;
        }
        this.ledPins[i].SetDriveMode(GpioPinDriveMode.Output);
        this.ledPins[i].Write(GpioPinValue.Low); // turn off
    }
```

```
    // Set the default to "IN" and "Available"
    this.ledPins[0].Write(GpioPinValue.High);
    Debug.Print("Good to go!");
}
```

Finally, we add a method to process the messages from the client. We will name this method OnMessageReceived. This method is responsible for deciphering the message from the client by converting the JSON string into an instance of the OfficeStatus class. Yes, you do that! That's part of what makes using JSON so nice—we can convert it to a C# class and call our get/set methods on it.

To convert the JSON string to the OfficeStatus class, we use the JsonSerializer. Deserialize() template method using the class as the type pass. The following is an example:

```
OfficeStatus = JsonSerializer.Deserialize<OfficeStatus>(message.ToString());
```

After the message is received and deserialized, we can then change the servo and LEDs to correspond with the data received. Listing 14-4 shows the complete OnMessageReceived() method.

Listing 14-4. Server Class OnMessageReceived Method

```
private void OnMessageReceived(object sender, Message message)
{
    // Here we want to take the message received, deserialize the JSON and
    store it
    // in our class.
    Debug.Print("Message received: " + message.ToString() + "\n");
    OfficeStatus officeStatus = JsonSerializer.Deserialize<OfficeStatus>(
    message.ToString());
    if (officeStatus.inOffice)
    {
        this.servoPin.Write(GpioPinValue.Low);  // turn off
    }
    else
    {
        this.servoPin.Write(GpioPinValue.High);  // turn off
    }
```

```
    // Set the LED pins
    for (int i = 0; i < 5; i++)
    {
        if (i == officeStatus.location)
        {
            this.ledPins[i].Write(GpioPinValue.High); // turn off
        }
        else
        {
            this.ledPins[i].Write(GpioPinValue.Low); // turn off
        }
    }
}
```

Notice we used a loop to loop over the array of LEDs. We set them on or off depending on the value of the location attribute. If location is 0, the first LED is turned on and the reset turned off. Cool, eh?

Well, that's the complete code for the Server class. Now, let's look at the code for the base application, which is a special background task class known as the StartupTask.

StartupTask

The StartupTask class was created for us when we created the project from the template. And, since we put all of the code for the GPIO in the Server class, we need only create a new instance of the Server class, initialize the GPIO, and start it. Listing 14-5 shows the complete code for the StartupTask class. That was easy, yes?

Listing 14-5. StartupTask Class

```
using Windows.ApplicationModel.Background;

namespace OfficeSignServer
{
    public sealed class StartupTask : IBackgroundTask
    {
        private static BackgroundTaskDeferral _Deferral = null;
```

```
        public void Run(IBackgroundTaskInstance taskInstance)
        {
            _Deferral = taskInstance.GetDeferral();
            Server server = new Server("192.168.42.15", 13001);
            server.InitGPIO();
            server.StartServer();
        }
    }
}
```

But there is a trick here. Do you see it? Yes, it's that deferral thing. It turns out, Windows 10 IoT Core will only allow a background task to run so long and then it gets stopped. To prevent that from happening, you can simply declare a variable of type BackgroundTaskDeferral and retrieve it with taskInstance.GetDeferral(). This will keep the operating system from closing the application after a timeout.

Tip   For more information about the background deferral, see https://docs.microsoft.com/en-us/uwp/api/windows.applicationmodel.background.backgroundtaskdeferral?view=winrt-19041.

Also, notice the use of the IP address 192.168.42.15 and port 13001. This information comes from your IoT device. It should match the IP address of where you plan to run the server (on your Raspberry Pi), **not** your PC. The port is largely arbitrary, but it should match the one you use in the client. In fact, we will see the same port and hostname (IP) when we look at the client application.

OK, now we're ready to look at the client application.

Writing the Client Application

The client application is a Universal Windows application also written in C#, so we will use the *C# Blank App (Universal Windows)* template. Use the name *OfficeSignClient* for the project name. You can save the project wherever you like or use the default location or save it with the server project.

This application is similar to the server application with the addition of the user interface. We will use the same libraries and create a client class named `Client`. However, unlike the server application, there is no GPIO code in the `Client` class because there is no GPIO! Rather, we will have a user interface that we will use to form a message that is sent to the server. Thus, the client portion of the networking code is quite basic.

Rather than repeat the steps for installing the NuGet packages, refer to the preceding sections named "EasyTcp" and "System.Text.Json" to install the packages. Once that is done, you're ready to code the client class.

Client Class

The client class is used to send messages to the server in response to actions (radio buttons selected) in the user interface. Recall, we will form a JSON string (document) that includes the state of the two radio button groups: the in- or out-of-office indicator and the location indicator. Recall, the server expects the JSON string to have two attributes: *inOffice* and *location*. We will use the actions in the user interface to generate the values. For example, if we want to indicate we are out of the office and at lunch, the JSON string would be as follows:

```
{
    inOffice: false,
    location: 3
}
```

To add the client class, right-click the project and choose *Add* ➤ *New Item....* Choose the class entry in the list and name it `Client.cs` and then press the *Add* button as shown in Figure 14-7.

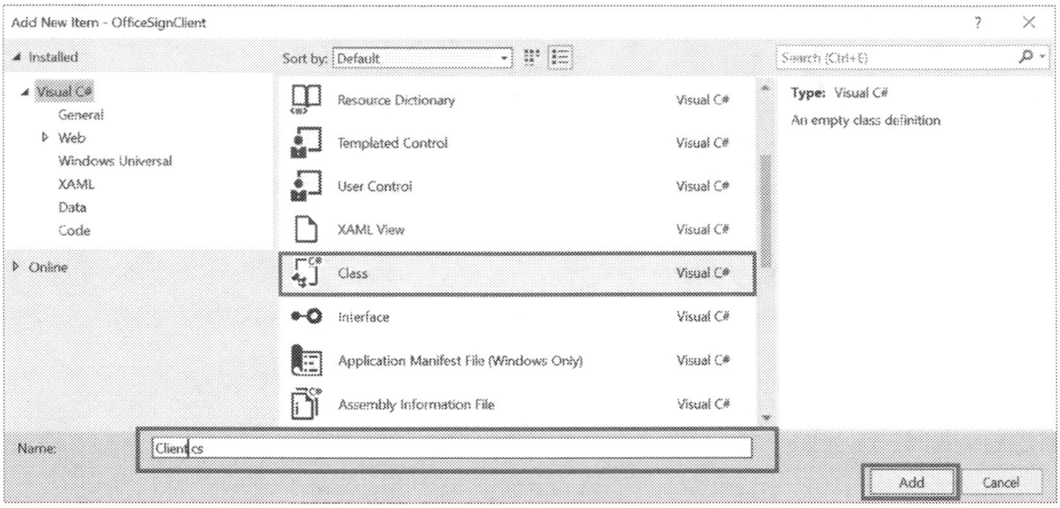

Figure 14-7. *Adding the Client class*

Once the dialog closes, double-click the Client.cs file. This is where you put most of the code for the project for working with the TCP socket. In this case, the class will simply send the message to the server. The user interface will be responsible for building the JSON string (message) to send to the server.

We'll begin with the namespaces. Add the following namespaces. Here, we see we need to use diagnostics for our debug statements, the JSON library, and several from the EasyTcp library. Go ahead and add those now.

```
using System;
using EasyTcp3.ClientUtils;
using System.Diagnostics;
using EasyTcp3;
```

Next comes the code for the Client class itself. We will need a number of constants and variables. We will use variables for storing the port and hostname (IP). We also need to define a variable for the EasyTcp client. The following shows the start of the class. Note that the class definition adds the public and sealed keywords.

```
namespace OfficeSignClient
{
    public sealed class Client
    {
        private ushort port;
```

```
    private string hostname;
    private EasyTcpClient client;
```
...

Next, we will need the constructor method to set the port and host variables on instantiation as shown in the following:

```
public Client(string hostname, ushort port)
{
    this.port = port;
    this.hostname = hostname;
}
```

We need only two other methods: one to connect to the server and another to send the message. We name the connect method simply Connect(). This is the method that has the most work. In this method, we must create an instance of the EasyTcp client, then attempt to connect to the server using the host and port passed when the client class is created. If we are not successful, we throw an exception, or print a debug message that we've connected successfully. Listing 14-6 shows the complete Connect() method.

Listing 14-6. Connect Method for the Client Class

```
public void Connect()
{
    // Create new easyTcpClient
    client = new EasyTcpClient();

    // Connect client
    // Make the app fail if you cannot connect.
    if (!client.Connect(hostname, port))
    {
        throw new Exception("ERROR: Cannot connect to the server. Check
        connections and IP address!");
    } else {
        Debug.Print("Connected to server!");
    }
}
```

The method where we send the message to the server is named `SendMessage()`. In this method, we check to see if the client is connected, and if not, we print a debug message and return. If it is connected, we send the message. Yep, it's that simple! Listing 14-7 shows the complete code for the method.

Listing 14-7. SendMessage Method for the Client Class

```
public void SendMessage(String message)
{
    if (!client.IsConnected())
    {
        Debug.Print("Cannot send the message. Client is not connected!");
        return;
    }
    client.Send(message);
    Debug.Print("Message " + message + " sent.\n");
}
```

Now that the client class is complete, let's look at the code for the user interface.

User Interface

The user interface for the application is a simple affair consisting of seven radio buttons. Two are grouped together in a series allowing you to choose one or the other. This will be used for the IN/OUT flag. The remaining five radio buttons are grouped together as well, allowing the user to choose one of five possible locations. Listing 14-8 shows the code for the XAML interface.

Listing 14-8. User Interface for the Client Class

```
<Page
    x:Class="OfficeSignClient.MainPage"
    xmlns="http://schemas.microsoft.com/winfx/2006/xaml/presentation"
    xmlns:x="http://schemas.microsoft.com/winfx/2006/xaml"
    xmlns:local="using:OfficeSignClient"
    xmlns:d="http://schemas.microsoft.com/expression/blend/2008"

    mc:Ignorable="d"
    Background="{ThemeResource ApplicationPageBackgroundThemeBrush}">
```

```
<Grid Background="{ThemeResource ApplicationPageBackgroundThemeBrush}">
    <StackPanel Width="400" Height="350">
        <TextBlock x:Name="Title" Height="40" TextWrapping="NoWrap"
        Text="Office Sign - Client" FontSize="20" Foreground="Blue"
        HorizontalAlignment="Center"/>
        <Line X1="0" Y1="0" X2="400" Y2="0" Stroke="White"
        StrokeThickness="2" />
        <TextBlock x:Name="Status" Height="20" TextWrapping="NoWrap"
        Text="1) Choose Status" FontSize="16" Foreground="Blue"
        HorizontalAlignment="Left"/>
        <StackPanel Orientation="Vertical" Width="347" >
            <RadioButton x:Name="radioIN" Content="IN"
            GroupName="Group1" Checked="Status_Checked" />
            <RadioButton x:Name="radioOUT" Content="OUT"
            GroupName="Group1" Checked="Status_Checked" />
        </StackPanel>
        <Line X1="0" Y1="0" X2="400" Y2="0" Stroke="White"
        StrokeThickness="2" />
        <TextBlock x:Name="Locations" Height="20" TextWrapping="NoWrap"
        Text="2) Choose Location" FontSize="16" Foreground="Blue"
        HorizontalAlignment="Left"/>
        <StackPanel Orientation="Vertical" Width="345">
            <RadioButton x:Name="radioAvail" Content="Available"
            GroupName="Group2" Checked="Locations_Checked"/>
            <RadioButton x:Name="radioDND" Content="Do not disturb"
            GroupName="Group2" Checked="Locations_Checked"/>
            <RadioButton x:Name="radioBBL" Content="Be Back later"
            GroupName="Group2" Checked="Locations_Checked"/>
            <RadioButton x:Name="radioLunch" Content="At lunch"
            GroupName="Group2" Checked="Locations_Checked"/>
            <RadioButton x:Name="radioGFTD" Content="Gone for the day"
            GroupName="Group2" Checked="Locations_Checked"/>
        </StackPanel>
    </StackPanel>
</Grid>
</Page>
```

Now, to wire in the client class to send a message, we use the same embedded class we had in the server code to contain the message and use the "on" events for the radio button groups to set the values in the embedded class. We then send that message to the server each time it is updated (the status is changed).

Let's begin with the libraries we need to include. Open the `MainPage.xaml.cs` file. The following shows the libraries we will need. These are the most basic libraries for a typical Universal Windows application. We also add the diagnostics, threading, and JSON libraries as shown in the following:

```
using Windows.UI.Xaml;
using Windows.UI.Xaml.Controls;
using System.Diagnostics;
using System.Threading;
using System.Text.Json;
```

Next, add the embedded class. Recall, we will use a small, embedded (sometimes called nested) class to hold the information about the office sign—specifically, the state of the flag (controlled by the servo) and the location (as indicated by the LEDs). We'll call this class `OfficeStatus` and place it after the `namespace` statement as shown in the following:

```
namespace OfficeSignClient
{
    public sealed class OfficeStatus
    {
        public bool inOffice { get; set; }
        public int location { get; set; }
    }

    public sealed partial class MainPage : Page
    {
...
```

Next, we will need two variables as shown in the following. We need one that is an instance of the `OfficeStatus` class and another for the `Client` class. Place these after the class declaration as shown:

```
public sealed partial class MainPage : Page
{
    public OfficeStatus officeStatus = new OfficeStatus();
    Client officeClient;
```

...

Next, we need to initialize the user interface, set the default settings for the OfficeStatus, initialize the client class with the host (IP) address and port of the server, connect to the server, and set the initial radio button values. Listing 14-9 shows the code for the main page constructor.

Listing 14-9. MainPage Constructor for the Client Application

```
public MainPage()
{
    this.InitializeComponent();
    this.officeClient = new Client("192.168.42.15", 13001);
    this.officeClient.Connect();
    this.officeStatus.inOffice = true;
    this.officeStatus.location = 0;
    radioIN.IsChecked = true;
    radioAvail.IsChecked = true;
}
```

Caution   Make sure the host (IP) and port match those values you used in the server application. If they do not match, the client will not connect, and nothing will happen when you change the radio buttons in the user interface.

We are almost there. We need only two more methods: one for when the IN/OUT office radio buttons are changed and another for when the location radio buttons are changed. These methods are also straightforward in that we simply change the data in the instance of the OfficeStatus class and send the message to the server. Let's look at the details of each method.

Recall from the XAML user interface, we named the IN/OUT group of radio buttons Status, so the event for when those buttons are changed will be named Status_ Checked(). This method is fired as an event (in a thread) when any of the radio buttons in the group are changed (checked). When the method is fired (run), we need to do several things.

First, if the officeClient (the variable that contains the instance of the client class) is null, we do nothing (we return). This is necessary because, during startup, we set the initial values of the radio buttons and thus this method is fired, but this can occur before the client is instantiated or connected to the client. Thus, we only want to execute this method if we have a connected, viable client instance.

Next, we check the event to determine which radio button was checked and use that information to update the OfficeStatus class instance. After that, we serialize the OfficeStatus class instance using the JSON utilities and request the client to send the information to the server. Finally, we call the sleep on the thread to ensure we give the client time to send the message to the server (otherwise, rapid changes on the client could appear delayed). That's it! Listing 14-10 shows the complete code for the method.

Listing 14-10. Status_Checked Method for the Client Application

```
private void Status_Checked(object sender, RoutedEventArgs e)
{
    // Wait until the client is initialized. Return if it is null.
    if (officeClient == null) {
        return;
    }
    Debug.Print("Status got checked!\n");
    // Get the instance of clicked RadioButton instance
    RadioButton rb = (RadioButton)sender;
    Debug.Print(rb.Name.ToString());
    Debug.Print("\n");
    if (rb.Name.Equals("radioIN"))
    {
        Debug.Print("Owner is IN!");
        this.officeStatus.inOffice = true;
    }
```

```
    else if (rb.Name.Equals("radioOUT"))
    {
        Debug.Print("Owner is OUT!");
        this.officeStatus.inOffice = false;
    }
    string message = JsonSerializer.Serialize(officeStatus);
    officeClient.SendMessage(message);
    Thread.Sleep(100);
}
```

The method for the location radio buttons is very similar. But since it is a different grouping, it has its own method. The group is named locations, and the method is named Locations_Changed. The complete code is shown in Listing 14-11.

Listing 14-11. Locations_Changed Method for the Client Application

```
private void Locations_Checked(object sender, RoutedEventArgs e)
{
    // Wait until the client is initialized. Return if it is null.
    if (officeClient == null)
    {
        return;
    }
    Debug.Print("Locations got checked! ");
    // Get the instance of clicked RadioButton instance
    RadioButton rb = (RadioButton) sender;
    Debug.Print(rb.Name.ToString());
    Debug.Print("\n");
    if (rb.Name.Equals("radioAvail"))
    {
        Debug.Print("Owner is in: available.");
        this.officeStatus.location = 0;
    }
    else if (rb.Name.Equals("radioDND"))
    {
        Debug.Print("Owner is DND!");
        this.officeStatus.location = 1;
```

```
    }
    else if (rb.Name.Equals("radioBBL"))
    {
        Debug.Print("Owner is BBL!");
        this.officeStatus.location = 2;
    }
    else if (rb.Name.Equals("radioLunch"))
    {
        Debug.Print("Owner is EATING!");
        this.officeStatus.location = 3;
    }
    else if (rb.Name.Equals("radioGFTD"))
    {
        Debug.Print("Owner is GFTD!");
        this.officeStatus.location = 4;
    }
    string message = JsonSerializer.Serialize(officeStatus);
    officeClient.SendMessage(message);
    Thread.Sleep(100);
}
```

There is one last thing you need to do. Recall from previous projects, we must modify the package manifest to include the *Internet (Client & Server)*, *Internet (Client)*, and (optionally) *Private Networks (Client & Server)* capabilities. You can do this by clicking Package.appxmanifest and then the *Capabilities* tab and selecting the items in the list.

OK, now you're ready to deploy the server application to your device and then run your client application on your PC. Go ahead and set everything up and power on your device.

Deploy and Execute

Since we have two applications, let's look at the specifics of deployment for each. You should begin by deploying the server application first. This will avoid potential connection problems on the client. That is, starting the client before the server may result in the client not connecting (timing out before the server is ready).

Deploy the Server

Recall from Chapter 7 (see the "Deploy and Debug" section and Figure 7-10), to deploy our application in debug mode, we must set the debug settings for our application, setting the following in the debug dialog:

- *Target device*: Remote Machine

- *Remote Machine*: The name of your device

- (Optional) *Uninstall and then re-install*: Checked

- (Optional) *Deploy optional packages*: Checked

- *Application process*: Managed Only

- *Background task process*: Managed Only

Once you make any changes in this dialog, be sure to save your solution. Building or deploying does not automatically save these settings.

We should also change the package name in the package manifest. Recall from Chapter 7 (see the "Deploying Your Application (No Debugging)" section and Figure 7-8), we can change the package name. Do that now so you can find your application easier in the Device Portal.

Since this is a startup task application, you can start and stop the application using the Device Portal. You won't see it in the list of applications in the default user interface on the device. In fact, you don't even need a monitor for the device to run this project. As the name suggests, the startup task can be set up (again, in the device manager) to start when the device is booted. Later, we will see one example of creating an enclosure for the project that allows you to install the office sign anywhere you want without the need of a keyboard and monitor. Cool!

Now you can deploy your application. Go ahead and do that now. You won't see anything appear on the device screen, but you may hear the servo initialize, and you should see the LEDs illuminate; the available LED should be lit. Nothing else is going to happen at this point because the client isn't connected. Let's deploy that now.

Deploy the Client

Deploying the client is easy. You simply set the platform to x86 or x64 and start the application in Visual Studio. It is best to start it in debug mode initially so you can see the debug statements and check that everything is working correctly (there are no errors). Be sure to fix any errors that occur and be sure to double-check that your device is connected to the network and you have the correct host (IP) and port specified in the client.

When the application starts, you should see the user interface in the initial settings as shown in Figure 14-8. Notice the default values are IN for the office status and available for the location.

Figure 14-8. *Client user interface*

Now, turn your attention to the server and try out the radio buttons. You should see the servo turn to one position when the IN input is clicked and another when the OUT input is clicked (and back again). This may not seem very interesting since you want to use the servo to raise and lower a flag, but you can use your imagination at least and see the wonder of remotely controlling mechanical devices. In fact, I show you how to build an enclosure complete with the mechanical flag apparatus in the next section.

Tip   You do not have to complete the enclosure task. I provide it as a mild diversion to show what is possible. As you will see, it requires a bit of mechanical aptitude (must be able to use a hobby knife, hand tools, etc.) as well as a bit of patience to get it going. If nothing else, it may be an interesting read that sparks your own ideas.

If you got this to work, congratulations! You have now been introduced to a whole new venue of IoT applications—making things move. If this interests you, I encourage you to look for more helpers and aides, such as the Servo Trigger board from SparkFun. There are an awful lot of cool projects you can imagine with just a servo or a dozen. But making things move in your IoT projects isn't limited to servos. There are stepper motors, continuous rotation motors, actuators, and much more that you can explore.

Now let's go over how to get started embellishing this project by prototyping an enclosure.

Prototyping the Out-of-Office Sign Enclosure

The following is a brief detour from your discovery of Windows 10 for the IoT. In this section, I present one possible way to take the preceding project and make it into something you could use every day. That is, one thing that separates an experiment from a practical application is how it is packaged.

You could leave your components plugged into a breadboard and use it, but some projects like the one in this chapter aren't very practical. In this case, it is because you have a mechanical element—the IN/OUT flag. Wouldn't it be nice to see this in an enclosure so you can see the flag move?

You may be wondering how to get started. Or perhaps you have doubts about how to build such a thing. Read on to see one approach to discovering how to build a permanent enclosure. You're going to use a technique called *prototyping*, where you build a solution with the intent to experiment on how best to solve the problem.

Most prototypes therefore are called *throwaway prototypes* because they often are of lower quality and makeshift at best. Moreover, they often have little or no aesthetic value. However, prototypes can also result in a near-complete form that can be used without modification. The example in this section is one of those—one way to build a functional enclosure. In this manner, prototypes are often used to prove a concept. That is what you will do in this section. The goal is to see how to design an enclosure, but you also build it so that you can dismantle it either to build a permanent solution or to recover the components for use in other projects.

As you can imagine, this task requires a bit of work, and of course you need to power down your device and partially disassemble the project. I try to keep it relatively easy by using 3D printed parts and as few tools and additional components as possible. That said, the following lists some additional parts you need if you want to follow along with this exercise:

- 3D printed parts from the source download
- Assorted M3 screws and nuts (or equivalent)
    - (1) M3x20 bolt for the flag
    - (4) M3x12 bolts for the Raspberry Pi
    - (4) M3x60 bolts for the enclosure
    - (8) M3 nuts
    - (1) M3 Nylock
    - (5) M3 washers
    - (4) M3 lock washers
    - (2) M2x10 bolts
- Mini breadboard with adhesive tape
- Additional male/female jumper wires
- Soldering iron and solder
- 6 to 8 inches of small gauge wire to connect servo to the flag

OK, so that's a lot more than a few items. And, yes, you are going to be soldering a bit. You could skip this part, but you need some way to attach wires to the LED legs and resistors to the LEDs. As you will see, the board that you will build isn't complicated. If you do not know how to solder, ask a friend to help you—now is a great time to learn![4]

The 3D printed parts are pretty simple and can be found in the source code archive for the book from the Apress website. The parts include the following:

- IN/OUT flag
- Servo mount
- Front plate

---

[4]Depending on the temperament of your friend, it could be a good time to watch someone solder. Be sure to take notes and pay attention so you can learn on your own.

- Rear plate

- (4) spacers

The enclosure forms an open-sided box and is designed to allow you to attach the parts (including the Raspberry Pi) to the front and rear panels using a minimum of hardware. In fact, you need only a few bolts as described earlier. The servo mount and servo controller boards can be attached to the rear of the front plate by melting the nubs that correspond to the mount points.

If you do not have a 3D printer or don't know anyone who does, you can visit a number of online 3D printing services, which you can upload designs to and (for a fee) they will print and ship them to you. I like `www.shapeways.com` as they have an excellent reputation and offer a variety of materials and finishes. However, you could also check out freelance 3D printing services, such as 3D Hubs (`www.3dhubs.com`), which allows you to search in your area for someone willing to print parts for you—sometimes at a much reduced price or even for only the price of the weight of filament (plastic) used. Enthusiasts are cool like that.

The 3D parts for the servo and flag are shown in Figure 14-9.

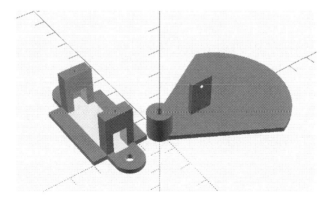

Figure 14-9. *3D printed parts for enclosure (flag, servo mount)*

The parts for the enclosure are shown in Figure 14-10. These include the front and rear panels as well as the four spacers. You connect the panels together using the (4) M3x60 bolts and the 3D printed spacers.

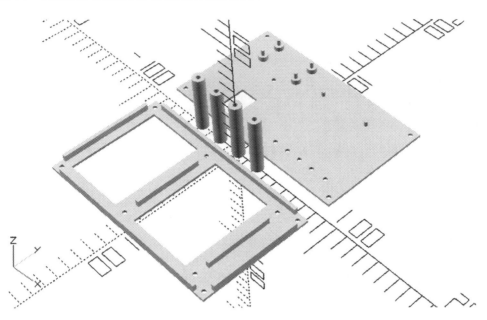

Figure 14-10. *3D printed parts for enclosure (front and back panels, spacers)*

WHAT ABOUT A CARDBOARD BOX?

Cardboard boxes make excellent prototyping material. They are plentiful, cheap (free), easy to cut, you can write on them, and more. Best of all, if you mess one up, just throw it in the fire and start over with a new one! You could use a cardboard box instead of the 3D printed parts if you'd like. However, you will need to use a double-sided tape to attach the servo mount and servo controller board. Of course, you'll need to do some experiments with cutting the holes. Hint: See the template overlap for guidance.

Now, let's see how to assemble the enclosure.

Assembling the Enclosure

The assembly of the enclosure isn't difficult, but it does require using either a hot glue gun or a 3D printer pen to secure the LEDs, servo mount, and servo controller board.

Front Panel

Begin by laying the front panel face down on a cutting mat or similar safe surface. We will start with the LEDs. Take each LED and spread the pins apart slightly. Then, one at a time (remember, the green LED goes in the topmost hole or the hole nearest the opening for the flag), press them in the holes and secure them with hot glue or a 3D pen as shown in Figure 14-11. Notice the longer leg (positive) is bent toward the center of the panel. We will connect these legs to the resistors mounted on a mini breadboard.

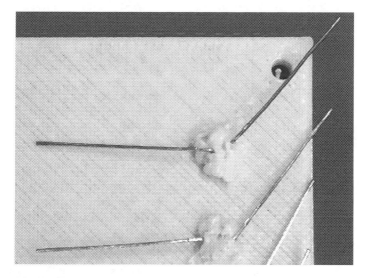

Figure 14-11. *Mounting the LEDs in the enclosure*

Note   The 3D print file is set to create holes for 3mm LEDs. If your printer settings differ or you use larger LEDs, you may need to drill out the holes or open the 3D printer source file in OpenSCAD and change the size of the holes.

Next, attach the mini breadboard next to the LEDs and bend the positive legs so that they are inserted into one of the rows in the breadboard. Then, bend all of the negative legs for the LEDs down in a vertical row and solder them together as shown in Figure 14-12. This may require some patience and trimming of some of the legs. Take your time doing this and ensure you have the breadboard oriented as shown.

Figure 14-12. *Soldering the LED negative pins*

Next, locate the raised pins for the servo mount and servo controller and then place them on the front panel. Use a hot glue gun, 3D pen, or an old soldering iron to melt the pins. Figure 14-13 shows the results of this step.

Figure 14-13. *Mounting the servo mount and servo controller*

Next, take a section of thin wire, such as the type used in auto and motorcycle racing. The wire I use is called *safety wire*, which is used to secure nuts and bolts from coming loose or worse falling out.[5] A length of about 6 to 8 inches should be enough. This wire is often called a *cable* or a *rod* when used with a servo. The small plastic piece that mounts to the servo is called the *servo arm*. Choose the longest of the arms and place it on the servo. Don't secure it yet. You need to make some modifications first. Take the wire and end it so that one end forms a vertical area that the servo can use to push and pull. The other end can be bent in a U shape for the same purpose when mounted to the flag.

---

[5]See https://en.wikipedia.org/wiki/Safety_wire.

Finally, you need to put another bend in the center so that the arm can align with the flag (the servo is taller than the flag mount). Figure 14-14 shows an illustration of how the arm should be shaped.

Servo   Flag

Figure 14-14. *Forming the servo rod/cable*

Before you mount the flag, place a piece of white paper over the flag, cut it to fit with some overlap, and then tape the ears down, as shown in Figure 14-15. This allows you to write on the flag from outside the box. Go ahead and mount the flag in the box with a long M3 bolt and nut or similar.

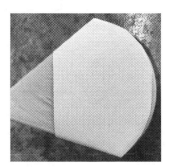

Figure 14-15. *Masking the flag*

Next, you can mount the servo and flag. Use the M3 Nylock nut and a washer to ensure the flag doesn't loosen from use. Go ahead and attach the rod to the servo and the flag. You may need to drill out the servo arm and the flag mount a bit depending on the thickness of the wire. Looser is better. While you're at it, go ahead and connect the servo to the servo controller pins at the top as shown. Figure 14-16 shows the completed front panel.

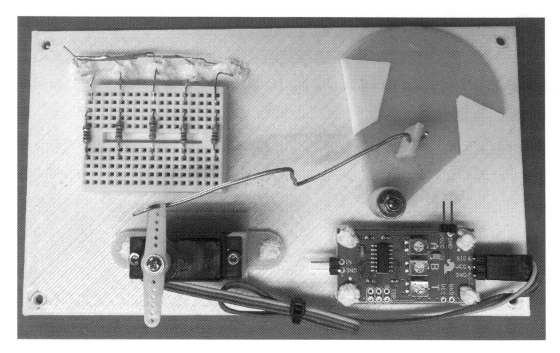

Figure 14-16. *Completed front panel*

Caution   Never rotate a servo when it is powered on. Also, never connect or disconnect a servo when the circuit is powered on.

Note that we haven't written anything on the flag yet. We save this until we get the project working. More specifically, we write "IN" on the flag when the flag is in the correct position and the client has sent the correct message. Similarly, we write "OUT" on the flag when the client sets the flag to the out position.

Now, let's work on the rear panel.

Rear Panel

The rear panel is where we mount the Raspberry Pi. Orient the rear panel with the raised sections pointing up and the smaller opening at the top. Then, orient the Raspberry Pi board so that the USB and Ethernet connectors are facing the right side. Use (4) M3 bolts to mount the Raspberry Pi to the rear panel as shown in Figure 14-17.

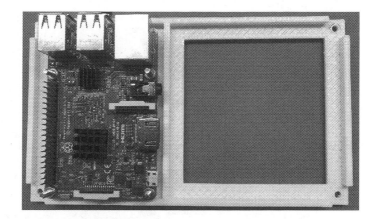

Figure 14-17. *Mounting the Raspberry Pi*

Now you can make all of the remaining wiring connections.

Putting It All Together

For this step, lay the front and rear panels face down on a flat surface with the rear panel on the right. Go ahead and place the long M3 screws through the front panel and slide the spacers over them. Use a GPIO map to locate all of the pins from Table 14-2.

Use jumper wires to make all of the connections for the LEDs. Then, use jumper wires to make the connection for the servo. You should now have what looks to be quite a mess of wires. If you are using longer jumper wires, there may be excess wire. You can bundle the wires at this point if you'd like, but you may need to adjust as we assemble the enclosure. Figure 14-18 shows the completed wiring connections.

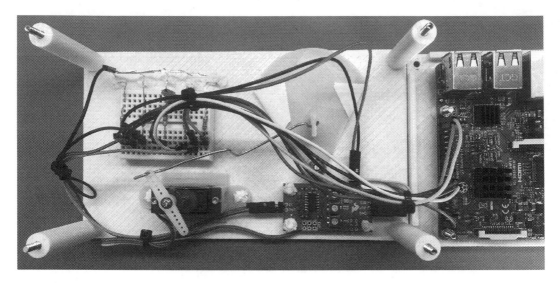

Figure 14-18. *Wiring completed*

Be sure the Servo Trigger is unobstructed or at least positioned so that you can get to the three potentiometers on the board to tune it should you need to do so.

Caution   Be sure to double-check and triple-check your wiring before powering on your device. There are a lot of wires in there, and it is very easy to get things mixed up. Use the tables earlier in the chapter as a guide to trace each and every wire from the Raspberry Pi GPIO pin to the device. For example, ensure that GPIO 20 is connected to the Servo Trigger signal pin.

OK, now you can carefully assemble the panels. Lift the rear panel and place it over the front panel and align the screws. Be careful not to disconnect any of the wires. Secure the panels with M3 lock washers and M3 nuts.

At this point, you have a nice looking box, but there's nothing written on the front. You can find a template in the form of a Word document you can use to print and glue (or tape) to the front panel. Simply download the file named officestatus_faceplate. docx from the chapter source code, change the text as you see fit, and print it out. You can then place the template over the front panel and mark the holes for the LEDs, servo bolt, and IN/OUT flag. Then, use a hole punch or hobby knife to cut out the holes. You can wait to do the flag opening after you've attached it to the front panel. Figure 14-19 shows what the template looks like.

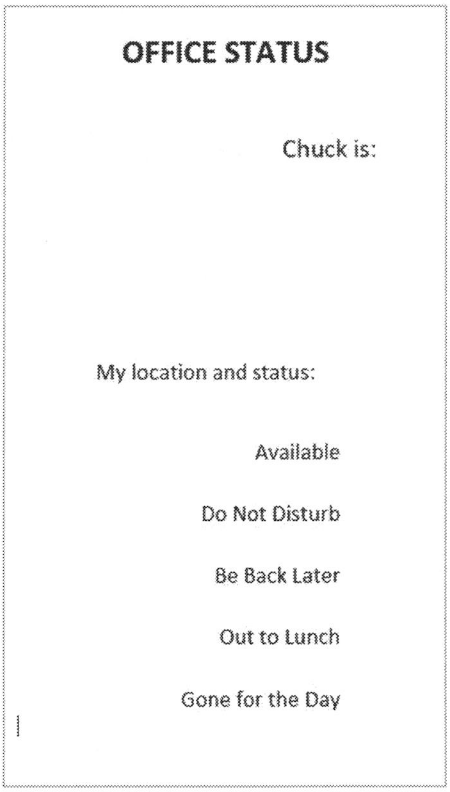

OFFICE STATUS

Chuck is:

My location and status:

Available

Do Not Disturb

Be Back Later

Out to Lunch

Gone for the Day

Figure 14-19. *Faceplate template*

There's just one more step to perform. You need to tune the servo so that the flag moves correctly, and it is shown in the window.

Tuning the Servo Trigger

The Servo Trigger has three small potentiometers on the board for adjusting the servo. Use these to change the off and on (triggered) positions of the servo so that your flag clearly shows in the cutout window and that it travels far enough to change from one value to another (from IN to OUT).

Use a small flat blade screwdriver to adjust the potentiometers. The A potentiometer adjusts the off position, and the B potentiometer adjusts the on (triggered) position. If you find this to be the opposite, you can either roll with it that way or simply rotate the entire servo mount 180 degrees.

You may find that you have to experiment a bit to get this right. This is really the hardest part of the prototype and the reason you're using temporary housing. You may have to adjust the wire that connects the servo arm to the flag shortening or bending it so that the action runs smooth. You may also have to detach and realign the servo arm to get it in the correct location.

Once you get the servo tuned correctly, take a moment to draw a small outline of the cutout on the flag in each position. You can then disassemble the flag and write IN and OUT in those locations.

OK, now you have your components assembled and the servo tuned. Let's test it out!

Testing the Prototype

OK, so now the real fun begins. Before you power on the device, if you haven't already done so, write the IN/OUT values on the flag and label the LEDs. I added my own comments here and there.

Now, fire up the device and start the application. Your prototype should look very similar to the one in Figure 14-20.

Figure 14-20. *Chuck is in his office!*

Now, try out your application and watch your sign change according to your commands. Isn't that cool? The sound of the servo cycling is a really powerful satisfying aspect of this prototype. Figure 14-21 shows the end state of my prototype for the day. I'm out of here!

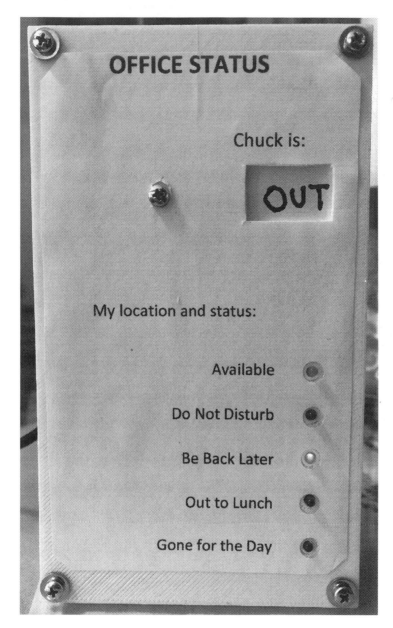

Figure 14-21. *Chuck has left the building...*

Now that everything is working, I am sad to say the coolness and jubilation eventually wears off.

Taking It a Step Further

I chose to use 3D printed parts for the enclosure for the project because it is easy to modify and reprint. However, you may want to consider implementing this project using a more robust enclosure. For example, you could purchase a plastic or aluminum enclosure and cut or drill the holes for the LEDS and flag mount. Another possibility is using a small piece of thin plywood to make the bezel and mount it in a deep photo frame or shadow box.

If you like this project and had as much fun as I did building it or better still if you want to take it to work and put it in service, I encourage you to consider taking the project a step further with a better enclosure. See Chapter 17 for ideas on where to publish your work. Be sure to reference this book as the origins of your idea.

Summary

I find those IoT projects (or any project) with mechanical elements really fun to design and implement. Mechanical movements allow you to bring a bit of whimsy and wonder to your projects. My interest in mechanical movements began as a child when I saw the early animatronics displays at Disney World and other amusement parks. Now that you have had a small taste of one method you can use to create such devices, you can begin to think about how to incorporate similar mechanisms in your projects.

In this chapter, you discovered how to build a nifty out-of-office sign with a mechanical flag and LEDs controlled across the network. This represents the fundamental building blocks for other remote controlled IoT projects.

In the next chapter, you explore an exciting new venue for Windows 10 IoT development—a brief introduction to how to take your IoT ideas into the enterprise, commercial realm. Yes, that's right. You can build consumer, enterprise, and even industrial solutions with Windows 10! How cool is that?

CHAPTER 15

Azure IoT Solutions: Cloud Services

Now that you've seen a number of projects ranging from basic to advanced, it is time to discuss how to make your IoT data viewable by others via the cloud. More specifically, you will get a small glimpse at what is possible with the Microsoft Azure cloud computing services and solutions.

I say a glimpse because it is not possible to cover all possible solutions available in Microsoft Azure for IoT solutions in a single chapter or even over two chapters.[1] Once again, this is a case where learning a little bit about something and seeing it in practice will help you get started. Like the other chapters where you've had a lightning tour, this chapter presents a few of the newer concepts and features of Microsoft Azure at a high level and in context of an example.

If you are just learning how to work with IoT solutions or have no plans to immediately host your solution in the cloud (or on Microsoft Azure), you can still learn quite a bit about the technology by following along and implementing the sample project. In fact, with a little patience, it's a fun alternative project.

Since the topic is quite large, we will discuss the general concepts of using Azure in this chapter and implement our project in Chapter 16.

---

[1]But we're going to give it a shot.

© Charles Bell 2021
C. Bell, *Windows 10 for the Internet of Things*, https://doi.org/10.1007/978-1-4842-6609-0_15

What Is Microsoft Azure?

That is a very interesting question. As it turns out, Microsoft Azure (hence Azure) is a cloud platform for all manner of cloud solutions. In fact, Azure[2] is a growing collection of cloud products, such as virtual machines, databases, networking, storage, reporting, and more. Furthermore, Azure is Microsoft's platform and infrastructure for building, deploying, and managing applications and services, including platform as a service (PaaS) and infrastructure as a service (IaaS).

WHAT IS CLOUD COMPUTING?

The term *cloud computing* is sadly overused and has become a marketing term for some. True cloud computing solutions are services that are provided to subscribers (customers) via a combination of virtualization, grid computing (distributed processing and storage), and facilities to support virtualized hardware and software, such as IP addresses that are tied to the subscription rather than a physical device. Thus, you can use and discard resources on the fly to meet your needs.

These resources, services, and features are priced by usage patterns (called *subscription plans* or *tiers*), in which you can pay for as little or as much as you need. For example, if you need more processing power, you can move up to a subscription level that offers more CPU cores, more memory, and so forth. Thus, you only pay for what you need, which means that organizations can potentially save a great deal on infrastructure.

A classic example of this benefit is a case where an organization experiences a brief and intense level of work that requires additional resources to keep their products and services viable. Using the cloud, organizations can temporarily increase their infrastructure capability and, once the peak has passed, scale things back to normal. Clearly, this is a lot better than having to rent or purchase a ton of hardware for that one event.

---

[2]Yes, the Azure websites are all colored an interesting shade of blue.

Sadly, there are some vendors that offer cloud solutions (typically worded as *cloud enabled* or simply *cloud*) that fall far short of being a complete solution. In most cases, they are nothing more than yesterday's Internet-based storage and visualization. Fortunately, Microsoft Azure is the real deal: a full cloud computing solution with an impressive array of features to support almost any cloud solution you can dream up.

If you would like to know more about cloud computing and its many facets, see `https://en.wikipedia.org/wiki/Cloud_computing`.

You can use Azure from a host of different operating systems, so it is not tied exclusively to Microsoft Windows. Together with programming APIs, frameworks, and drivers in Visual Studio, you can build solutions to run as a web application in Azure. Better still, you can write applications for Windows, Android, and even iOS that integrate with Azure through JavaScript, Python, Node.js, and other web development tools and languages.

There are many tools you can use to interact with Azure including the following:

- Azure Portal (`https://portal.azure.com`).

- Azure IoT Explorer application that replicates some of the features of the portal (`https://docs.microsoft.com/en-us/azure/iot-pnp/howto-use-iot-explorer`).

- Azure command-line interface (CLI) available for a host of platforms that uses a set of commands to interact with Azure (`https://docs.microsoft.com/en-us/cli/azure/install-azure-cli`). These commands can also be run in the Azure Shell that is part of the portal.

- The Cloud Explorer in Visual Studio.

There are other ways beyond those, but we will keep to these few for simplicity.

Tip   Do not forsake the Azure documentation! This site provides a complete albeit terse at times look into all of the features of Azure. Be sure to visit there often as you grow beyond the examples in this chapter. For more information about Azure, see `https://docs.microsoft.com/en-us/azure/?product=featured`.

Getting Started with Microsoft Azure

As part of your tour, you will see how to use Azure with IoT solutions by writing a sample project to send messages to and from Azure. Consider it a Hello, World for Microsoft Azure. Naturally, Azure has features that fully support IoT solutions, and you will concentrate on those features in this chapter. And, Visual Studio has a number of project templates designed to make starting Azure projects easier.

More specifically, there are tools built into the Azure Portal and Visual Studio extensions that work with your source code project files to aid in setting up applications to interact with Azure features. In fact, you will see how to gather sensor data on your Raspberry Pi and send it to Azure to be stored, processed, and displayed and hosted as a website. Yes, Azure completes your IoT solutions by providing the Internet-facing systems and features to share your data.

Although you will only scratch the surface of what is possible with Azure, the projects in this chapter give you a taste of what is possible for IoT solutions using Azure. Let's continue the lightning tour and see how to get started using Azure.

> **Note**   This lightning tour is only one path to learn how to use Azure. The presentation is tailored for using Azure with IoT solutions. As such, I cover only a small portion of the features of Azure. Also, there may be more than one way to achieve your goals using other features in Azure. Indeed, you could spend a lot of time learning everything there is to know about Azure.[3]

Sign Up for an Azure Account

The very first thing you need to do to use Azure is to sign up and create an account. While Azure is a paid service platform, Microsoft permits you to create a temporary free account that is active for 30 days. After the 30 days, you can continue to use Azure, but any resources you created or used during your trial period may become billable, and you can always upgrade to a paid account. Fortunately, if you build your IoT solutions using an Azure free account, you can continue to use them as paid resources once you upgrade your account.

---

[3]While I consider myself a veteran of cloud computing, learning Azure has proven to be like peeling an onion. Every time I use a new feature, I find another cool thing I want to explore.

The process to create an Azure account is quite easy but requires a valid Microsoft account to get started. If you do not have a Microsoft account, you need to create one by following the steps on `https://signup.live.com`. You do not have to have a Microsoft email account—any valid email account is OK to use for setting up your Microsoft account.

You also need to enter a credit card to complete the Azure account signup. Microsoft bills you $1 when you set up your account, but they refund you within a few days. This is Microsoft's way of verifying your account and is quite common for monthly subscription services that offer a trial period.

You may be wondering how Microsoft can offer a free account when all the resources are billable. They do this by crediting your account $200. This is normally more than enough to get you through the 30-day trial period. Having the credit card on file allows you to keep your resources going without interruption should you choose to continue your subscription.

Now, let's see how to create your Microsoft Azure account. Visit `https://azure.microsoft.com/en-us/free/` and click *Start free*. Be sure to read the conditions of signing up on the bottom of the page. As you will see, there are some services that may be free beyond the 30-day trial, including a feature called an IoT Hub.

Signing up is easy. The process starts with logging in to Microsoft Live with your Microsoft account. If you are running on Windows, you have the option of using Windows Hello to log in. After that, you must enter your billing information—including your name, email, and phone number—and then click *Next*.

You are then asked to agree to the various agreements including subscription, offer details, and the privacy statement. Once you tick the box that you agree, you can click *Sign up* to proceed. Once the registration process is complete, you can submit feedback, register to watch tutorial videos, and explore the products. When you're ready, click Go to the portal to launch the Azure Portal. Or you can navigate to `https://portal.azure.com` later to log in. Figure 15-1 shows the initial portal screen that you see once logged in.

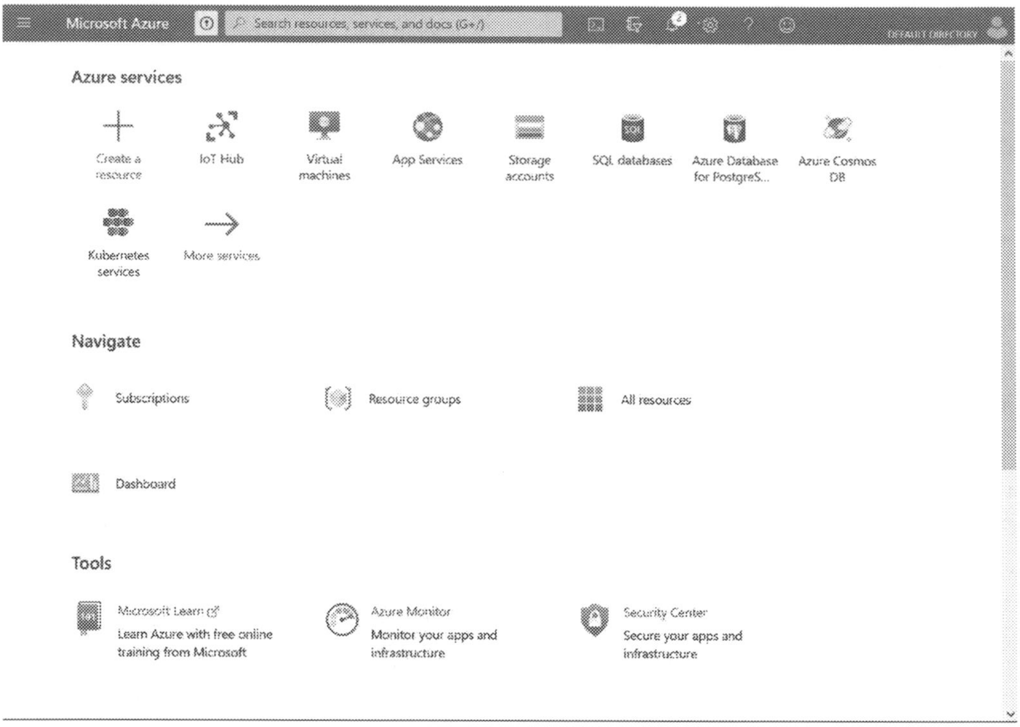

Figure 15-1. *Azure Portal (initial login)*

While you see only a portion of the features of Azure in this chapter, there are a few features, terminology, and concepts that you should understand in order to complete the projects. The following lists the Azure features and concepts used in this chapter. I discuss the more complex topics in more detail in the following sections. You will see the others in action as you explore the projects.

- *Azure IoT Hub*: A relatively recent addition to Azure, IoT Hubs allow you to create a service that allows you to send and store messages to the cloud as well as send messages from the cloud to a device. The IoT Hub can be connected to other Azure services to complete your IoT solution.

- *Azure IoT Accelerators*: A dedicated solution for IoT projects that provides preconfigured templates for creating several types of common IoT solutions. It combines a number of existing Azure features with a streamlined setup process. Note that this was formerly named "Azure IoT Suite."

- *JavaScript Object Notation (JSON)*: An open-standard format that uses human-readable text to transmit data objects consisting of attribute-value pairs.[4]

Azure IoT Hub

The Azure IoT Hub is a service designed to allow bidirectional message passing between IoT devices and the cloud (the IoT Hub or connected services). Thus, the IoT Hub can be considered the main Azure feature for connecting your IoT solution to Azure. There are a number of features and benefits to using the IoT Hub, including the following:

- Reliable device-to-cloud and cloud-to-device messaging.

- Includes provisions for security credentials and access control.

- Permits monitoring for device connectivity.

- Microsoft provides a number of libraries to support a host of devices, languages, and platforms.

For more information about the Azure IoT Hub, see `https://docs.microsoft.com/en-us/azure/iot-hub/`.

IoT Accelerators

The Azure IoT Accelerators is a combination of existing Azure features, IoT services, and preconfigured solutions that allow you to get started quickly. The preconfigured solutions are complete implementations of common enterprise-grade IoT solutions. This allows you to get started experimenting with Azure and linking your IoT solution to the cloud much more quickly and with far less effort than configuring the Azure features yourself individually.

The list of capabilities of the preconfigured solutions in Azure is quite impressive. The following is a list of the features and benefits of the preconfigured solutions:

- Easy-to-set up devices to collect data

- Built-in analytics to capture the data

---

[4]`https://en.wikipedia.org/wiki/JSON`

- Sample web application to display the data in real time

- Sample alarms for watching the data for major events

- Visualize real-time and historical data

If that list isn't impressive enough, consider that you can take the preconfigured solution and modify it as your needs change. That is, you can start out with the basic solution and then add more capability in the form of faster processing, more devices, more storage, and so forth. It really is the best way to get your enterprise-grade IoT solution up and running in the cloud and even mature it as you continue to develop.

For more information about the IoT Accelerators, see `https://docs.microsoft.com/en-us/azure/iot-accelerators/`.

Now that you have an Azure account and know a bit more about the technologies that you will use, let's now walk through creating a sample IoT application to demonstrate how to set up resources for storing IoT data in Azure.

Building IoT Solutions with Azure

Building IoT solutions that use Azure can be rather challenging. This is mostly due to the rather steep learning curve in understanding Azure and the Azure features needed. Rather than jump directly into a complex Azure-based IoT solution, let's set the stage for the chapter project, where you see a simple demonstration on how messages are exchanged with Azure. Essentially, you will implement the project in two stages: one to see how to write sample data to Azure and another to see how to connect real sensors to store and view data in the cloud using the Azure IoT Suite.

The project in this section is designed to learn how to set up an IoT Hub and connect the application running on your device to send and receive messages. This teaches you how the data is sent to the cloud and prepares you for writing code to send live data to Azure.

Thus, the application is simplistic in the sense that you will not use a sensor; rather, you will use a user interface to get sample values and send those to the IoT Hub. Once again, this is so that you can learn the basics of working with the IoT Hub, and not make the project more difficult from the complexity of reading data from sensors and viewing them in the cloud.

The steps in this section require several steps, taking you from setting up your IoT Hub to configuring your PC and, finally, seeing the cloud messages in action, as shown next. I discuss each in more detail in the upcoming sections.

1. Set up an IoT Hub in Azure.

2. Set up your PC for working with the Azure IoT Hub.

3.   Create the example application.

4.   Monitor messages sent from the device to the IoT Hub message queue.

5.   Send a message from the IoT Hub to the device.

These steps sound daunting,[5] but if you follow along, you should be able to go through the steps in only a few minutes. Some of the steps you perform in this project can be reused in the chapter project.

Set Up an IoT Hub

An IoT Hub is a feature in Azure for connecting devices to other features in Azure. Most notably, the IoT Hub provides a message queue for you to send messages to the cloud. These messages can then be consumed by other features (or other applications) in Azure or by another application connecting to Azure. Let's begin by logging in to the Azure Portal (`www.portal.azure.com`). Once again, there are many ways to interact with Azure, but the easiest is using the portal.

If you're not on the home page, click *Home* in the upper left. Then, hover over *IoT Hub* and then click *Create* or search for IoT Hub in the search bar at the top of the portal. Figure 15-2 shows the selections made.

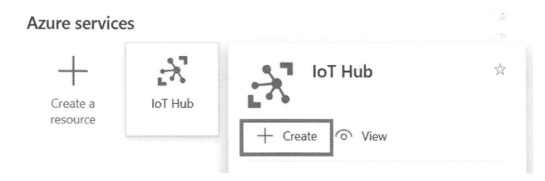

Figure 15-2. Adding an IoT Hub

---

[5]I've condensed the essentials to make it easy to follow.

Tip   The Azure Portal page is best viewed on a wide screen because it spreads things out quite a bit by opening sliding forms (called *blades*) for settings, selections of features, and so forth.

Once you initiate the operation to create a new IoT Hub, you will need to fill out some information in a series of pages (or blades). Let's explore the process. Begin by completing the following on the first page. Sample values are in italics. Feel free to use whatever you like for names, but keep them consistent as you follow along.

- *Subscription*: Leave this set to *Free Trial*.

- *Resource group*: Recall, a resource group is a container that we can use to group resources. I recommend creating a new one for this project—*CABWeatherRG1*.

- *Region*: Choose your (general) location—*East US*.

- *IoT Hub name*: Choose a name that is unique (Azure tells you if it is not). You have only a few characters to work with, so keep it short—*CABWeatherHub1*.[6]

Since we are starting our first IoT Hub, we likely don't have a resource group yet, so you will need to create one by clicking the *Create* link under the text box. Figure 15-3 shows how to create the resource group listed earlier. Click *OK* to complete the creation.

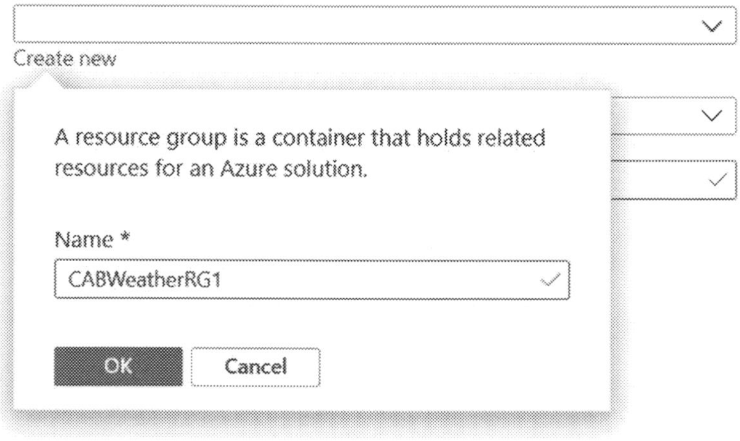

Figure 15-3.   *Create resource group*

---

[6]I append "1" to the name here but drop it later in the chapter project.

Once the resource group is created, you can fill in the rest of the information as shown in Figure 15-4.

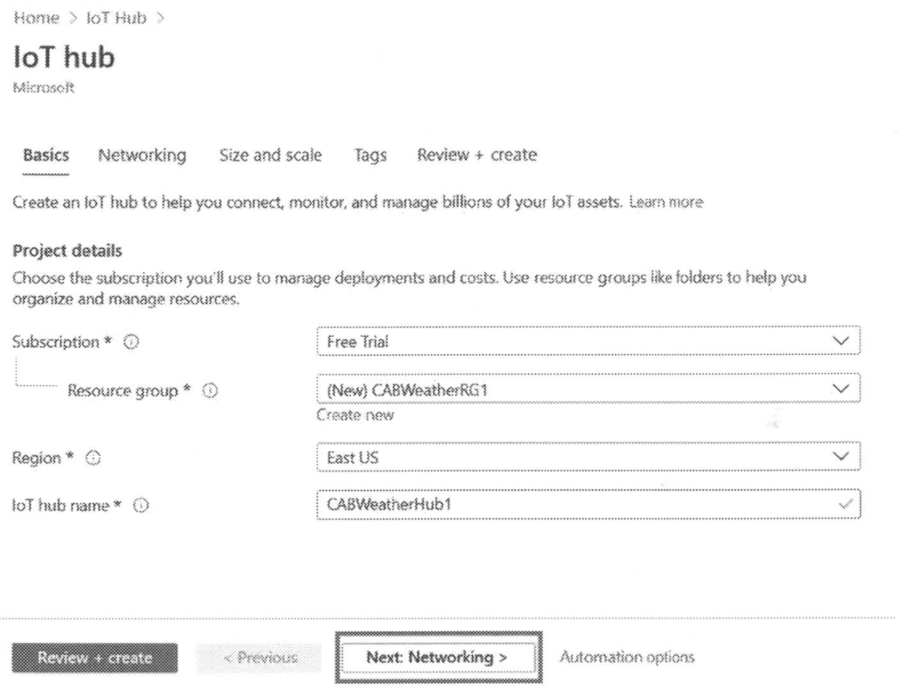

Home > IoT Hub >

IoT hub
Microsoft

Basics   Networking   Size and scale   Tags   Review + create

Create an IoT hub to help you connect, monitor, and manage billions of your IoT assets. Learn more

Project details

Choose the subscription you'll use to manage deployments and costs. Use resource groups like folders to help you organize and manage resources.

Subscription * ⓘ	Free Trial
Resource group * ⓘ	(New) CABWeatherRG1
	Create new
Region * ⓘ	East US
IoT hub name * ⓘ	CABWeatherHub1

Review + create   < Previous   Next: Networking >   Automation options

Figure 15-4. *Create IoT Hub—basics*

When ready, click the *Next: Networking* ➤ button to proceed to the next step. In this step, you will keep the defaults as shown in Figure 15-5. There are several links to more information about each of the options. I encourage you to explore those, but for this project we will use the defaults.

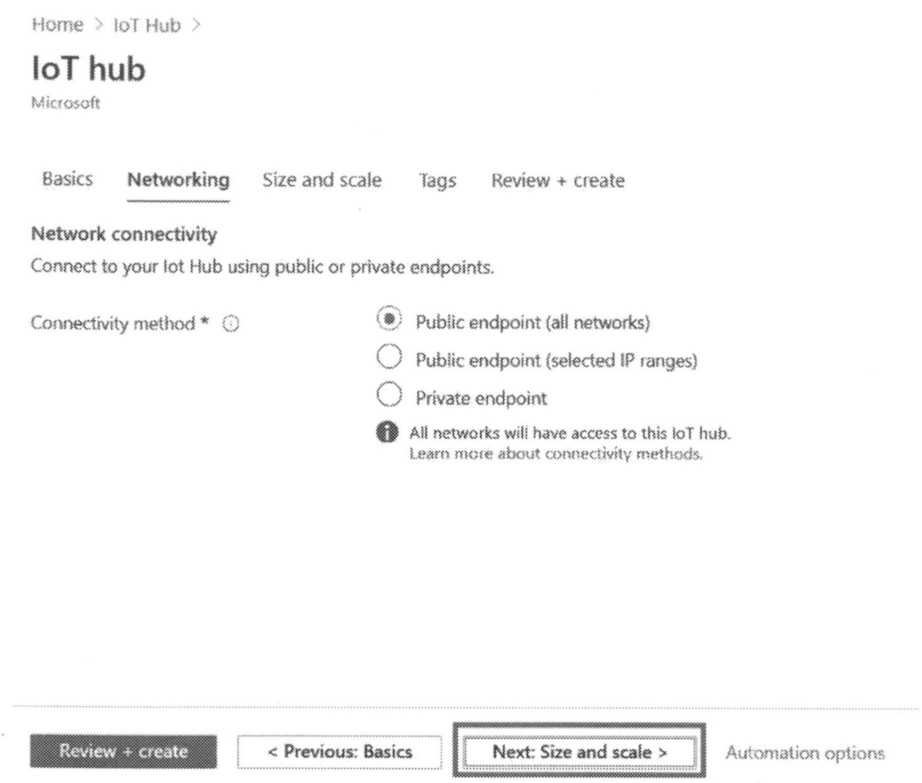

Figure 15-5. Create IoT Hub—networking

When ready, click the *Next: Size and scale* ➤ button to proceed to the next step. In this step, you will keep the defaults as shown in Figure 15-6. There are several links to more information about each of the options. I encourage you to explore those, but for this project we will use the defaults.

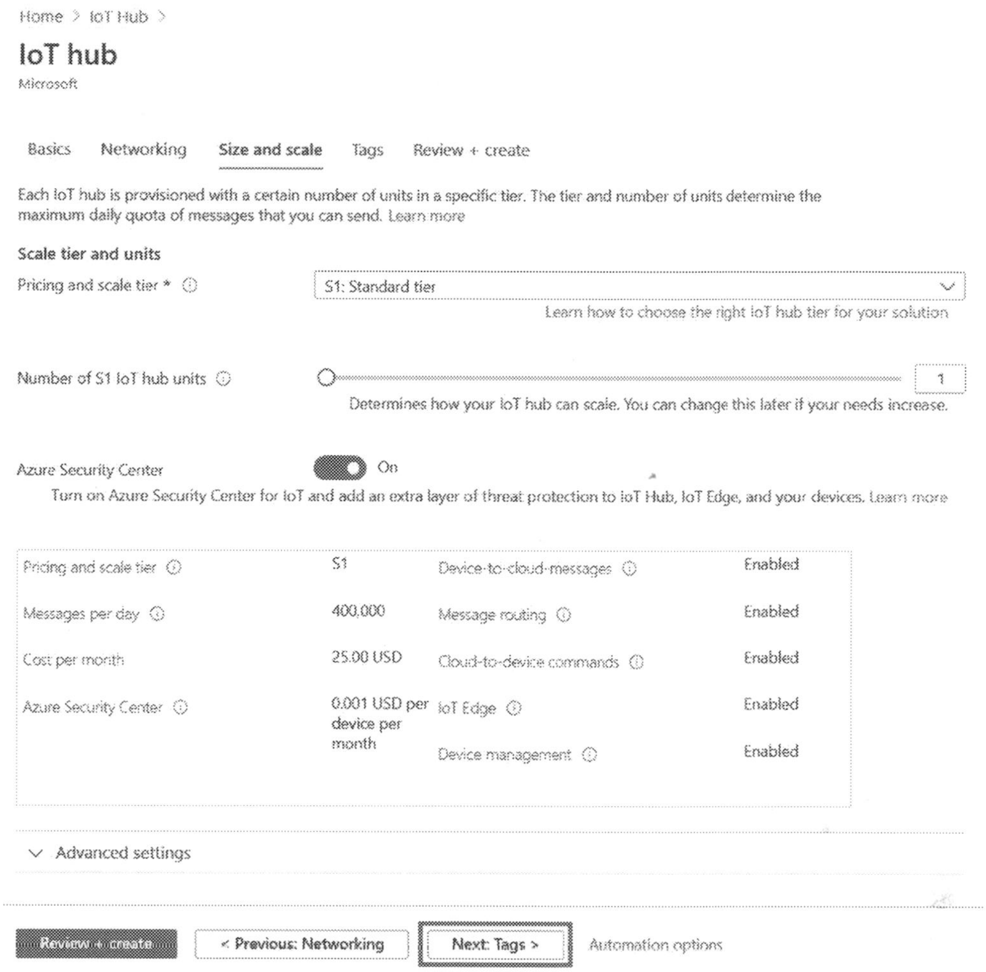

Home > IoT Hub >

IoT hub
Microsoft

Basics   Networking   **Size and scale**   Tags   Review + create

Each IoT hub is provisioned with a certain number of units in a specific tier. The tier and number of units determine the maximum daily quota of messages that you can send. Learn more

Scale tier and units

Pricing and scale tier * ⓘ

S1: Standard tier ⌄

Learn how to choose the right IoT hub tier for your solution

Number of S1 IoT hub units ⓘ

1

Determines how your IoT hub can scale. You can change this later if your needs increase.

Azure Security Center   ⬤ On

Turn on Azure Security Center for IoT and add an extra layer of threat protection to IoT Hub, IoT Edge, and your devices. Learn more

Pricing and scale tier ⓘ	S1	Device-to-cloud-messages ⓘ	Enabled
Messages per day ⓘ	400,000	Message routing ⓘ	Enabled
Cost per month	25.00 USD	Cloud-to-device commands ⓘ	Enabled
Azure Security Center ⓘ	0.001 USD per device per month	IoT Edge ⓘ	Enabled
		Device management ⓘ	Enabled

⌄ Advanced settings

Review + create   < Previous: Networking   Next: Tags >   Automation options

Figure 15-6. *Create IoT Hub—size and scale*

When ready, click the *Next: Tags* ➤ button to proceed to the next step. In this step, you can assign one or more tags to help you locate the resource with searches. For this example, we'll keep it simple and add only the one tag *Source* with the value *Windows 10 for the IoT* as shown in Figure 15-7. Feel free to add any tags you'd like.

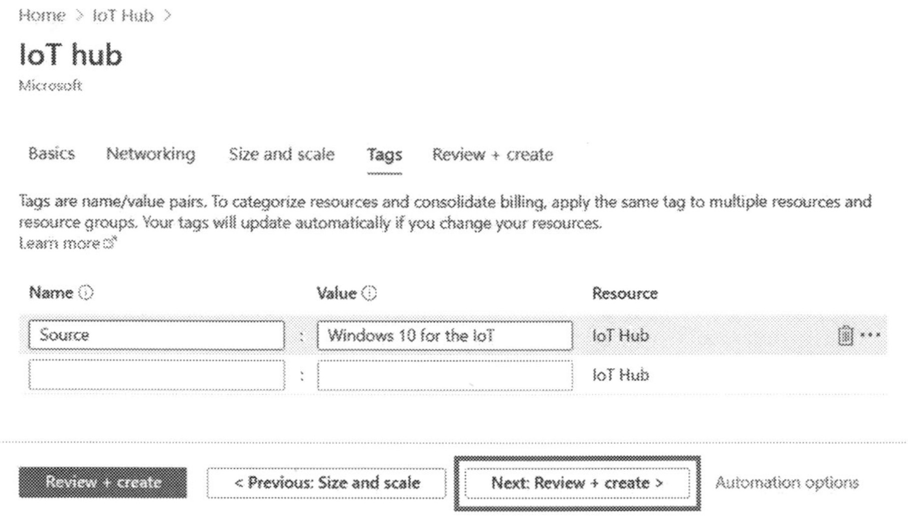

Figure 15-7. *Create IoT Hub—tags*

Once you have the tags you want, click *Next: Review + create* ➤ to review the information before creating the resource as shown in Figure 15-8.

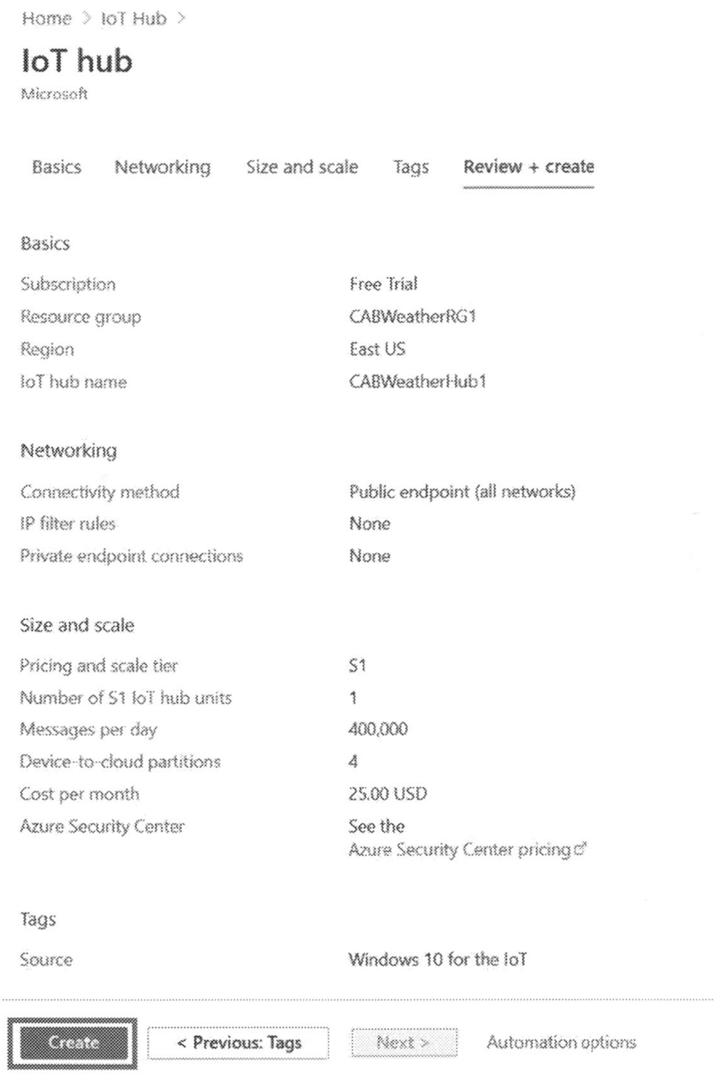

Home > IoT Hub >

IoT hub
Microsoft

Basics   Networking   Size and scale   Tags   **Review + create**

Basics

Subscription	Free Trial
Resource group	CABWeatherRG1
Region	East US
IoT hub name	CABWeatherHub1

Networking

Connectivity method	Public endpoint (all networks)
IP filter rules	None
Private endpoint connections	None

Size and scale

Pricing and scale tier	S1
Number of S1 IoT hub units	1
Messages per day	400,000
Device-to-cloud partitions	4
Cost per month	25.00 USD
Azure Security Center	See the Azure Security Center pricing ↗

Tags

Source	Windows 10 for the IoT

[Create]   [< Previous: Tags]   [Next >]   Automation options

Figure 15-8. *Create IoT Hub—review*

Once you have reviewed the information, click *Create* to create the IoT Hub or use the buttons provided to go back and make changes.

Note   You can have only one free IoT Hub subscription per account.

The next step is the registration process. Azure merrily creates your IoT Hub for you in the background. An occasional pop-up message keeps you updated of the status. I should note that, depending on resources being allocated, the registration could take some time. Now is a good time to refresh that beverage, pay some attention to the family, pet your dog, and so forth, and check back in about 15 to 30 minutes.[7] You will see the deployment screen after a few moments as shown in Figure 15-9. There's nothing to do here as it is purely informational.

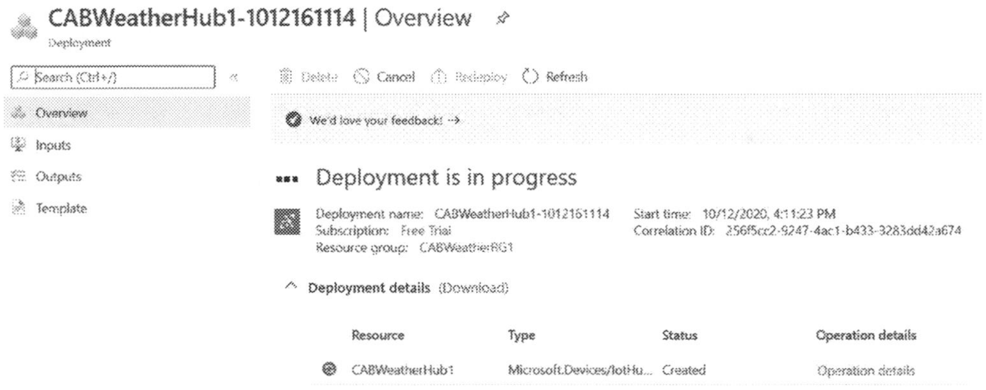

Figure 15-9. *Deploying the IoT Hub*

Caution   Remember, any resource you create will be charged toward your account. If you're using the free account, the cost will be subtracted from your initial credit. While the resources don't cost much on a daily basis, you can accrue quite a bill over time. It is always best to delete any resources you aren't using so you don't get billed for them.

Once created, you see a message stating Deployments Complete as shown in Figure 15-10. If at this point you do not see your IoT Hub in the dashboard on the portal, you can try refreshing the screen or click *Resource Groups* to refresh the dashboard.

---

[7]It usually only takes about 5 to 15 minutes, but since you're relaxing, take your time!

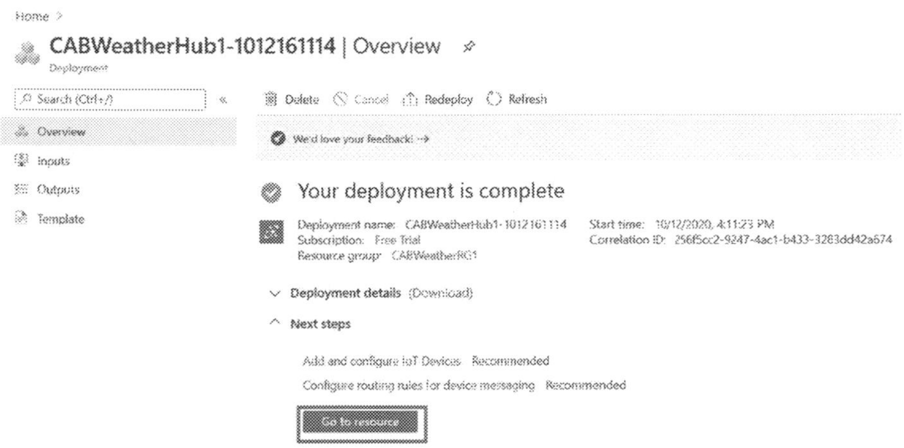

Figure 15-10.   *IoT Hub deployed*

Click the *Go to resource* button to open the IoT Hub. This opens the IoT Hub details page. This is where it is helpful to have a wide screen as the blades can fan out after a few clicks. It is also a long page. Figure 15-11 shows an excerpt.

Figure 15-11.   *IoT Hub details page (excerpt)*

Note   The banner shown is typical of how Microsoft communicates changes to their services. For now, you can ignore the banner.

There is one very important bit of information you will need about your IoT Hub for this project. You will need to know the connection strings for connecting a device to the hub. To find them, click *Shared Access Policies* in the list on the left, then select the user *iothubowner* as shown in Figure 15-12. Notice the connection strings are on the blade to the right complete with buttons to allow you to copy them (for copy and paste).

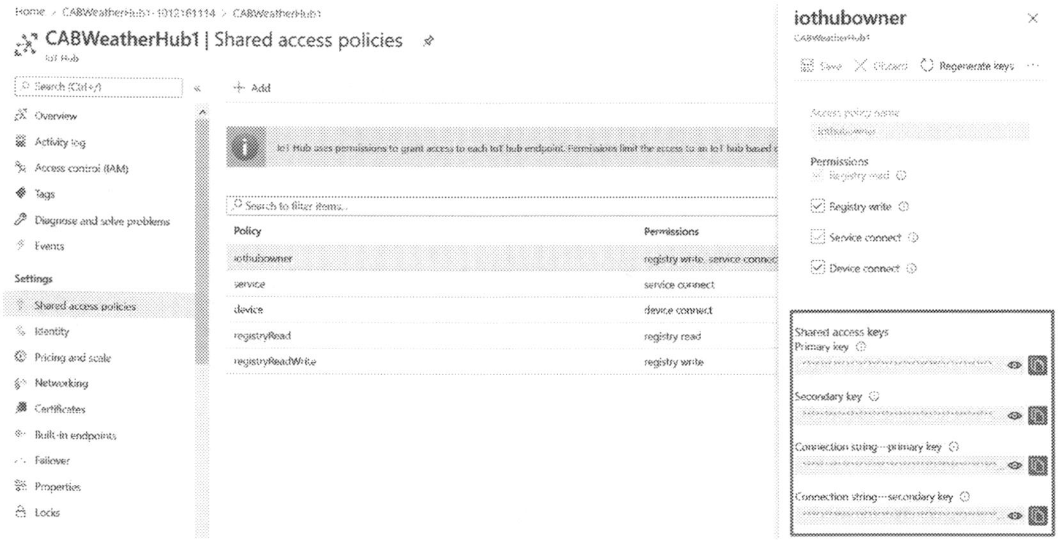

***Figure 15-12.** Finding the connection strings*

Caution   Your keys, Ids, and subscription-related tokens should be well protected. Someone with access to this data can create applications that can connect to your Azure account. Be sure to strip them or redact them before publishing any of your own code.

OK, the IoT Hub is ready for messages. Next, let's set up your PC.

Set Up the PC

There are a few things that you need to do on your PC to create the sample project. You must install the Azure CLI (if you haven't already). We will use the Azure CLI to test the sample application. Instructions on how to install the Azure CLI (use the current version, not the beta) can be found at *https://docs.microsoft.com/en-us/cli/azure/install-azure-cli*. You will also want to install the Azure command-line extension, which we will see in the demonstration.

Once the Azure CLI is installed, open a terminal (command window). Notice, we first add the Azure IoT extension using the az extension add --name azure-iot command, then log in with the az login command, which prompts for your Azure user account and password via your browser, demonstrate how to get help for a command using -- help after the command, show a sample informational command to list the IoT Hubs we've created using the az iot hub list command, and finally log out with the az logout command. Listing 15-1 shows these commands in action (excerpted for brevity— the Azure CLI is "chatty" that way).

Listing 15-1. Starting the Azure CLI

```
C:\Users\cbell>az extension add --name azure-iot
 - Installing ..
C:\Users\cbell>az login
You have logged in. Now let us find all the subscriptions to which you have
access...
[
...
]

C:\Users\cbell>az iot hub --help


Group
    az iot hub : Manage entities in an Azure IoT Hub.

...


Commands:
    create                            : Create an Azure IoT hub.
    delete                            : Delete an IoT hub.
    generate-sas-token                : Generate a SAS token for a target
                                        IoT Hub, device or
                                        module.
    invoke-device-method              : Invoke a device method.
    invoke-module-method              : Invoke an Edge module method.
    list                              : List IoT hubs.
...
```

```
C:\Users\cbell>az iot hub list
[
  {
...
    "location": "eastus",
    "name": "CABWeatherHub1",
...
    "tags": {
      "Source": "Windows 10 for the IoT"
    },
    "type": "Microsoft.Devices/IotHubs"
  }
]

C:\Users\cbell>az logout
```

Tip   See https://docs.microsoft.com/en-us/cli/azure/ for a complete reference manual on the Azure CLI.

You should also ensure you have installed the Azure development tools for Visual Studio. If you followed the example installation of Visual Studio from Chapter 4, you should have already checked this option. If not, you can add it by running the Visual Studio Installer and selecting *Modify...* and then choosing the *Azure development* workflow as shown in Figure 15-13.

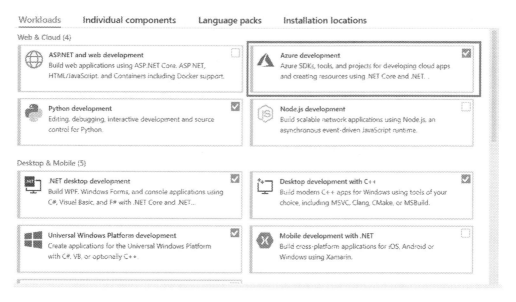

Figure 15-13. *Installing the Azure development tools (Visual Studio)*

One of the tools that this gives you is the Cloud Explorer, which allows you to connect to and work with your Azure components and, for this chapter, the Azure IoT Hub and devices. Cool!

Now, let's create the sample Hello, World Azure style project.

Hello, World! Azure Style

In this section, you will see how to leverage the concepts and features you've seen in the previous sections (plus a couple of new ones) to create a sample Azure IoT solution. We will not be using our IoT device for this example; rather, we will be running an application on our PC and simulating a sensor through a user interface. We will enter some values in the user interface and then verify the data is sent to our IoT Hub using the Azure CLI.

The project covered all the basic things that you need to know to get an IoT solution talking to Azure. You will see the code for creating the connections to Azure as well as how to properly format and send and receive messages using JSON strings.

As I mentioned previously, this project lays the groundwork for the chapter project. Indeed, much of what you develop in this project is used later. The only exception is the user interface. You use it as a means to test the Azure connection and message passing in this sample project, but you will not need it for the chapter project because it is a headless application.

Finally, this project runs on your PC, so there is no need for additional library references (but you can run it on your device if you want). Just make sure that you are connected to the Internet.[8]

The steps we will use for this project mirror those of previous chapters, so the explanations will be brief. In this project, we will need two supporting classes: one to manage the data we will be sending to the cloud and another to contain the code for communicating with the Azure IoT Hub.

Let's start with creating the project.

Create a New Project

You will use a C# project template for this project—the Blank App (Universal Windows) template. Use the name `HelloAzure` for the project name. You can save the project wherever you like or use the default location.

Install Devices.Client NuGet Package

Next, we need to install the *Microsoft.Azure.Devices.Client* NuGet package. Recall, we do this by using the *Project ➤ Manage Nuget Packages...* menu and then search for the name of the package and install it. Refer to the previous chapters if you need more specific instructions or examples.

Now, let's build a user interface, and then you'll add a class to simulate collecting data.

User Interface XAML Code

Let's build a simple user interface to send messages to the cloud as well as display the last message received from the cloud. Click the `MainPage.xaml` file and add the code shown in Listing 15-2.

Listing 15-2.   User Interface XAML Code

```
<Grid Background="{ThemeResource ApplicationPageBackgroundThemeBrush}">
    <StackPanel Width="400" Height="400">
        <TextBlock x:Name="title" Height="60" TextWrapping="NoWrap"
```

---

[8]It's easy to forget this step. Guess how I know.

```
                    Text="Hello, World Azure Style!" FontSize="28"
                    Foreground="Blue"
                    Margin="10" HorizontalAlignment="Center"/>
            <TextBlock Text="Temperature:" />
            <TextBox Name="InTemperature" Text="25.00" />
            <TextBlock Text="Pressure:" />
            <TextBox Name="InPressure" Text="10.00" />
            <Button x:Name="send_button" Content="Send Data" Width="75"
            ClickMode="Press"
                    Click="send_button_Click" Height="50" FontSize="24"
                    Margin="10" HorizontalAlignment="Center"/>
            <TextBlock Text="Message from Azure:" />
            <TextBlock Name="OutAzure" Text="" />
        </StackPanel>
    </Grid>
```

Tip   If you're using a PC that is set up with the dark theme and want to see the application run with a white background, you can add `RequestedTheme="Light"` to the lines above the `<Grid>` declaration to switch to the light theme.

You create a number of labels as well as two text boxes to allow you to choose values for temperature and barometric pressure and a button to send the data to Azure cloud. Figure 15-14 shows an example of what the user interface looks like when the application is running.

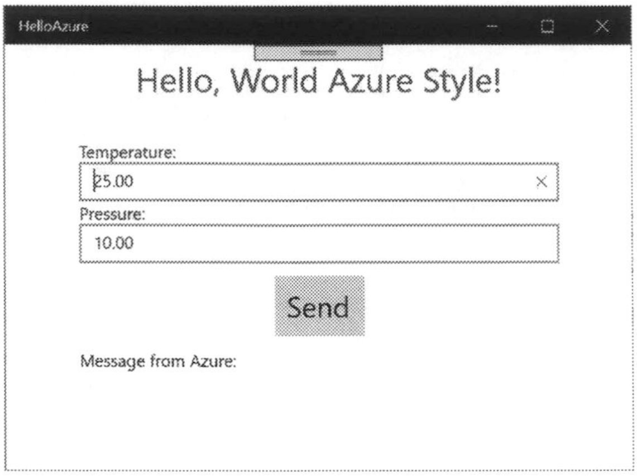

Figure 15-14.   *Sample user interface*

Next, we need a class to simulate encapsulation of capturing data from the sensor.

Sensor Data Class

To create a repository for storing the simulated sensor data, add a new class named SensorData.cs. You can add the new class by right-clicking the project in the Solution Explorer and choosing *Add* ➤ *Add New Item...* menu and then choosing the *Class* entry in the list. Use the name *SensorData* and click the *Add* button.

Now, open that file and add the code to create a new setup attributes for the data values. Since you are simulating data from a sensor that reads temperature and barometric pressure, you create the two data attributes, as shown next:

```
namespace HelloAzure
{
    class SensorData
    {
        public double Temperature { get; set; }
        public double Pressure { get; set; }
    }
}
```

You use this class to fill in the data from the simulated sensor (the user interface). But first, we need code to communicate with our Azure IoT Hub.

Azure IoT Hub Class

To create a class to manage communication with Azure IoT Hub, add a new class named AzureIoTHub.cs. You can add the new class by right-clicking the project in the Solution Explorer and choosing *Add ➤ Add New Item...* menu and then choosing the *Class* entry in the list. Use the name *AzureIoTHub* and click the *Add* button.

In this class, we will need a string to store your connection string, a variable to contain an instance to the DeviceClient class, and two methods—one to send data to the cloud and another to receive messages from the cloud. These are asynchronous methods because we will be calling them from our user interface code.

To send messages to the cloud, we must convert the sensor data we specified in the user interface to a JSON string, formulate a new message, and send it to the cloud. Simple, yes?

Receiving a message from the cloud is almost as easy. In this case, we check to see if a message is available and, if it is, retrieve its body and then retrieve any properties that may be in the payload. For simplicity, we formulate a string so we can paste it in the user interface as an example. We conclude with acknowledging receipt of the message.

Listing 15-3 shows the details of the code for this class. Take a few moments and look it over to ensure you understand how it works.

Listing 15-3. Azure IoT Hub Class

```
using Microsoft.Azure.Devices.Client;
using Newtonsoft.Json;
using System.Text;
using System.Threading.Tasks;

namespace HelloAzure
{
    class AzureIoTHub
    {
        const string deviceConnectionString = "YOUR KEY GOES HERE";
        static DeviceClient deviceClient = DeviceClient.CreateFromConnectio
        nString(deviceConnectionString, TransportType.Amqp);

        public static async Task SendDeviceToCloudMessageAsync(SensorData
        data)
```

```
        {
            var jsonMsg = JsonConvert.SerializeObject(data);
            var message = new Message(Encoding.ASCII.GetBytes(jsonMsg));
            await deviceClient.SendEventAsync(message);
        }

        public static async Task<string> ReceiveCloudToDeviceMessageAsync()
        {
            while (true)
            {
                Message receivedMessage = await deviceClient.
                ReceiveAsync();
                if (receivedMessage == null) continue;
                var messageData = Encoding.ASCII.GetString(receivedMessage.
                GetBytes());
                messageData += "{";
                foreach (var item in receivedMessage.Properties)
                {
                    messageData += " " + item.Key + ":" + item.Value + ",";
                }
                messageData += "}";
                await deviceClient.CompleteAsync(receivedMessage);
                return messageData;
            }
        }
    }
}
```

Caution   Remember to copy in your Azure IoT Hub connection string that includes the device in the string. We will do this in a later step.

Now, we can return to the user interface and complete the code there.

User Interface Code

Return to the user interface MainPage.xaml.cs and add the code shown in Listing 15-4 to add the data from the user interface to the SensorData. We also add a timer to periodically check for messages from the cloud. Once again, the code is presented without explanation as an exercise.

Listing 15-4. Code for Reading Sensor Data from the User Interface

```
using System;
using System.Diagnostics;
using Windows.UI.Xaml;
using Windows.UI.Xaml.Controls;

// The Blank Page item template is documented at https://go.microsoft.com/
fwlink/?LinkId=402352&clcid=0x409

namespace HelloAzure
{
    public sealed partial class MainPage : Page
    {
        public MainPage()
        {
            this.InitializeComponent();
            DispatcherTimer timer = new DispatcherTimer();
            timer.Tick += timerTick;
            timer.Interval = TimeSpan.FromSeconds(3);
            timer.Start();
        }

        private async void send_button_Click(object sender, RoutedEventArgs e)
        {
            // Read data from the simulated sensor
            SensorData data = new SensorData();
            data.Temperature = Convert.ToDouble(this.InTemperature.Text);
            data.Pressure = Convert.ToDouble(this.InPressure.Text);

            // Send data to the cloud
```

```
        await AzureIoTHub.SendDeviceToCloudMessageAsync(data);

        Debug.WriteLine(String.Format("Data sent: {0}, {1}",
            data.Temperature, data.Pressure));
    }

    private async void timerTick(object sender, object e)
    {
        this.OutAzure.Text = await AzureIoTHub.
        ReceiveCloudToDeviceMessageAsync();
    }
  }
}
```

OK, we now have all of the code we need, and we're ready to test it. At this point, if you're following along, it would be a good idea to compile the code to ensure everything is correct in the code. If not, be sure to fix it before you proceed.

Next, we need to create a test device in our Azure IoT Hub. For this, we will use the Cloud Explorer feature in Visual Studio.

Add a Test Device

For this step, we could use the Azure Portal (and you're welcome to do so) or the Azure CLI to create a device. However, let's look at another cool feature in Visual Studio—the Cloud Explorer.

The Cloud Explorer is a small tool in Visual Studio that lets you connect to your Azure IoT Hub and perform a number of simple operations such as monitoring messages and creating devices. This last feature is what we will demonstrate here.

To open the Cloud Explorer, click the *View* ➤ *Cloud Explorer* menu. This will open a new pane on (typically) the right side of Visual Studio. If you haven't associated an Azure account, you will be asked to log in. Visual Studio will remember this account going forward. You may need to log out and back in again if your account doesn't show in the list. Figure 15-15 shows an example of the Cloud Explorer connected to a trial account (mine).

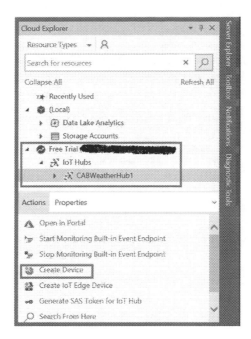

Figure 15-15. *Cloud Explorer (Visual Studio)*

Notice the marked areas in the figure. First, we see a tree control we can use to expand to reveal the list of resources in our account. In this example, there is only one Azure IoT Hub named CABWeatherHub1.

Once you click a resource, a context menu opens at the bottom to show you which functions are available. You can also get this menu by right-clicking an item in the tree.

In this case, we want to select the Azure IoT Hub, then click the *Create Device* menu item as shown. Once you do that, a small text box will appear to allow you to name the device. Go ahead and do that now and name the device *TestDevice*. Figure 15-16 shows the results of adding the device.

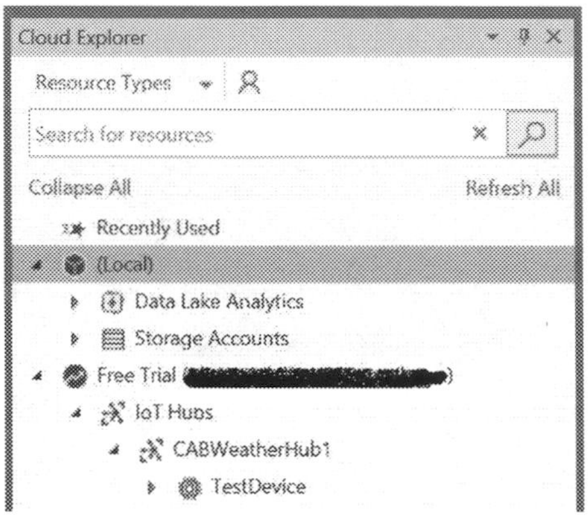

Figure 15-16. *Cloud Explorer with device (Visual Studio)*

That's it! Quick and easy—no forms to fill out or anything. What this means is you will get a device associated with that Azure IoT Hub using all of the defaults. You can, of course, go into your Azure Portal and make any changes you want to the device later.

OK, we've got our code. We've got a device. Let's test it out.

Testing the Application

To test the example project, we're going to use two tools: first, we will use the Azure CLI to monitor events for our device to see if the data is reaching the cloud; second, we will use the Azure Portal to send a message from the cloud to the device to ensure it can receive messages. The first is known as device-to-cloud and the second cloud-to-device, common terms used in the Microsoft Azure documentation.

Testing Device-to-Cloud Messages

To ensure our device is successfully sending messages (our simulated sensor telemetry) to the Azure IoT Hub, we will use the Azure CLI to connect to the hub's event endpoint and echo any messages received. If we see the messages here, we can verify they are received by Azure.

To begin, you must first copy your connection string for your device into your code. Recall, we will be placing the string in the AzureIoTHub.cs class module variable. To locate your connection string, log in to the Azure Portal and navigate to our hub and then click the TestDevice. This will show the main page (blade) for the device that includes all of your connection strings. You will want to click the Copy button for the primary connection string as shown in Figure 15-17. Now, paste that string into the deviceConnectionString variable in the AzureIoTHub.cs file.

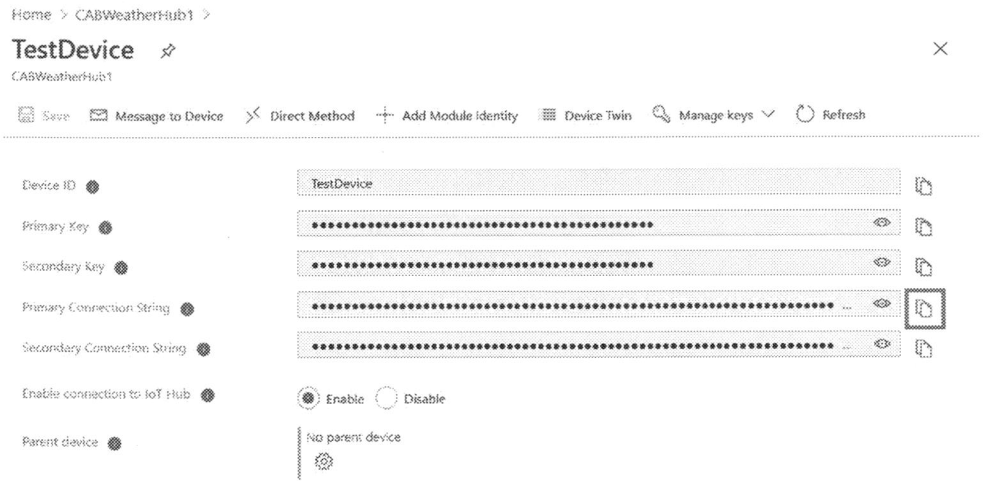

Figure 15-17. *Locating the primary connection string for a device*

OK, now we can launch the application. Once the application is started, open a new terminal window and log in to Azure using the command `az login`. Recall, this will open your browser and request your login.

Once logged in, we need to create a temporary shared access signature (SAS[9] token). Use the following command to create the credential with a duration of 3600 seconds for the Azure IoT Hub named `CABWeatherHub1`. This will be cached and active while we test our application.

```
az iot hub generate-sas-token --duration 3600 -n CABWeatherHub1
```

Next, we need to start monitoring the default event endpoint for the hub. Use the following command to start the monitor for the specific hub as before:

```
az iot hub monitor-events --hub-name CABWeatherHub1
```

Next, return to your application and change the values in the text boxes and click Send. Do this a couple of times. What you should see in the terminal are those messages you sent to the cloud. Depending on your Internet speed, you may see them momentarily, or it may take a few seconds for them to appear. Listing 15-5 shows a transcript of a test session.

---

[9]`https://docs.microsoft.com/en-us/azure/storage/common/storage-sas-overview`

Notice the payload section of the messages found. This contains the properties you sent for the telemetry. When you are finished monitoring messages, click *Ctrl+C* and then *Y* to end the monitor command.

Listing 15-5. Monitoring Messages Using the Azure CLI

```
C:\Users\cbell>az login
You have logged in. Now let us find all the subscriptions to which you have
access...
[
  {
    "cloudName": "AzureCloud",
...
  }
]

C:\Users\cbell>az iot hub generate-sas-token --duration 3600 -n
CABWeatherHub1
{
  "sas": "SharedAccessSignature sr=CABWeatherHub1..."
}

C:\Users\cbell>az iot hub monitor-events --hub-name CABWeatherHub1
Starting event monitor, use ctrl-c to stop...
{
    "event": {
        "origin": "TestDevice",
        "module": "",
        "interface": "",
        "component": "",
        "payload": "{\"Temperature\":23.3,\"Pressure\":101.22}"
    }
}
{
    "event": {
        "origin": "TestDevice",
        "module": "",
        "interface": "",
```

```
        "component": "",
        "payload": "{\"Temperature\":23.55,\"Pressure\":101.33}"
    }
}
{
    "event": {
        "origin": "TestDevice",
        "module": "",
        "interface": "",
        "component": "",
        "payload": "{\"Temperature\":24.91,\"Pressure\":102.43}"
    }
}
Stopping event monitor...
Terminate batch job (Y/N)? y
```

OK, now we know our application can send data; let's see if it can receive data.

Testing Cloud-to-Device Messages

To test sending messages to the device, we will go back to the Azure Portal and navigate to the device as we did previously. On the device page, click the Message to Device link at the top. This will open a new page that will permit you to specify a message and optionally any properties you want to send.

Let's send a simple message and add two properties. We first must add the properties one at a time with a key and a value. Once they are set, click *Add Property* to add the property. Do that again for another property. The keys and values do not matter as we are simply testing receipt of the payload. Finally, type a message in the body. Figure 15-18 shows the panel (blade) prior sending the message to the device.

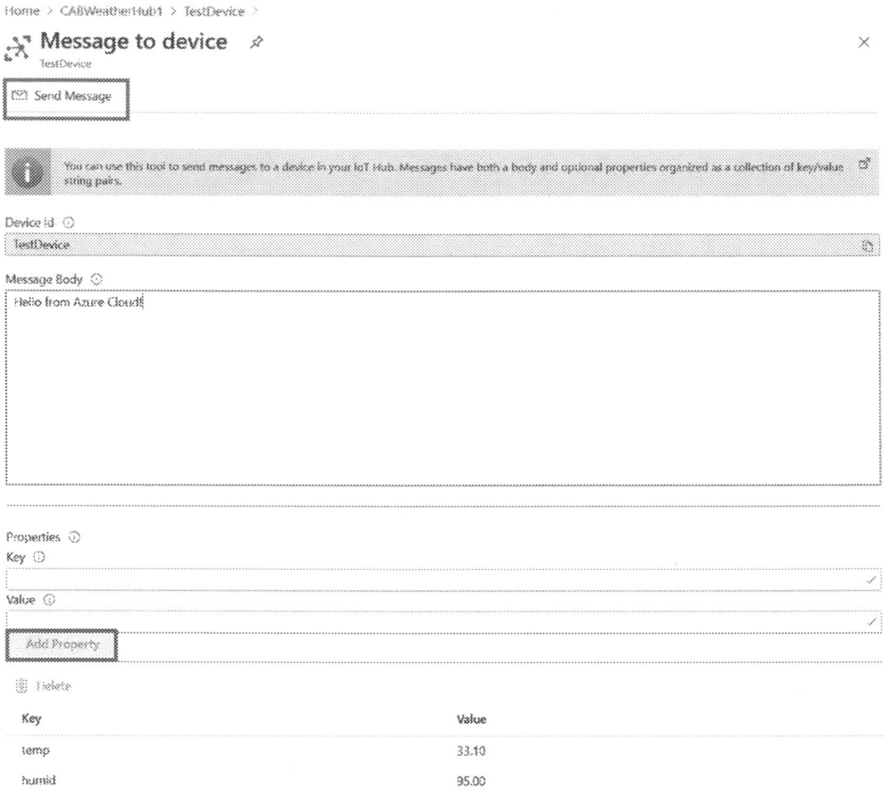

Figure 15-18. *Sending a message to the device (Azure Portal)*

Before you send the message, make sure your application is running. When you are ready, click Send Message to send the message to the device. After a moment or so, you should see the message captured by the application as demonstrated in Figure 15-19. Notice we've captured the properties from the message.

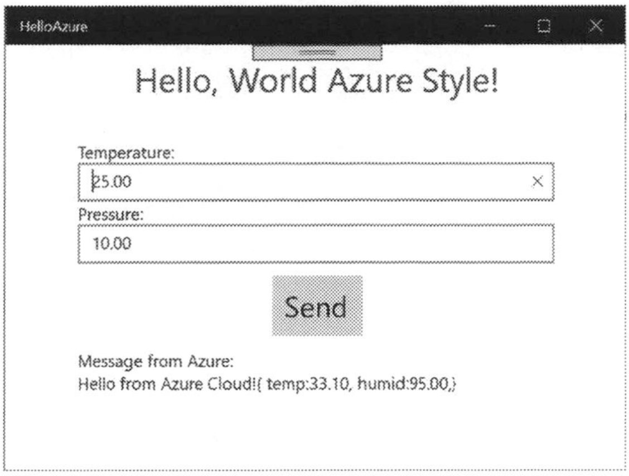

Figure 15-19. *Application captures message from cloud*

So, what just happened? The portal sent the message to the device message queue, and then the application (in the timer tick event) connected to Azure and retrieved the message. How fantastic is that?

If you got all of this to work, take a victory lap around your house, apartment, or office. You've just created a sample project that stores data in the cloud.

Caution   Be sure to delete your resource group **and** the cluster when you no longer need them to prevent unexpected charges on your account.

Summary

Using cloud computing services, especially those that provide tools and features specifically for IoT solutions, such as Microsoft Azure, can be a steep learning curve. Fortunately, Microsoft has worked very hard to streamline the process to get you going quickly and is supported with an extensive documentation library. While the setup process is easy and the code to interact with the services is not difficult, there is nothing simple or basic about the features included in the Azure IoT Suite.

In this chapter, we went on a whirlwind tour of a part of the Azure IoT products and witnessed how messages are sent to and from the cloud. In the next chapter, we will take these concepts a step further and build our Azure IoT weather project from scratch. We will also see an advanced product from Azure to store and display the data. Cool!

Azure IoT Solutions: Building an Azure IoT Solution

Now that you've gotten a taste of what working with Azure is like, let's put those new skills to work building a real project that uses an IoT device and a sensor to send data to the cloud where we will store and display them. The examples in the last chapter showed how to display the data on the device (or your PC), but we didn't see any way to save the data.

In this chapter, we will take the example project a step further. You will reuse the weather sensor that you used in Chapter 13. However, rather than saving the data in MySQL, you will send it to Microsoft Azure. You will also see how to present the data via a website connected to the data in Microsoft Azure, but it's going to be a bumpy ride. The technology and code is rather straightforward but can be tedious to set up. Never fear, I will walk you through every step.

Getting Started

Now that you have seen how to get started writing IoT solutions with Azure, including how to set up an account, an IoT Hub, and a sample application that sends and receives messages from the cloud, let's now see how to send real data to Azure, process the data, and view it in a website—all of which is hosted in Azure. While it may seem complex, the solution is still quite small yet provides a demonstration of what is possible.

© Charles Bell 2021
C. Bell, *Windows 10 for the Internet of Things*, https://doi.org/10.1007/978-1-4842-6609-0_16

Note   The following demonstrates one of the many ways you can utilize Azure to build a cloud-based IoT solution. For larger enterprise-grade solutions, check out the Azure IoT Suite management website at `www.azureiotsolutions.com/Accelerators`.

To implement this solution, you are going to discover several new tools and features. It's a bumpy ride through the many screens and a bit of a minefield with respect to terminology, but if you take your time and follow along, you can get this project running and see your IoT data in the cloud.

Before we dive into the code and start working with Azure, let's look at an overview of the Azure components we will use. The following lists the services and products we will use in Azure. They are presented in the order in which you would typically create and deploy them in Azure. While the products used may differ, this process is a good baseline to use for developing Azure IoT solutions.

1. *Configure the IoT device*: Create an application deployed to the IoT device that generates data from one or more sensors or other inputs.

2. *Test the IoT device reading data from a sensor*: Test to ensure the IoT device is capturing data from the sensor correctly.

3. *Create an Azure IoT Hub and device*: Create the gateway for our devices to push data to the cloud.

4. *Test the IoT device sending to the cloud*: Test to ensure the data captured by the IoT device is being sent to the IoT Hub.

5. *Create and configure an Azure Data Explorer Cluster*: Create the data storage product composed of multiple Azure resources including a database storage component to route data from the IoT Hub for storage.

6. *Test the data stream*: Test to ensure the data captured by the IoT device is being routed to the Data Explorer Database.

7. *Set up data visualization in the Data Explorer Dashboard*: Generate queries against the database to display the data using visualization tools.

8. *Test the visualization*: Test the entire process to ensure the data captured by the IoT device is being visualized in the Data Explorer Dashboard.

As you can see, we'll be using three major components in Azure. If we view them in reverse skipping the intermediate testing steps, it is easy to see how the solution works; we visualize data in the Data Explorer Dashboard, which is connected via queries to the database in the Data Explorer Cluster, which is in turn pulling the data from the IoT Hub, which queues the data sent from our IoT device.

As you study how to build more sophisticated solutions in Azure, you will learn that, while simplified, this is a general pattern most solutions use to process data through Azure components. In essence, we use the components in Azure to connect the pieces we want until we reach our goal routing the data as we go.

In this section, we will see how to build this solution from the IoT device up through the chain of Azure components.

Configure the IoT Device

In this step, we will create an application and deploy it to our IoT device. We will write the code so that it retrieves data from the temperature sensor and prepares to send it to the cloud. The project resembles the project from Chapter 13, so if you have completed that project and the other projects in this book, this section should be familiar to you.

Note   Some of this section is taken from Chapter 13 and intentionally repeated here for clarity and completeness.

Required Components

The following lists the components that you need. You can get the BMP280 sensor from Adafruit (`www.adafruit.com`) either in the Microsoft IoT Pack for Raspberry Pi or purchased separately, SparkFun (`www.sparkfun.com`), or any electronics store that carries electronic components. However, if you use a sensor made by someone other than Adafruit, you may need to alter the code to change the I2C address. Since this solution is a headless application, you do not need a monitor, keyboard, and mouse.

- Adafruit BMP280 I2C or SPI barometric pressure and altitude sensor

- Jumper wires: (4) male-to-female

- Breadboard (full size recommended but half size is OK)

- Raspberry Pi 2 or 3

- Power supply

Set Up the Hardware

Although there are only four connections needed for this project, you will plan for how things should connect, which is good practice to hone. To connect the components to the Raspberry Pi, you need four pins for the BMP280 sensor, which requires only power, ground, and two pins for the I2C interface. Refer to Chapter 13 for how to wire the sensor. I include Figure 16-1 as a reminder of how things are connected.

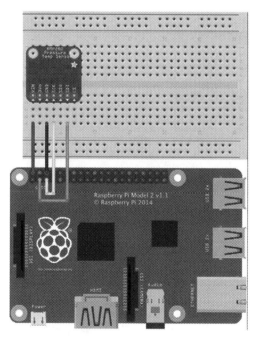

Figure 16-1. *Connections for the weather sensor project*

If you are following along with this chapter working on the project, go ahead and make the hardware connections now. Don't power on the board yet but do double- and triple-check the connections.

Write the Code

Now it's time to write the code for the project. You will use the same BMP280 sensor from Chapter 13, but instead of writing the data to MySQL, you will send it to the cloud. We will also use C# instead of Visual Basic. You won't use a user interface, so this is a headless solution.

For this portion of the project, we will create a class that we can use to store and read the data from the sensor. We will name the class WeatherSensor, which incorporates the WeatherData structure from Chapter 13, and use it in the main code module to read data using a timer callback. Let's get started.

New Project

You will use a C# project template for this project—the C# Background Application (IoT) template. Use the name AzureWeather for the project name. You can save the project wherever you like or use the default location.

We use this project type because IoT solutions like this example typically do not use a user interface. Plus, it's a good example to show you some coding alternatives for your own project.

Note   The book source code stores the code for this section in a folder named AzureWeatherBase since this is the starting point for the chapter project.

Before we go any further, let's add the additional packages we will need for this project. We need to add two references from NuGet to the solution. You need the *Glovebox.IoT.Devices* library, which in turn requires the *Units.NET* library. Recall, we do this by using the *Project ➤ Manage Nuget Packages...* menu and then search for the name of the package and install it. Refer to the previous chapters if you need more specific instructions or examples.

Caution   Be sure to select version 3.34 when installing the library. Other versions are not compatible.

Now, let's add a class to manage the sensor for us.

Add the Sensor Class

We use a new class named WeatherSensor to manage interaction with the BMP280. The class will contain a struct named WeatherData that is the same structure we used in Chapter 13 to contain the data read. We will also add a method to read data from the sensor. Later, we will expand the class with additional functionality.

Right-click the project and choose *Add* ➤ *New Item....* Choose *Class* in the list and name the file WeatherSensor.cs. Then, double-click the file in the project to open the file in the editor (if not already open).

Listing 16-1 shows the complete code for this class. Since it uses code we're familiar with, I leave the details of how the code works as an exercise, but it is the same code from Chapter 13 albeit a bit refactored. Notice I had to move the using clause for the Glovebox library to the class.

Listing 16-1. The WeatherSensor Class

```csharp
using Glovebox.IoT.Devices.Sensors;
using System;

namespace AzureWeather
{
    class WeatherData
    {
        public double degreesCelsius { get; set; }
        public double degreesFahrenheit { get; set; }
        public double bars { get; set; }
        public double hectopascals { get; set; }
        public double atmospheres { get; set; }
        public string eventTime { get; set; }
    }

    class WeatherSensor
    {
        private BMP280 tempAndPressure;      // Instance of BMP280 class
        public WeatherSensor()
        {
            // Instantiate a new BMP280 class instance
```

```
            tempAndPressure = new BMP280();
        }
        public WeatherData ReadData()
        {
            WeatherData weatherData = new WeatherData();
            weatherData.degreesCelsius = tempAndPressure.Temperature.
            DegreesCelsius;
            weatherData.degreesFahrenheit = tempAndPressure.Temperature.
            DegreesFahrenheit;
            weatherData.bars = tempAndPressure.Pressure.Bars;
            weatherData.hectopascals = tempAndPressure.Pressure.
            Hectopascals;
            weatherData.atmospheres = tempAndPressure.Pressure.Atmospheres;
            weatherData.eventTime = DateTime.Now.ToString();
            return weatherData;
        }
    }
}
```

You may be wondering why the code was moved to a class. This is simply to demonstrate how to build your projects in a more modular way. Sure, you could place all of the code for the project in the main code file, but it would not be as reusable and may suffer from maintenance issues should the code continue to evolve.

OK, now let's finish the code for the project.

Complete the Code

Now we will complete the code for the project. Double-click the StartupTask.cs file in the project tree. There are a number of namespaces you need to include. Go ahead and add those now, as shown next:

```
using System;
using System.Diagnostics;
using System.Timers;
using Windows.ApplicationModel.Background;
```

Next, we will need some variables. We will need a timer to use as a timing loop to set the sampling frequency (how often we take samples)—we will set this very high for testing, but you should change it to something more appropriate later. We also will need a variable to hold an instance of the WeatherSensor class that we created. Finally, we will need a variable to capture a task deferral to keep the background application running. The following shows an excerpt with these variables added (shown in bold):

```
namespace AzureWeather
{
    public sealed class StartupTask : IBackgroundTask
    {
        private static Timer timer;  // Timer for reading from sensor
        WeatherSensor weatherSensor = new WeatherSensor();
        private static BackgroundTaskDeferral _Deferral = null;
...
```

The next modification to the StartupTask.cs file is to add code in the Run() method to set up the timer and its callback. This is very similar to how we used the DispatcherTimer in XAML except we will use the system timer. The following shows the code for the Run() method. Notice I used a timer interval of 10 seconds. You will want to increase that for typical weather data acquisition.[1]

```
public void Run(IBackgroundTaskInstance taskInstance)
{
    _Deferral = taskInstance.GetDeferral();
    timer = new System.Timers.Timer();
    timer.Interval = 10000;

    timer.Elapsed += this.BmpTimer_Tick;
    timer.AutoReset = true;
    timer.Enabled = true;
}
```

---

[1]Think about it this way; how often do you need to check temperature? Is it likely to change much in seconds, minutes, or hours, and at what interval is such a change pertinent to your analysis and goals?

Finally, we add the callback method named BmpTimer_Tick(). In this method, we will read data from the WeatherSensor class. Listing 16-2 shows the completed method. Notice we simply get the data from the sensor using the ReadData() method, which returns a struct named WeatherData that we use to print the values read using the debugging methods.

Listing 16-2. The BmpTimer_Tick Callback Method

```
private void BmpTimer_Tick(Object source, System.Timers.ElapsedEventArgs e)
{
    // Read new values from the sensor
    WeatherData data = weatherSensor.ReadData();

    // Record the data to debug
    Debug.WriteLine("Data Read at: " + DateTime.Now.ToString("MM/dd/yyyy
    HH:mm:ss"));
    Debug.WriteLine("--------------------------------");
    Debug.WriteLine(data.degreesCelsius.ToString("F2") + " C");
    Debug.WriteLine(data.degreesFahrenheit.ToString("F2") + " F");
    Debug.WriteLine(data.bars.ToString("F2") + " Pressure (Bars)");
    Debug.WriteLine(data.hectopascals.ToString("F2") + " Pressure
    (Hectopascals)");
    Debug.WriteLine(data.atmospheres.ToString("F2") + " Pressure
    (Atmospheres)");

    // Send data to Azure IoT Hub
}
```

Now that we've coded the application, we can test it to ensure we have the code working to read data from the sensor. For now, we're just printing it to a debug statement, but as you can see in the preceding code, we will add the code to send the data to Azure in a later step.

Test the IoT Device Reading Data from Sensor

In this section, we will test to ensure the IoT device is capturing data from the sensor correctly. Begin by deploying the application to your IoT device.

Deploy and Execute

For this project, we will use debug mode in order to see the debug messages; you will need to be using the debug build. Recall from Chapter 6 (see the "Deploy and Debug" section and Figure 6-11), to deploy our application in debug mode, we must set the debug settings for our application, setting the following in the debug dialog:

- *Target device*: Remote Machine

- *Remote Machine*: The name of your device

- (Optional) *Uninstall and then re-install*: Checked

- (Optional) *Deploy optional packages*: Checked

- *Application process*: Managed Only

- *Background task process*: Managed Only

Once you make any changes in this dialog, be sure to save your solution. Building or deploying does not automatically save these settings.

We should also change the package name in the package manifest. Recall from Chapter 6 (see the "Deploying Your Application (No Debugging)" section and Figure 6-9), we can change the package name. Do that now so you can find your application easier in the Device Portal.

Now you can deploy your application. Go ahead and do that now. You can run the deployment from the Debug menu, and when you do, you see the debug statements in the output window. Listing 16-3 shows an excerpt of the data you should see. When you start the debugger, the output window automatically opens and displays the debug output. If it does not, you can open the window manually and select debug from the drop-down menu.

Listing 16-3. Example Debug (Output Window)

```
Data Read at: 11/01/2020 17:12:37
--------------------------------
19.49 C
67.08 F
1.01 Pressure (Bars)
1006.35 Pressure (Hectopascals)
0.99 Pressure (Atmospheres)
```

```
...
Data Read at: 11/01/2020 17:12:46
---------------------------------
23.28 C
73.90 F
1.01 Pressure (Bars)
1006.30 Pressure (Hectopascals)
0.99 Pressure (Atmospheres)
...
Data Read at: 11/01/2020 17:12:56
---------------------------------
20.80 C
69.44 F
1.01 Pressure (Bars)
1006.27 Pressure (Hectopascals)
0.99 Pressure (Atmospheres)
```

OK, that's it! We have now completed two of the eight steps in our Azure IoT solution. We have written an application to read data from a sensor and tested it to ensure it works. Now, let's move on to the first phase of interactive with Azure.

Create an Azure IoT Hub and Device

In this step, we will create the gateway for our devices to push data to the cloud. Specifically, we will create an Azure IoT Hub and a device associated with the hub. The steps in this section are identical to the preceding section entitled "Set Up an IoT Hub" and repeated here in overview form. Refer to the previous section if you need additional details on how to create the Azure IoT Hub.

Create the IoT Hub

If you're not on the home page, click *Home* in the upper left. Then, hover over IoT Hub and then click *Create*. Unlike the previous section, we will take a shortcut and populate only the data on the first page. Values for this project are shown in italics:

- *Subscription*: Leave this set to *Free Trial*.

- *Resource group*: Recall, a resource group is a container that we can use to group resources. I recommend creating a new one for this project—*CABWeatherRG*.

- *Region*: Choose your (general) location—*East US*.

- *IoT Hub name*: Choose a name that is unique (Azure tells you if it is not). You have only a few characters to work with so keep it short— *CABWeatherHub*.

Figure 16-2 shows the completed form for creating the IoT Hub for this project.

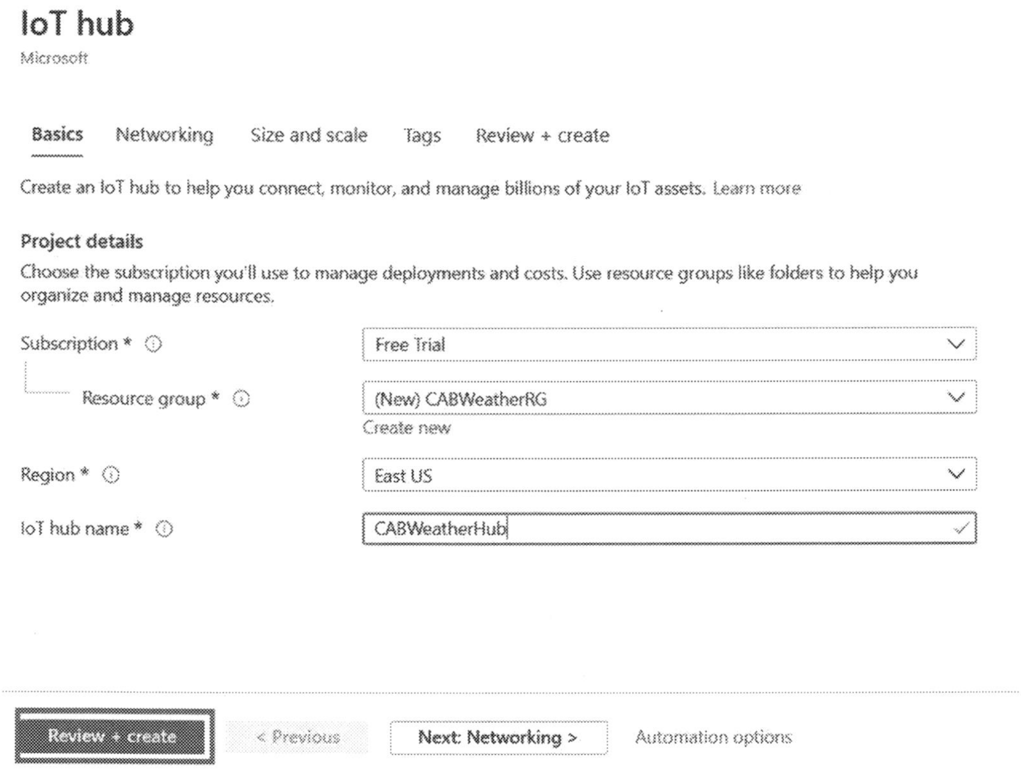

IoT hub
Microsoft

Basics   Networking   Size and scale   Tags   Review + create

Create an IoT hub to help you connect, monitor, and manage billions of your IoT assets. Learn more

Project details

Choose the subscription you'll use to manage deployments and costs. Use resource groups like folders to help you organize and manage resources.

Subscription * ⓘ	Free Trial ▾
Resource group * ⓘ	(New) CABWeatherRG ▾
	Create new
Region * ⓘ	East US ▾
IoT hub name * ⓘ	CABWeatherHub ✓

Review + create	< Previous	Next: Networking >	Automation options

Figure 16-2.   *Create the IoT Hub*

Once you have the data entered, click *Review + create* and then *Create* on the summary page to create the resource.

> **Note**   If you don't have a resource group yet, you will need to create one by clicking the *Create* link under the text box.

The next step is the registration process, which can take some time to complete. Be sure to wait until the resource is deployed before proceeding.

> **Caution**   Be sure to delete any resources you aren't using so you don't get billed for them!

Once you see the Deployments Complete message, click the *Go to resource* button to open the IoT Hub. Next, let's set up your device in the IoT Hub.

Create the Device in the IoT Hub

To create a device in the hub, scroll down and select the entry *IoT devices* in the list and then click the *New* button as shown in Figure 16-3.

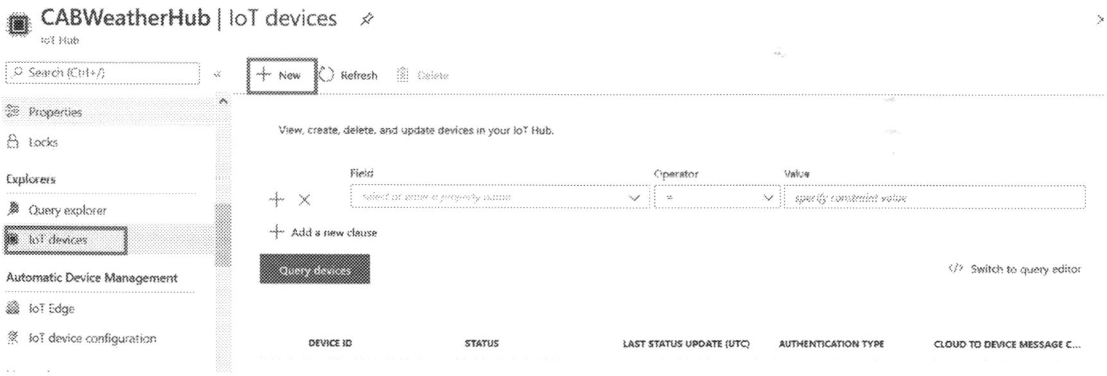

Figure 16-3. *Manage devices in IoT Hub*

On the next page, enter a name for the device (use RasPi_BMP280_1) and leave the default values for keys. Make certain the *Auto-generate keys* tick box is ticked so keys are generated for you. Then click *Save* to create the device as shown in Figure 16-4.

Figure 16-4. *Create new device in IoT Hub*

Once the device is created and listed in the devices blade for the hub, click it to open. Recall, here is where we go to find (and copy) the keys we use to connect our IoT device to the device in the hub. Figure 16-5 shows where you can find the keys.

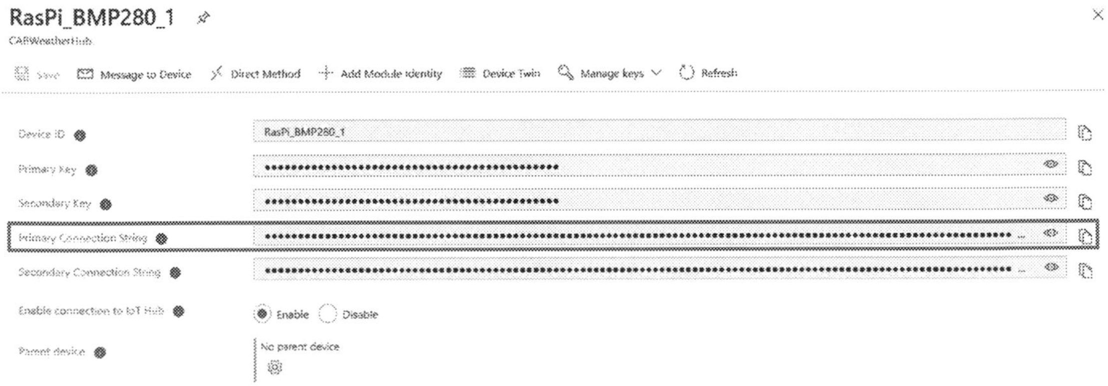

Figure 16-5. *IoT Hub device details—keys*

Now that we have the IoT Hub and device created and we've located the correct keys, we can write the next portion of the project to send data to the hub.

WHAT ABOUT USING THE AZURE CLI?

If you've been following along with the chapter but prefer to work with command-line clients like the Azure CLI, you can! In fact, you can do all manner of things with the client including creating the IoT Hub. The following shows examples of the commands you will need to use. These are the commands I used to create the IoT Hub for this example. Some commands may require installing certain extensions to the CLI, but these are also well documented.

```
# Login to Azure
az login

# Create resource group
az group create --name CABWeatherRG --location eastus

# Create IoT Hub
az iot hub create --name CABWeatherHub --resource-group CABWeatherRG --sku S1
```

If you'd prefer using the CLI, you'll find it has a very nice help system. Just use `--help` with any command to see what options are available or visit the online documentation at `https://docs.microsoft.com/cli/azure/`.

Test the IoT Device Sending to the Cloud

In this section, we will connect our application to the Azure IoT Hub and ensure the data captured by the IoT device is being sent to the IoT Hub. Let's start with modifying the project.

Write the Code

If you haven't already, open the AzureWeather project in Visual Studio. Fortunately, all of the work we did in the tutorial from earlier in the chapter can be reused. We just need to add one more resource and some code to format the messages for the BMP280.

Install Devices.Client NuGet Package

First, we need to install the *Microsoft.Azure.Devices.Client* NuGet package. Recall, we do this by using the *Project ➤ Manage Nuget Packages...* menu and then search for the name of the package and install it. Refer to the previous chapters if you need more specific instructions or examples.

Azure IoT Hub Class

Recall from the tutorial, we want to create a class to manage communication with Azure IoT Hub. We add a new class named `AzureIoTHub.cs`. You can add the new class by right-clicking the project in the Solution Explorer and choosing *Add ➤ Add New Item...* menu and then choosing the *Class* entry in the list. Use the name *AzureIoTHub* and click the *Add* button.

In this class, we will need a string to store your connection string, a variable to contain an instance to the `DeviceClient` class, and a single method to send data to the cloud.

To send messages to the cloud, we must convert the sensor data we specified in the `WeatherData` struct to a JSON string, formulate a new message, and send it to the cloud. Listing 16-4 shows the complete code for the class.

Listing 16-4. AzureIoTHub Class

```
using Microsoft.Azure.Devices.Client;
using Newtonsoft.Json;
using System.Text;
using System.Threading.Tasks;

namespace AzureWeather
{
    class AzureIoTHub
    {
        const string deviceConnectionString = "YOUR KEY GOES HERE";
        static DeviceClient deviceClient = DeviceClient.CreateFromConnectio
        nString(deviceConnectionString, TransportType.Amqp);

        public async Task SendData(WeatherData data)
        {
```

```
            var jsonMsg = JsonConvert.SerializeObject(data);
            var message = new Message(Encoding.ASCII.GetBytes(jsonMsg));
            await deviceClient.SendEventAsync(message);
        }
    }
}
```

We also need to change the connection string. Go back to your Azure Portal and copy the primary connection string from the IoT Hub device and paste it into the deviceConnectionString variable declaration.

Next, we must add a call to the class in the timer callback. Open the StartupTask. cs file and add the following using statement and code to the callback method (shown in bold) as shown in Listing 16-5.

Listing 16-5. Completed StartupTask.cs Code

```
using System;
using System.Diagnostics;
using System.Threading.Tasks;
using System.Timers;
using Windows.ApplicationModel.Background;

namespace AzureWeather
{
    public sealed class StartupTask : IBackgroundTask
    {
        private static Timer;  // Timer for reading from sensor
        WeatherSensor = new WeatherSensor();
        private static BackgroundTaskDeferral _Deferral = null;
        public void Run(IBackgroundTaskInstance taskInstance)
        {
            _Deferral = taskInstance.GetDeferral();
            timer = new System.Timers.Timer();
            timer.Interval = 10000;

            timer.Elapsed += this.BmpTimer_Tick;
            timer.AutoReset = true;
            timer.Enabled = true;
```

```
        }

        private void BmpTimer_Tick(Object source, System.Timers.
        ElapsedEventArgs e)
        {
            // Read new values from the sensor
            WeatherData data = weatherSensor.ReadData();

            // Record the data to debug
            Debug.WriteLine("Data Read at: " + DateTime.Now.ToString("MM/
            dd/yyyy HH:mm:ss"));
            Debug.WriteLine("--------------------------------");
            Debug.WriteLine(data.degreesCelsius.ToString("F2") + " C");
            Debug.WriteLine(data.degreesFahrenheit.ToString("F2") + " F");
            Debug.WriteLine(data.bars.ToString("F2") + " Pressure (Bars)");
            Debug.WriteLine(data.hectopascals.ToString("F2") + " Pressure
            (Hectopascals)");
            Debug.WriteLine(data.atmospheres.ToString("F2") + " Pressure
            (Atmospheres)");

            // Send data to Azure IoT Hub
            AzureIoTHub hub = new AzureIoTHub();
            var t = Task.Run(() => hub.SendData(data));
        }
    }
}
```

That's it! The code is complete. Be sure to compile the code to ensure it builds before moving to the next step.

Deploy and Execute

Once the code compiles successfully, go ahead and deploy it to your device like we did earlier. You can use debug to step through the code if you'd like or run it in debug mode but remove any breakpoints you may have added—whatever you'd like.

Recall we can start and stop our application from the Device Portal. Figure 16-6 shows the entry in the application list and the Actions drop-down for starting, stopping, and uninstalling the application.

Figure 16-6. *Actions for application in Device Portal*

Verify Data Received by the IoT Hub

Recall, we won't see much in the Azure Portal about our device, but we can see if it is sending messages. If you navigate to the IoT Hub overview page, you can scroll down and see some graphs that indicate devices have connected and the number of messages sent as demonstrated in Figure 16-7. Note that you may have to change the setting to the current hour to see recent updates.

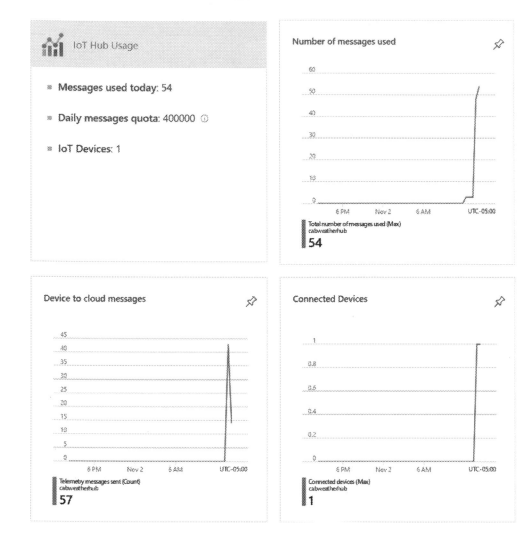

Figure 16-7. *IoT Hub metrics*

However, none of the charts will show you what messages were received. For that, we can use the Azure CLI. Recall, we will first log in, then create an access token, and finally monitor the events. Listing 16-6 shows an excerpt of such a session taken while the application is running. Refer to the previous section if you need more details about any of the commands.

Listing 16-6. Monitoring Events Using Azure CLI

```
C:\Users\cbell>az login
You have logged in. Now let us find all the subscriptions to which you have
access...
[
  {
    "cloudName": "AzureCloud",
...
}
]

C:\Users\cbell>az iot hub generate-sas-token --duration 3600 -n
CABWeatherHub
{
  "sas": "SharedAccessSignature sr=CABWeatherHub.azure-devices.net&sig=..."
}

C:\Users\cbell>az iot hub monitor-events -d RasPi_BMP280_1 --hub-name
CABWeatherHub
Starting event monitor, filtering on device: RasPi_BMP280_1, use ctrl-c to
stop...
{
    "event": {
        "origin": "RasPi_BMP280_1",
        "module": "",
        "interface": "",
        "component": "",
        "payload": "{\"degreesCelsius\":18.990000000000009,\"degreesFahren
        heit\":66.182,\"bars\":1.0247225,\"hectopascals\":1024.7225,\"at
        mospheres\":1.0113224771773994,\"eventTime\":\"11/2/2020 9:51:18
        AM\"}"
    }
}
{
    "event": {
        "origin": "RasPi_BMP280_1",
```

```
        "module": "",
        "interface": "",
        "component": "",
        "payload": "{\"degreesCelsius\":18.98000000000018,\"degreesFahren
          heit\":66.164000000000016,\"bars\":1.024747,\"hectopascals\":1024
          .747,\"atmospheres\":1.0113466567974341,\"eventTime\":\"11/2/2020
          9:51:08 AM\"}"
    }
}
{
    "event": {
        "origin": "RasPi_BMP280_1",
        "module": "",
        "interface": "",
        "component": "",
        "payload": "{\"degreesCelsius\":18.98000000000018,\"degreesFahr
          enheit\":66.164000000000016,\"bars\":1.0247321,\"hectopascals\"
          :1024.7321000000002,\"atmospheres\":1.01133195164076,\"eventTi
          me\":\"11/2/2020 9:50:59 AM\"}"
    }
}
{
    "event": {
        "origin": "RasPi_BMP280_1",
        "module": "",
        "interface": "",
        "component": "",
        "payload": "{\"degreesCelsius\":18.990000000000009,\"degreesFahrenh
          eit\":66.182,\"bars\":1.0247424,\"hectopascals\":1024.7424,\"atmosp
          heres\":1.0113421169504071,\"eventTime\":\"11/2/2020 9:51:28 AM\"}"
    }
}
Stopping event monitor...
Terminate batch job (Y/N)? y
```

Take a moment and look at the data in the monitor stream. Notice we have the origin (which device) and the payload, which is the JSON string for the data collected. If you are watching your Visual Studio output window, you should be able to find those messages. In this example, the data doesn't vary much because the sample rate is too high (not enough time for significant changes).

Notice also we used the same commands as we did previously, only for the monitoring; we specified an additional parameter to limit display to the RasPi_ BMP280_1 device. Once you start adding multiple devices, this parameter can make remote debugging a little easier.

Well, that's it! We have written an application that reads data from a sensor and sends it to Azure. You may be thinking that the project thus far isn't different from the one in the tutorial earlier in this chapter. While that is somewhat true, it does differ in how the code is written, and it uses a real sensor. Further, we saw how to use the Azure Portal to set up the resources.

In fact, we now have a fully functional application that captures data from a real sensor and sends the data via a message to the Azure IoT Hub. But what do we do with it? In the next section, we will explore one of the many ways we can capture data messages sent to the hub and store them.

Tip   From here on, the project becomes much more complex using many more resources in Azure. Be sure to destroy any resources you create to avoid using up your trial credit (or getting a larger than expected invoice).

Create and Configure an Azure Data Explorer Cluster

In this step, we will leverage another Azure product to store the data from our IoT devices. So, what is the Data Explorer and why is it being used in this chapter? The Azure Data Explorer Cluster is a product composed of multiple Azure resources including a database storage component to route data from the IoT Hub for storage and retrieval.

The Data Explorer Cluster is designed to be a fast, scalable data storage and retrieval system specifically tailored to telemetry data. In fact, it can be used from related tasks such as data warehousing and IoT solutions like remote monitoring. While it is simple in concept and the setup isn't overly difficult, it does consume a lot of resources.

Tip   See `https://docs.microsoft.com/en-us/azure/data-explorer/` for more information about the Data Explorer Cluster.

Here, we will see how to create a Data Explorer Cluster and connect it to our IoT Hub. We'll test the stream in the next step.

Configuring the cluster properly for data ingestion requires two steps: creating the cluster and creating a database to store the data. In this section, we will also see how to stop and start the cluster.

Note   This product generates a number of resources in Azure, so be sure to monitor your usage and don't forget to stop the cluster if you don't need it running or delete it if you don't have any important data stored.

Create the Data Explorer Cluster

To create a Data Explorer Cluster, click *Create a Resource* in the Azure Portal and then search for *Azure Data Explorer*. You should see a number of tiles appear, and one of them is the Azure Data Explorer. This is because the search returns all related products. Figure 16-8 shows the tile we want. Just click the tile.

Azure Data Explorer
Microsoft
Azure Data Explorer is a big-data, interactive analytics platform that provides ultra-fast telemetry search

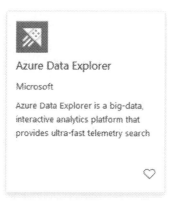

Figure 16-8. *Azure Data Explorer tile (from search results)*

Once you click the tile, you will see the overview page for the data cluster product as shown in Figure 16-9. Take a moment to read more about the product.

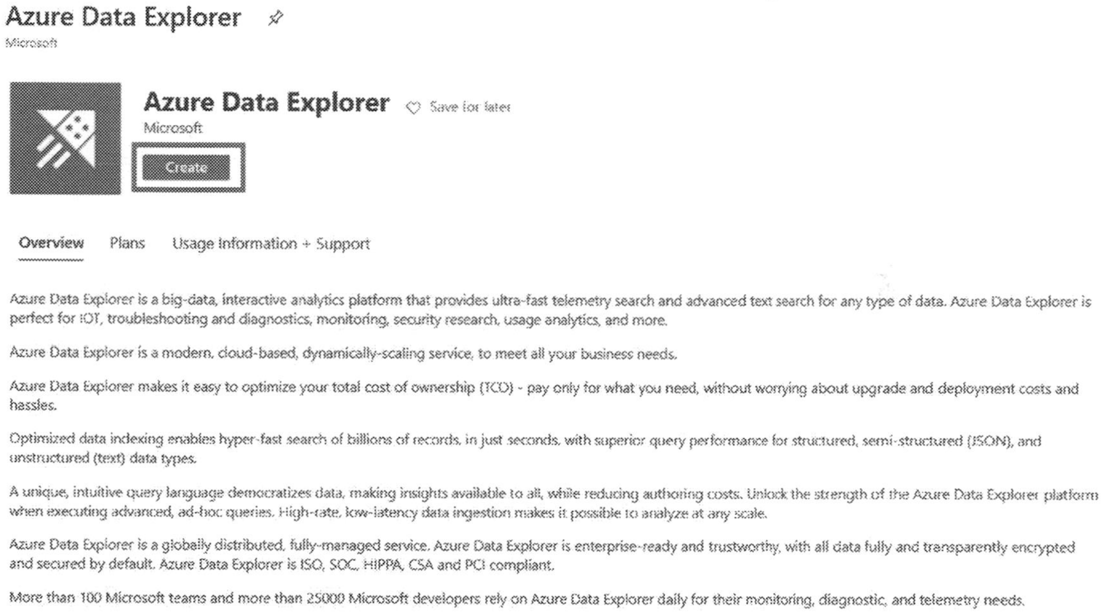

Azure Data Explorer
Microsoft
Azure Data Explorer
Save for later
Microsoft
Create
Overview
Plans
Usage Information + Support
Azure Data Explorer is a big-data, interactive analytics platform that provides ultra-fast telemetry search and advanced text search for any type of data. Azure Data Explorer is perfect for IOT, troubleshooting and diagnostics, monitoring, security research, usage analytics, and more.
Azure Data Explorer is a modern, cloud-based, dynamically-scaling service, to meet all your business needs.
Azure Data Explorer makes it easy to optimize your total cost of ownership (TCO) - pay only for what you need, without worrying about upgrade and deployment costs and hassles.
Optimized data indexing enables hyper-fast search of billions of records, in just seconds, with superior query performance for structured, semi-structured (JSON), and unstructured (text) data types.
A unique, intuitive query language democratizes data, making insights available to all, while reducing authoring costs. Unlock the strength of the Azure Data Explorer platform when executing advanced, ad-hoc queries. High-rate, low-latency data ingestion makes it possible to analyze at any scale.
Azure Data Explorer is a globally distributed, fully-managed service. Azure Data Explorer is enterprise-ready and trustworthy, with all data fully and transparently encrypted and secured by default. Azure Data Explorer is ISO, SOC, HIPPA, CSA and PCI compliant.
More than 100 Microsoft teams and more than 25000 Microsoft developers rely on Azure Data Explorer daily for their monitoring, diagnostic, and telemetry needs.

Figure 16-9. *Azure Data Explorer overview (portal)*

When you're ready to create a cluster, click *Create*. This opens a new details page that requires the following information. Sample values are in italics. Feel free to use whatever you like for names, but keep them consistent as you follow along.

- *Subscription*: Leave this set to *Free Trial*.

- *Resource group*: Recall, a resource group is a container that we can use to group resources. Use the same one you created earlier—*CABWeatherRG*.

- *Cluster name*: Choose a unique cluster name. Use only 4–22 lowercase letters and numbers—*cabweathercluster*.

- *Region*: Choose your (general) location—*East US*.

- *Workload/compute specifications*: Choose the compute configuration for the cluster. This describes the "shape" or "size" of the virtual computer(s) the cluster will use. Choose *Dev/test* for the workload and *Dev(No SLA)_Standard_E2a_v4* for the compute specification (this is one of the lower cost and lower performance options). You can also click *Select from all options* to view "larger" shapes.

- *Availability zones*: Choose 1, 2, and 3 from the drop-down list (tick each).

Figure 16-10 shows an example of how to fill out this form to create the cluster.

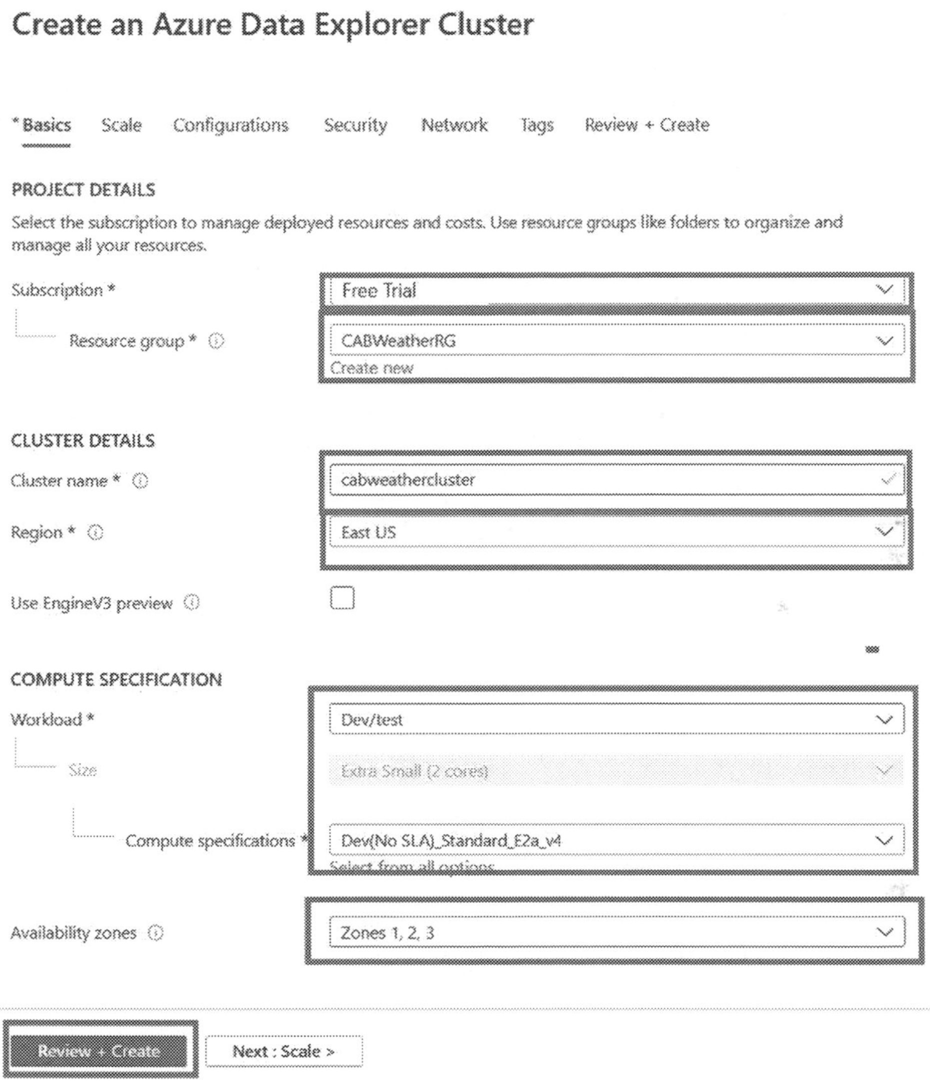

Create an Azure Data Explorer Cluster

*Basics   Scale   Configurations   Security   Network   Tags   Review + Create

PROJECT DETAILS

Select the subscription to manage deployed resources and costs. Use resource groups like folders to organize and manage all your resources.

Subscription * | Free Trial   ∨ |

Resource group * ⓘ | CABWeatherRG   ∨ |
                                         | Create new |

CLUSTER DETAILS

Cluster name * ⓘ | cabweathercluster   ✓ |

Region * ⓘ | East US   ∨ |

Use EngineV3 preview ⓘ   ☐

COMPUTE SPECIFICATION

Workload * | Dev/test   ∨ |

Size | Extra Small (2 cores)   ∨ |

Compute specifications * | Dev(No SLA)_Standard_E2a_v4   ∨ |
                                         | Select from all options |

Availability zones ⓘ | Zones 1, 2, 3   ∨ |

[**Review + Create**]   [Next : Scale >]

Figure 16-10. *Create Azure Data Explorer Cluster*

When ready, click *Review + Create* and then wait for the options to be validated. Once validated, you can click the *Create* button to create the cluster.

Creating the cluster and all of its resources is not a simple operation. In fact, it could take some time for the deployment to complete (e.g., 30–45 minutes). Wait until the deployment is complete and then click *Go to resource* to open the overview page.

Create a Database

Next, we need to create a database to store our data. This isn't a detailed table creation; rather, it is a step in the configuration process to establish a database that you can later modify by adding a table. For now, we just create the empty database.

To create a database, open the Data Explorer overview page and click *Create database* as shown in Figure 16-11.

Figure 16-11. *Data Explorer overview*

On the next page, you will complete a name for the database as well as retention and cache periods in days. The default values are fine for these options, but use the database name weather as shown in Figure 16-12. When ready, click the *Create* button.

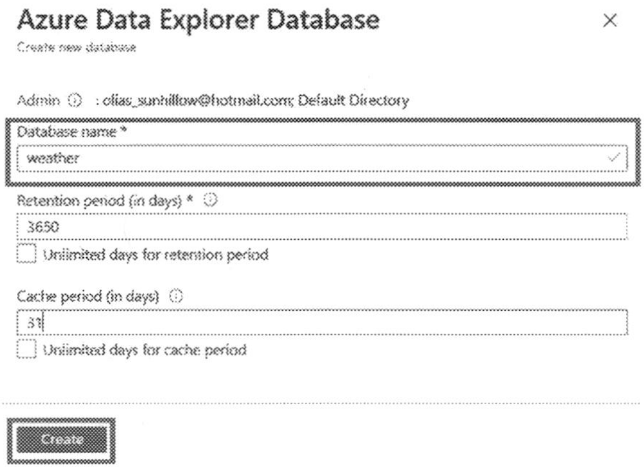

Figure 16-12. *Create new database (Data Explorer)*

We will see how to work with the database in a later step.

Start and Stop the Cluster

Finally, you should understand how to stop and start a cluster. When you create a cluster, it is in the started state (by default). However, if you are not using the cluster, you can save some resources (money) by stopping the cluster. Stopping the cluster may also be needed for some maintenance operations.

To stop a cluster, click the *Stop* button on the overview page as shown in Figure 16-13. If the cluster is stopped, the *Stop* button will change to *Start*, which you can click to start the cluster.

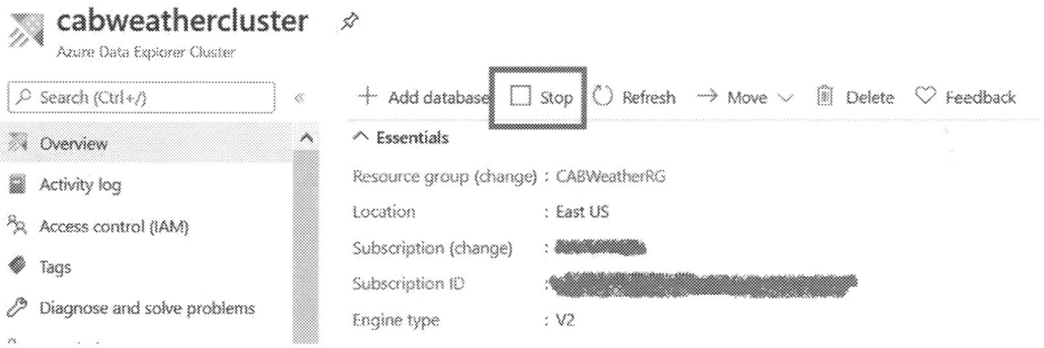

Figure 16-13. *Start and stop the Data Explorer Cluster*

Now that we know a little more about the Data Explorer Cluster, let's send some data and discover how to store the data in the cluster.

Test the Data Stream

In this section, we will ensure the data captured by the IoT device is being routed to the Data Explorer Database. We will be using the same application on our device and the same IoT Hub, so there is nothing to change there.

However, we must create a table in the database to store the data and then connect the cluster (table) to the IoT Hub. Once we do that, we can start our application and watch the data flow to the table. Let's begin by creating a table.

Create a Table

To create a table to store our data, navigate to the Data Explorer page and then choose Databases as shown in Figure 16-14.

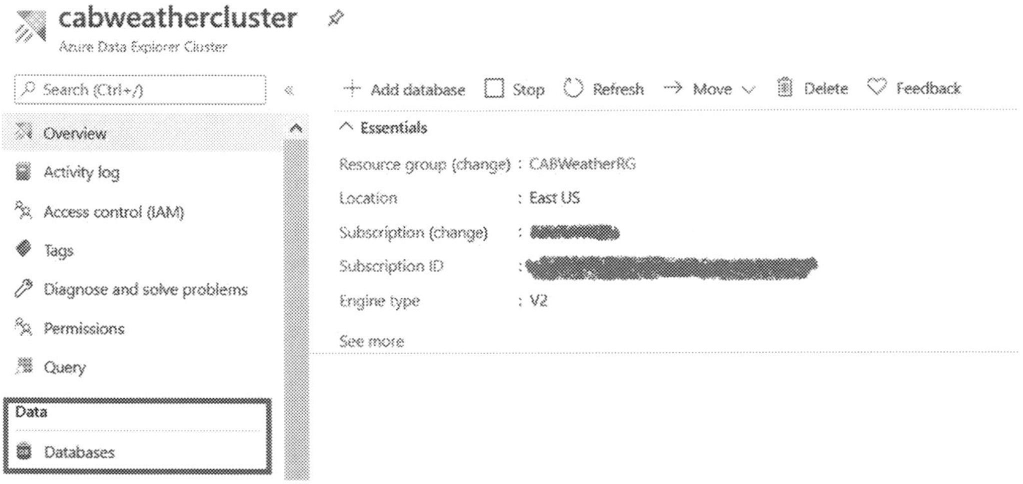

Figure 16-14. *Databases (Data Explorer Cluster)*

Click the database we created (weather) to see the details, then click *Query* as shown in Figure 16-15.

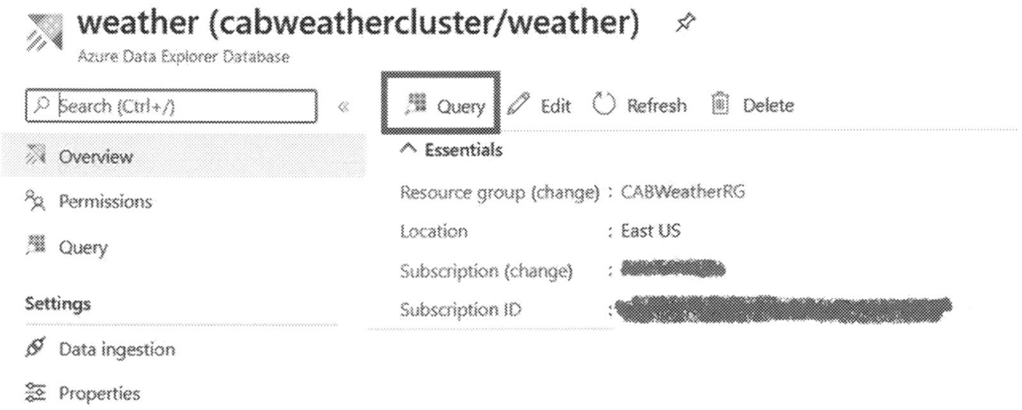

Figure 16-15. *Launch query window (Data Explorer Cluster)*

If you've used database systems before, the following will seem somewhat familiar in that we will issue queries on the database to create resources and show (select) data. However, rather than use SQL, Azure uses a different query language called the Kusto Query Language (KQL). You can read more about this language at `https://docs. microsoft.com/en-us/azure/data-explorer/kusto/query/`.

To create a table, issue the following command in the query window. Type this into the query area and then click *Run*. This will create the table with the columns we need to store the data. Here, we see we created one column for each of the data items using the data type `real` for all numeric values and `datetime` for the event time.

```
.create table WeatherData (
eventTime: datetime,
degreesCelsius: real,
degreesFahrenheit: real,
bars: real,
hectopascals: real,
atmospheres: real)
```

Figure 16-16 shows the results of executing the query. Note that I've expanded some of the results. Feel free to poke around in the interface to learn more.

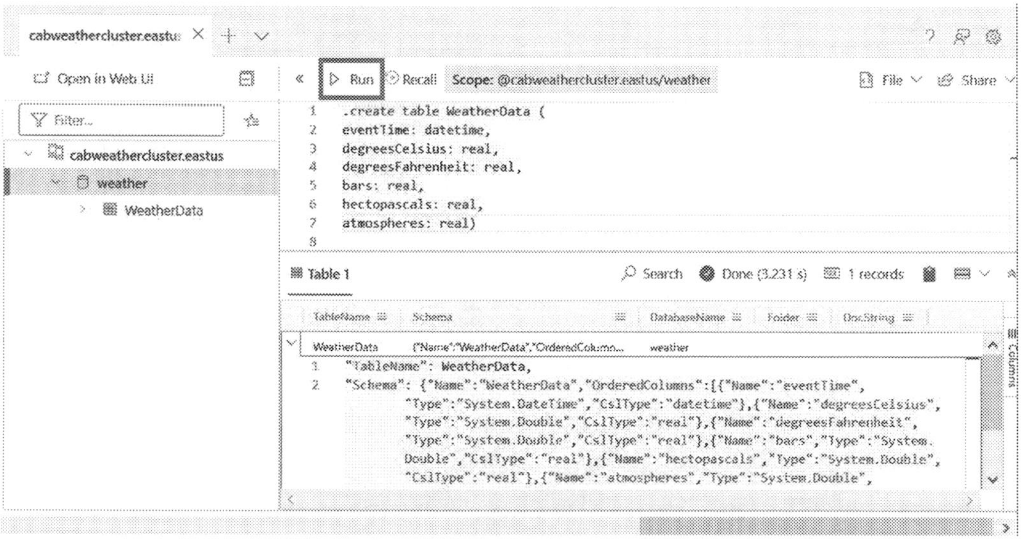

Figure 16-16. *Create the table (Data Explorer Cluster)*

However, to do so, we will need a way to map our JSON data (from the payload key) into the table. We do this using a table mapping. The following shows the table mapping (called data mappings) we need for this project. Type this command over the previous query and click *Run*:

```
.create table WeatherData ingestion json mapping 'WeatherMapping'
'['
'{"column":"eventTime","path":"$.eventTime"}, '
'{"column":"degreesCelsius","path":"$.degreesCelsius"}, '
'{"column":"degreesFahrenheit","path":"$.degreesFahrenheit"}, '
'{"column":"bars","path":"$.bars"}, '
'{"column":"hectopascals","path":"$.hectopascals"}, '
'{"column":"atmospheres","path":"$.atmospheres"} '
']'
```

Figure 16-17 shows the results of executing the query.

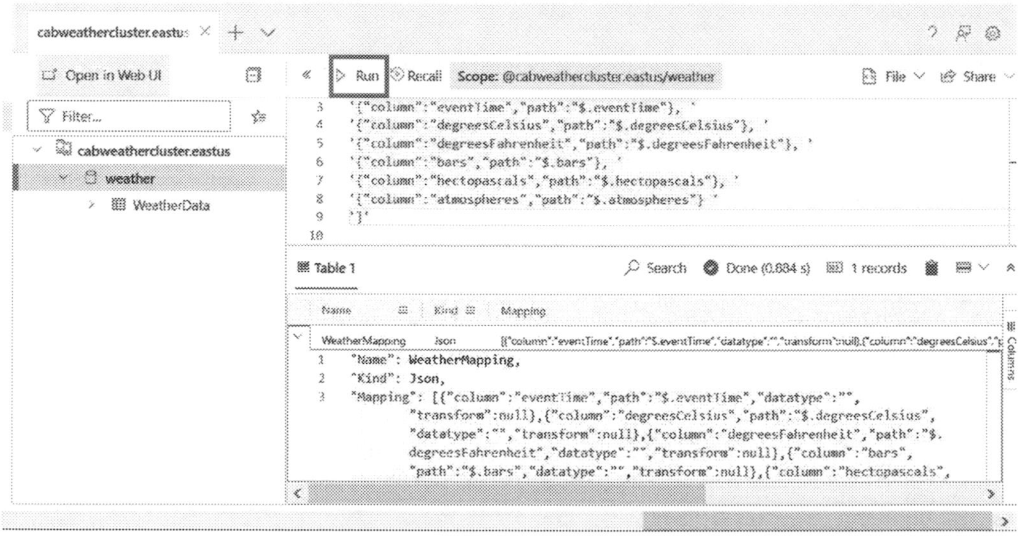

Figure 16-17. *Create table mapping (Data Explorer Cluster)*

Azure provides a vast array of table mapping options. If you plan to use Data Explorer for other projects, visit https://docs.microsoft.com/en-us/azure/data-explorer/kusto/management/mappings to learn more about creating data mappings.

Now that we have our cluster, database, and table created, we need one more thing; we need to connect our cluster to our IoT Hub.

Connect to the Azure IoT Hub

To connect the cluster to the IoT Hub, click Data ingestion on the query page as shown in Figure 16-18.

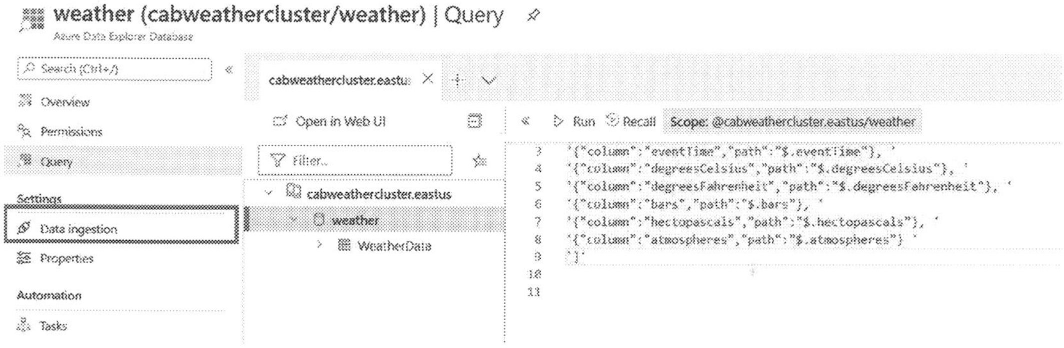

Figure 16-18. *Select Data ingestion (Data Explorer Cluster)*

On the data ingestion page, click *Add data connection* as shown in Figure 16-19.

Figure 16-19. *Create data ingestion (Data Explorer Cluster)*

Next, choose *IoT Hub* from the *Select connection type* drop-down box and fill in the following information. Note that, depending on the value chosen, you may see a different set of information you will need to provide. Sample values are shown in italics.

- *Select connection type*: Choose the connection type which can be a number of different resources in Azure. Use *IoT Hub*.

- *Data connection name*: Name the connection. Use *weatherconnection*.

- *Subscription*: Choose the name of your subscription. Use *Free Trial*.

- *IoT Hub*: Choose the name of the IoT Hub created earlier (*CABWeatherHub*).

- *Shared access policy*: Choose the access policy you want to use. For now, select the hub owner (*iothubowner*). You can explore different policies if you plan to make your application available to other users.

- *Consumer group*: Choose the consumer group (if you've created any, choose *$Default* otherwise).

- *Table name*: Enter the name of your table (*WeatherData*).

- *Data format*: Choose the data type mapping method you used earlier (*JSON*).

- *Mapping name*: Enter the name of the table mapping created earlier (*WeatherMapping*).

Figure 16-20 shows an example of how the form should be completed. When ready, click *Create*.

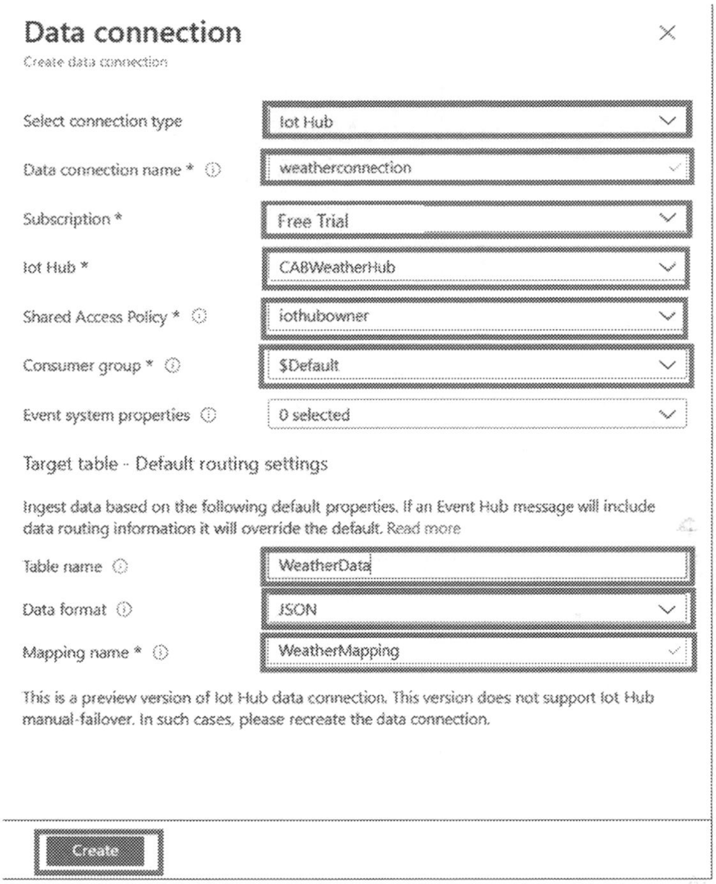

Figure 16-20. *Create connection (Data Explorer Cluster)*

Once the connection is created, we can go to our device and launch the application.

Start the IoT Device

Now, we can launch our application. If you have had your device running previous to creating the connection, the data will not show in the table. Only data that is sent to Azure after the connection is created will be saved to the table.

Show the Data

Once the application has run for about 5–10 minutes (we're still using the short 10-second sample rate), navigate back to the Data Explorer query window and issue the following query. Figure 16-21 shows an example output.

WeatherData

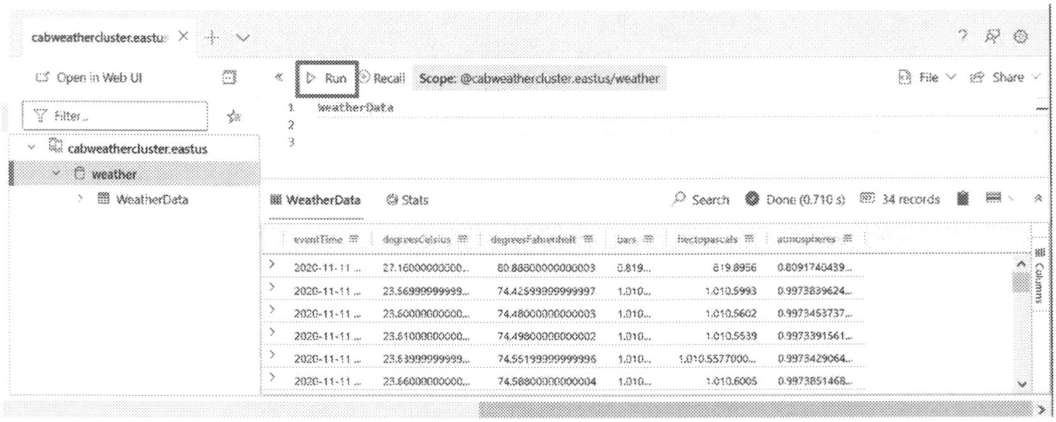

Figure 16-21. *Example query for data (Data Explorer Cluster)*

You may not see the exact number of rows that have been sent because the data ingestion takes a few seconds to route the data.

Tip   If you do not see any data in your table, try leaving your application running and then delete and recreate the connection in Data Explorer. You can also check your IoT Hub to ensure messages are being received.

OK, we've now successfully connected our Data Explorer Cluster to our IoT Hub and have seen how to store and query for data. Cool. Now if only we could see interesting graphics. Fortunately, there are Azure products to do just that!

Set Up Data Visualization in the Data Explorer Dashboard

There are many ways (products) in Azure you can use to generate visuals of your data. In this section, we will see how to use queries against the database to display the data using a visualization tool called *dashboards* in Data Explorer Cluster.

Tip   To learn more about dashboards, see `https://docs.microsoft.com/en-us/azure/data-explorer/azure-data-explorer-dashboards`.

While your application is running, navigate to the default page for your cluster and note the URL on the right side of the screen. Copy this into your browser and navigate to the site. This opens the dashboard view of your cluster, which is a separate web application. You will notice you have fewer options to choose from on the left side of the screen. Click *Dashboards* and then *Create new dashboard* as shown in Figure 16-22.

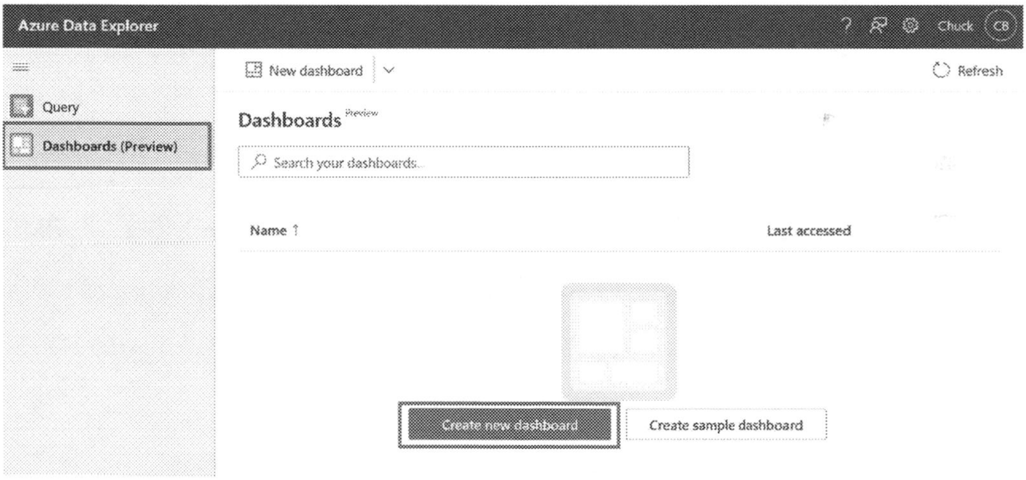

Figure 16-22.   *Dashboard view (Data Explorer Cluster)*

WAIT! IT'S NOT WORKING

You may see an error stating your user account does not exist in "Microsoft Services" when navigating to the query page. If you see a similar error, you can open the database in the Azure Portal and run the following command to grant access to your Azure account. This is needed due to recent changes to the Data Explorer Database access.

```
.add database weather users('msauser=<your Microsoft email>')
```

It is also possible to see errors about your account not existing as a tenant in "Microsoft Services" when navigating to the dashboards page. Should this occur, you can append the following to your URL to access the query or dashboards pages in the Data Explorer. There is a recent bug in the dashboards access that is not forwarding the tenant id. Passing it via the URL will solve that problem.

```
https://dataexplorer.azure.com/dashboards?tenant=<YOUR TENANT ID>
```

You can find your tenant id on the Azure Active Directory page in the Azure Portal.

Enter Weather Line Graph in the pop-up text box to name the dashboard and then click *Create*. At this point, we have a dashboard created but nothing to see. The first thing we need to do is create a data source for the data we want to display. Click Data sources at the top of the page to create the first data source. In the dialog that pops up, we want to paste the URI for our cluster (found on the cluster main page) and then click Connect as shown in Figure 16-23.

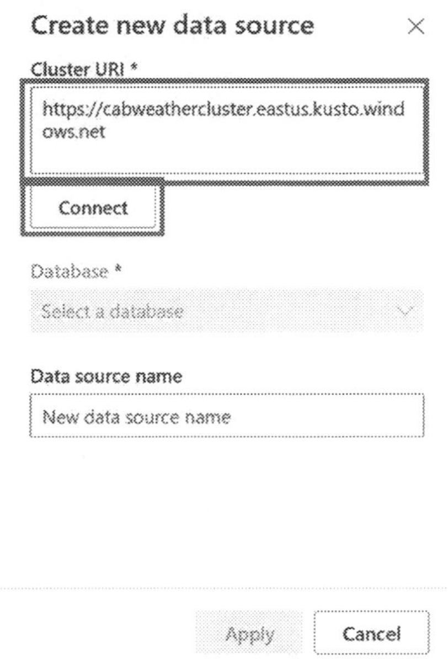

Create new data source   ×

Cluster URI *

https://cabweathercluster.eastus.kusto.wind
ows.net

Connect

Database *

Select a database

Data source name

New data source name

Apply   Cancel

Figure 16-23. *Add a new data source (Data Explorer Dashboard)*

Once the connection succeeds, choose the name of your database, and click *Apply* as shown in Figure 16-24.

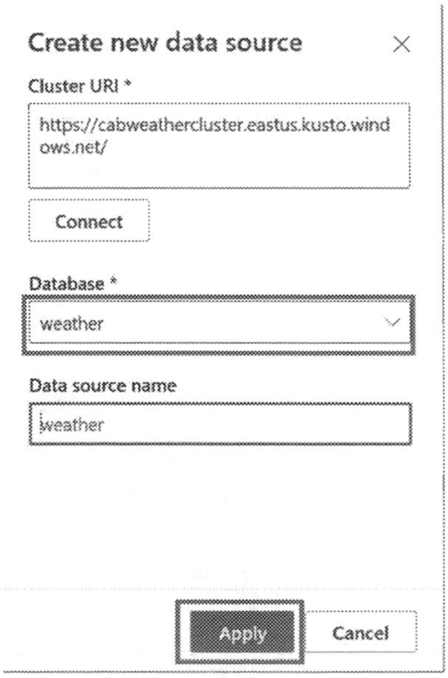

Create new data source   ×

Cluster URI *

https://cabweathercluster.eastus.kusto.wind
ows.net/

Connect

Database *

weather

Data source name

weather

Apply   Cancel

Figure 16-24. *Selecting the database for the data source (Data Explorer Dashboard)*

Now that we have a dashboard created with a data connection, let's test it by creating a graphic to display the data Azure calls the graphics *tiles*.

Test the Visualization

In this step, we will test the entire data ingestion process to ensure the data captured by the IoT device is being visualized in the dashboard. Begin by creating your first tile by clicking + *Add tile* as shown in Figure 16-25. Notice we need to first create the query, and then we can add the visual. The interface is very similar to the query interface in the Azure Portal for the Data Explorer Cluster complete with an area to enter queries and a *Run* button to execute queries.

Figure 16-25. Creating a new tile (Data Explorer Dashboard)

Use the following query to isolate only the temperature (Celsius and Fahrenheit) and event time. Enter the query and then click *Run*.

```
WeatherData
| project eventTime, degreesFahrenheit, degreesCelsius
```

What you should see is the query returns results in a typical grid format. However, we want a line graph! To add a visual, click the + *Add visual* button as shown in Figure 16-26.

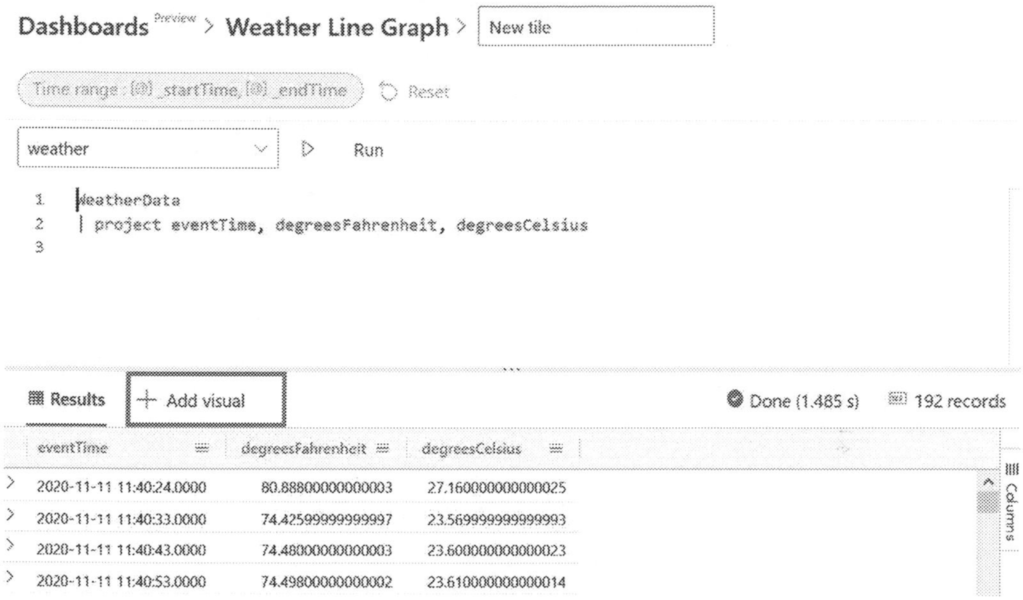

Figure 16-26. *Add visual after query execution (Data Explorer Dashboard)*

When the dialog opens, choose *Line Chart* for the *Visual type* as shown in Figure 16-27. You can also name the axis and label the chart however you'd like. I've done a little of that.

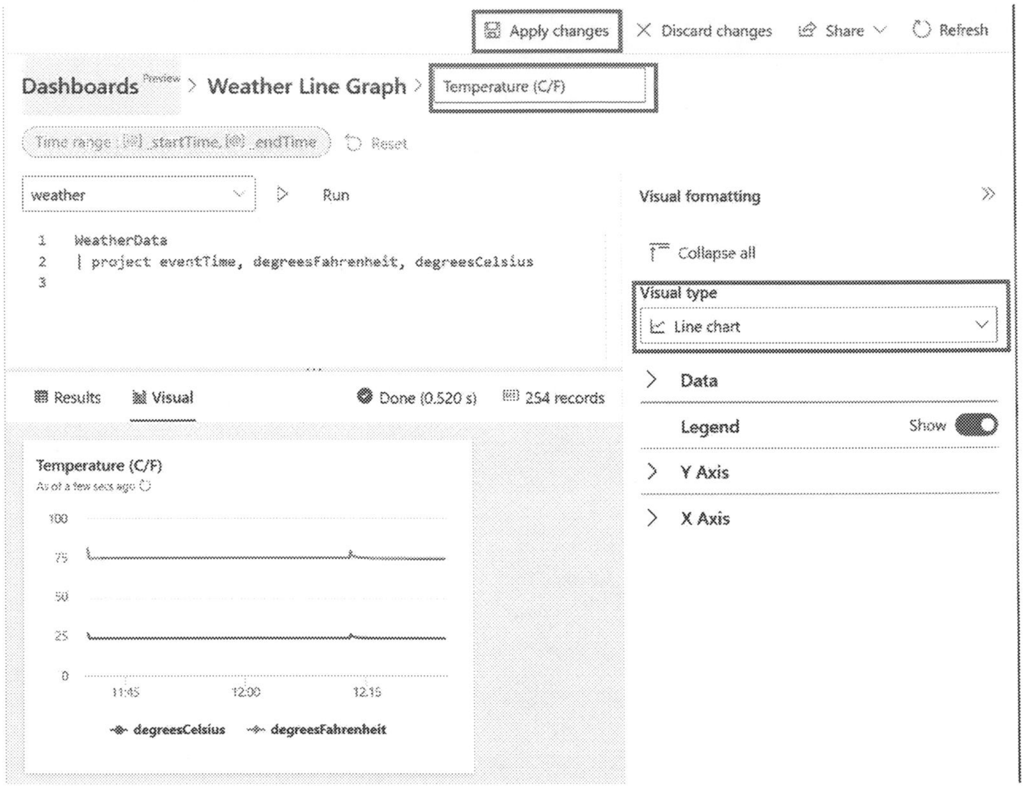

Figure 16-27. *Line chart visual (Data Explorer Dashboard)*

Once you're satisfied with the changes, click *Apply changes* and then click *Save changes*. This adds the tile to your dashboard as shown in Figure 16-28.

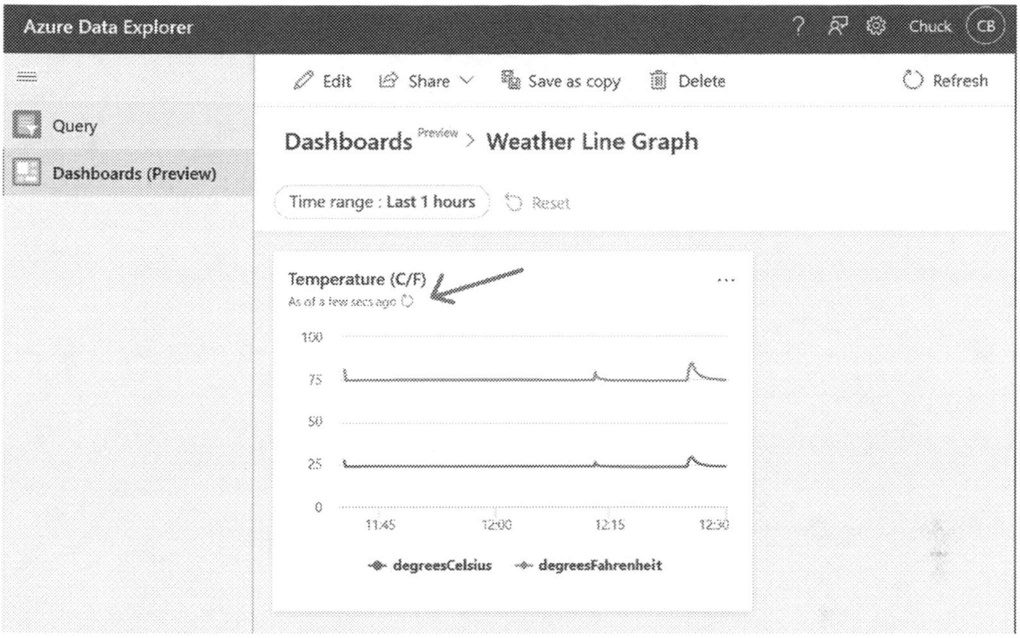

Figure 16-28. *Completed dashboard (Data Explorer Dashboard)*

Congratulations! You've created your first Azure IoT application! You now have a dashboard that you can visit anytime you want (while your cluster is running) and see the temperature sampled by the device. You can also click the small circle (see arrow) next to the status of each graphic to refresh the data. This can be handy if you want the latest data added.

I leave as an exercise adding the other data columns in other charts. Just follow the same steps to add new graphics. Go ahead and spend some time exploring how you can make visuals of your data.

If things don't work for you or you aren't seeing similar results as shown in this chapter, be sure to go back and diagnose each step along the way. You can always delete everything in Azure (by deleting the resource group) and start over. While this may seem drastic, it serves two purposes; it removes any mistakes you may have made along the way, and it stops Azure from charging your account while you diagnose, which can be helpful if you decide to leave it for another day!

Caution   Be sure to delete your resource group **and** the cluster when you no longer need them to prevent unexpected charges on your account.

Summary

As you can see, it is possible to build all manner of cloud-based solutions and services with Microsoft Azure. In this regard, Azure is very powerful, which is exactly what you want when developing IoT solutions that you want to put into service as either a reliable enterprise-grade product or even as a product you sell to others.

In this chapter, you close your journey of discovering how powerful Azure's IoT products are and how you can use them to build sophisticated IoT solutions from simple projects that you run for fun to full cloud-based IoT solutions. Now it's time for you to engage your own imagination and put the tools and techniques you learned in this book toward building your own IoT solutions.

I hope you enjoyed the ride and that reading about and working on the projects in this book were as much fun for you as I had writing them.

Where to Go from Here?

Now that you have had a thorough introduction to using Windows 10 IoT Core, which included the types of projects that you can create, and tutorials, as well as examples in three programming languages, it is time to consider what you can do beyond the pages of this book.

This chapter explores what you can do to continue the craft of building IoT solutions. Most people want to simply develop projects for themselves—for fun or to solve problems around the home or office. However, some of you may want to take your skills to the next level. Whatever the case, there are a few things that you should consider. The following sections look at sources for more example projects, discuss how to join the community of Windows 10 IoT enthusiasts through social media and other Internet resources, and explain how to become a contributing member of the growing throng of makers. We will also discover how to get started making your IoT solution more than a small, home or school project.

More Projects to Explore

If you want to work on more Windows 10 IoT projects, you will be happy to learn that there are many, perhaps hundreds, of examples that you can explore. The vast majority of examples are contributions from the community, ranging from a high-level overview to detailed instructions on how to complete the project. Sadly, most of the examples are presented with little or no instruction.[1] However, now that you have had detailed instructions on working with Windows 10 IoT projects and Visual Studio, you should be able to complete examples with little or no documentation.

---

[1]This is one of my major motivations for writing this book.

© Charles Bell 2021
C. Bell, *Windows 10 for the Internet of Things*, https://doi.org/10.1007/978-1-4842-6609-0_17

There are two major repositories for Windows 10 IoT example projects: the Microsoft Windows 10 IoT website (`https://github.com/microsoft/Windows-iotcore-samples`) and the Hackster.io general hardware community forum (`www.hackster.io/search?i=projects&q=iot%20core`).

Microsoft Windows 10 IoT Samples

The Microsoft sample page is really handy because you can filter the list of samples by tags. For example, you can find samples that are written in C#, C++, Visual Basic, headless, headed, or others. Many of the samples are presented in a number of languages, making them much easier to use. For example, you can visit the site and find a beginning project named Hello Blinky! implemented in C# or C++.

Fortunately, most of the projects on the site are documented and can be followed easily enough. That is, the documentation for some examples does not explain everything that you need to know to use the samples. In some cases, steps are missing or a familiarity with the tools and techniques are assumed. In a few cases (thankfully), the documentation is old and not applicable to the newer releases. Fortunately, this is rare, and I have seen some improvements for the older samples.

That aside, the site is the best and most complete set of Windows 10 IoT samples available. There are very basic samples, samples that demonstrate how to work with hardware, samples for use with the third-party libraries, and even some samples for creating utilities to run on your devices.

What makes this site even more useful is that the samples are stored in GitHub, allowing you to navigate directly to the source files and see the code in your browser, skipping the documentation or demonstration page. This makes it really nice if you just want to see how to implement some feature rather than walk through a long page of text. Of course, the best way to use the samples is to download the entire set of samples. Simply visit the GitHub site (`https://github.com/microsoft/Windows-iotcore-samples`) and click the *Download ZIP* button on the link. Once you download and unzip the file locally, you are able to see all the sample projects. How cool is that?

Most (if not all—I haven't checked all) of the samples are licensed under the MIT license, which makes it very convenient for everyone since the MIT license permits you to use and even publish the code. (See `https://opensource.org/licenses/MIT` for a sample of the MIT license.) This is really great because I have many times wanted to use a sample or demonstration of a project, only to discover the license doesn't permit it. Kudos to Microsoft for embracing the open source community.

Hackster.io

The Hackster site is a community dedicated to learning hardware. You can find all manner of hardware sample projects, including many of those for Windows 10 IoT, Raspberry Pi, Arduino, and more! There is a small but growing Windows 10 category. When you visit the parent site (`www.hackster.io`), you can choose the category by typing a subject. If you enter Windows 10, the site suggests the Windows IoT Core category. Just accept that and you will be presented with the list of samples for Windows 10 IoT.

On the category page, you see all of the projects you can explore. Each is marked with a relative level of difficultly, ranging from easy to advanced. You also see a count of the number of times the sample project was viewed and the number of thumbs-up ratings the project received from others in the community (you have to join Hackster to rate a project). Best of all, there is a comments section that you can use to encourage the designer or ask the designer for help with the project.

Tip   Use the golden rule when posting comments or questions in online forums. Resist the temptation to post opinions, fan flames of dissent or ridicule, and stick only to the facts.

What I like most about the Hackster site is that the samples are generally well documented and often include a number of photos of the project. Due to the unique structure of the site, the samples are organized in parts that make it easy to follow. For example, the web page for "Character LCD over I2C" (`www.hackster.io/dzerycz/character-lcd-over-i2c-ba8ee9`) contains an overview section that presents a short description of the project, associated tags, difficulty rating, publication date, and even the license. This makes reviewing the project very easy.

The Microsoft site has only a very terse introductory section that jumps directly into the documentation. For those that like to browse projects, the overview page is a great asset. Figure 17-1 shows the "Character LCD over I2C" overview. As you can see, it is a sample project for using a character-based LCD panel over the I2C bus.[2]

---

[2]This sample is an excellent example of a properly named project. Too often I find projects that are named differently or suggest a different implementation than what it really is. Be sure to use an accurate title when posting your own projects.

Jaroslav Živný
Published September 19, 2015 © MIT

Character LCD over I2C

Connect HD44780 LCD character display to Windows 10 IoT devices via I2C
and PCF8574

⚲ Beginner   ♡ Protip   ⊚ 8,055

Figure 17-1.   *Hackster.io sample project*

If you scroll down from the overview, you find sections that demonstrate how to connect the hardware (similar to how I introduce the projects in this book), a short walk-through of the code, and descriptions and demonstrations on how to use the project.

Some samples include short videos to demonstrate or explain the project. At the end of the page, you find the comments section, which you can use to read what others have said about the project, as well as the questions that others have had regarding the project. If you get stuck on a sample, be sure to read all of the comments—there is a good chance that someone has already asked the question or solved the problem.

Now that you have seen a couple of resources for more sample projects, let's discuss how you can join the community and contribute to the growing repository of all things Windows 10 IoT.

Join the Community

Once you have mastered the sample projects in this book, as well as a few from other resources, it is time to take your hobby a bit further by joining the community of Windows 10 IoT developers and enthusiasts.

In this section, I discuss some of the reasons that you may want to share your knowledge. I also discuss etiquette for sharing and contributing and suggest a few example communities that you may want to join or monitor. As you will see, not all of these are strictly Windows 10 (or even Microsoft) friendly, but they can be an excellent source for ideas.

Let's begin with why you would want to share.

Why Contribute?

As more and more free thinkers drive hobbies like IoT, the more prevalent the concept of sharing becomes. This is no accident. Many of the founders and pioneers of the IoT are open hardware and open source advocates. This applies not only to hardware and software but also to other intellectual products, such as the source code and documentation for IoT projects.

Many feel that their code should be free for anyone to use and modify, with reciprocal expectations. For example, if you modify someone else's design or code, you should share not only the improved design but also credit the originator. In some cases, this is as simple as listing the original author, but other times it may mean giving the original author your modifications. So long as you follow the guidelines of the license, all is fair and well in sharing.

However, depending on how the sample code was written (licensed), there may be some limitation to what can be shared. For example, it may not be possible to share code from a proprietary library. Although you may be the creator of the code that uses the library, you do not own the library and cannot share it. You most likely can share your code with others, but publication may be restricted.

Sharing your projects also means placing them someplace where others can find them. You may want to make them freely available to anyone, or you may want to limit what people can do with your project. Fortunately, there are websites that can handle either quite well.

So, why contribute your project? There are many reasons, including the fact that it can be a really good feeling to see one of your projects liked, used, and made by others. Perhaps the most important reason for contributing is to help others learn what you have or, better, learn how to avoid pitfalls or problems. In this way, we all benefit by learning best practices or simply better ways to implement our ideas. Finally, your own projects and experiences, when shared, inspire others to create other projects or perhaps improve yours.

I've had this happen with my own projects. People have taken what I have made and improved it. Since they, in turn, shared their project, I was able to incorporate a lot of their improvements in my projects—making them even better than I envisioned.

Now, let's focus on how to go about sharing your projects.

How We Share

You may be wondering why anyone would want to give away for free something that they have worked on for hours. While it is true that the expectation is that you should share your cool projects with others, it isn't a hard-and-fast rule. In fact, there are some who have made their projects available for a fee as a precursor to selling the IoT solution in a commercial avenue. However, the vast majority of enthusiasts share their ideas and projects for free.

There are several communities where you can share your projects. You will see some of these in the next section. But first, there are some things that you must understand about sharing objects. Believe it or not, there is a set of rules—some written, some not—that you are expected to follow should you decide to embrace the IoT or any similar community. The following section lists some guidelines (rules) that you would do well to heed when sharing your ideas, projects, and commentary with the community.

Keep Your Designs Original

Nobody likes a copycat. You didn't like it when you were five years old, and you won't appreciate it when you see something that you designed and shared for free being presented as the "design of the month" but credited to someone else.

Thus, you must do your homework to make sure that your design is unique. You don't have to purposefully alter your design so that it doesn't resemble someone else's, but you should do due diligence and at least search for similar projects. Remember, it is OK if you develop a similar project, but it generally is bad form (or perhaps a violation of the license) to simply reproduce something that someone else has published.

In the rare event that your project is nearly identical to another, so long as your work is your own, there shouldn't be a problem. In fact, this happened to me once. My response and that of the other designer were something like, "Cool project. Like minds, eh?" Once again, there is nothing wrong with this provided that you both acknowledge the resemblance and that there are no licensing issues.

If the other project is truly the same design but was licensed differently, you may have to negotiate with the other designer. This can happen when projects are licensed for ownership (e.g., commercial property), but it is rare given that most IoT sample repositories are sites where people share their projects for free.

Let's look at another non-source code example. What is the likelihood that a dozen different cases for a Raspberry Pi will be similar in size, have the same port openings, and perhaps even assemble the same way (snap together)? Very likely, yes? Does this mean that there is one original and 11 copies? No, certainly not. This is not what I am talking about.

What I mean by *unique* is that of those 12 cases, you should be able to identify some differences among them—be that how they are constructed, if they are made from several parts, whether they have ventilation, and so forth. Even if all 12 designers started at the same time, there will be some minor differences. More importantly, each is its own work. That is, no one used the design of another to pass off as their own.

In a software project, the source code is most likely going to be a little different among the examples. While the degree of differences is something for lawyers to sort out, suffice it to say, if your code and another's are nearly identical, but created without knowledge of the other, it is OK to share your code—provided that there are no licensing conflicts.

Finally, when you share your project, and it is based on the work of another, you must annotate your code, documentation, and project website and give credit to the original designer. That is, you state unequivocally that your project is a derivation of the original. It is also good form to include a link to the original design along with a list of your modifications. Once again, this assumes that the license permits it.

Check the License

I have mentioned licensing under the aspect of downloading and using sample projects. Recall that most repositories require you to specify a license for your project. This permits the repository to host your project and communicate to everyone what your intentions are regarding ownership, permissions to use, and so forth.

As I stated previously, you need to check the license before using any design. If you plan to modify the design, you need to pay close attention to the license. The vast majority of licenses allow you to use the design and most allow you to modify it.

However, where some licenses differ is in regard to the ownership of the modifications. Some open source licenses, like GPL, permit modifications but require you to surrender those modifications to the original owner (the person or organization that created and licensed it) if you plan to distribute those changes. That is, you can modify it at will for personal use, but once you distribute those changes, you have to give them to the owner of the license.

I have only run into this a couple of times, but in those cases, the designer was prototyping designs for a commercial product. The license and indeed the text of the project made it clear she was looking for help with the design but that the design would not be made public. Watch out for this and tread lightly. Any work you do could be for the benefit of the owner and not yours to keep or profit.

Tip   When in doubt about a license, contact the originator and ask them directly.

Since most sample IoT projects are licensed for sharing and free modification, you normally don't have to worry too much. However, I recommend that you check the license before using any project, especially if you intend to share or publish your derivation.

Keep It Appropriate

Believe it or not, there are hobbyists and enthusiasts with impressively vivid imaginations who have come up with IoT projects that some may consider inappropriate or even obscene. No matter what your own views are, you should strive to tolerate the views of others. That doesn't mean you must compromise your own views—just be aware that yours may offend and strive to minimize the offense.

More specifically, don't upload projects with inappropriate themes to sites that are viewable by everyone. It is (or may be) fine to upload some project that promotes a theme, ideal, and so forth (provided there are no copyright violations); just don't upload projects or commentary that are clearly offensive or intended to cause harm.

For example, if you consider the fact that IoT projects are being used in schools to teach children the technology and techniques of working with hardware and designing software, you shouldn't upload projects with themes that parents may deem as inappropriate. The most obvious, of course, are offensive language, adult themes, and slanderous images.

You should check the usage and user agreement for the site that hosts your chosen repository as part of the post-no-post decision. Make sure that you read the section about what is and is not appropriate and adhere to that.

There is another angle to consider. You should avoid uploading sample projects that are or could potentially be illegal or unlawful. This may be difficult to discern considering the IoT community includes the entire globe. However, most sites have language to suggest what is and is not permitted. And some have language in the agreement that gives them (the site) the right to remove things that they deem inappropriate.

For example, I once saw a project for a *radio frequency identification* (RFID) reader that could be used to read RFIDs from a distance. This sounds harmless, but consider the number of things that use RFID, such as security badges, identification, and even credit cards. Clearly, reading RFID from things you own is fine (indeed, that's what the project demonstrated), but the project could be (and most likely has been) used for evil. Fortunately, others noticed this, and the project site has been removed (the URL results in a 401 error).

So, before you upload a design or sample project, make sure that you understand and agree to the terms of the user agreement, particularly what is and is not appropriate. Most times, a misunderstanding is not something that will get you into trouble, but if you do it more than once, it's likely that someone at the site will want to speak with you or restrict your access. This brings me back to the opening of this section: be sure to respect the views of others—especially the intended audience of the site. If you disagree with the restrictions, find another site.

Annotate Your Work

I can tell if a sample IoT project is good or of high quality by the way that it is annotated and documented, that is, how well the designer described the project on its site. If I encounter a project that looks appealing only to discover that the designer didn't bother to describe how to connect the hardware, or explains the code in no more than seven words, or didn't provide any instructions, or worse didn't present any photos of the actual implementation, I won't use it.

Thus, you should strive to provide as full a description as possible. You don't have to write a novel, novella, or a dissertation, but you should provide enough information to describe the intended use, the problem that it solves, as well as a set of instructions on how to write the source code, compile it, and deploy it.

The only exceptions are when you are still working on a project or you plan to make changes before finalizing it. Here, you should mark (annotate) the project with some verbiage about a work in progress, being experimental, and so forth. If your repository has a feature to mark the project as such, use that. This way, others will know that your project isn't quite ready for general adoption. One reason for doing this is to get feedback from others. I've done this myself with mixed results. Mostly, people are happy to comment that they like it, but don't comment, or if they do, however encouraging, don't suggest any changes or improvements.

I would also suggest you provide some level of contact information so that others who have questions can contact you. Typically, viewers can easily contact you through the site, but you may want to provide other forms of contact (e.g., email). You don't want to provide your home address and phone number, but an email address is a nice way to make yourself open to the community.

For example, I have seen blogs, sample projects, and tutorials where people have posted their IRC handle, email address, and even in one case their business phone number. Although I may not go quite that far, I suggest providing an email address so that you can communicate with people who like your project. Plus, it's nice to connect 1:1 with someone to discuss your work!

Be a Good Citizen

Suppose you run across a sample project that not only isn't high quality but is also (in your opinion) designed or implemented incorrectly. Should you immediately comment and crush the designer's ego with a flippant remark about how dumb their code is? No, certainly not!

What I would do (most likely) is ignore the project altogether. I mean, why make things worse by pointing out the defects? I have found that the community at large (there are some exceptions) will likely do the same and not comment. Remember that one of the keys to determining whether a project is well designed (good) is the number of people who have used it. Typically, there is a counter you can check for this. If no one has liked it or even downloaded it, you can be sure it won't make it to the top of any search lists or as sample project of the month.

On the other hand, if you feel compelled to comment, be sure to either contact the designer privately or be as constructive as you possibly can. The goal should be to help the designer improve the project, not challenge their intellect (or pride).

When I do comment on projects that I find strange and perhaps flawed (and it is rare), I generally phrase my comments in the form of a question. A question normally doesn't put someone on the defensive, and if worded properly, it should also not offend.

For example, I may ask, "Have you found the code may hang if the user presses the button more than once?" This is a nice way of asking if the designer has tested their project under the conditions that you expect it to fail. This is good, constructive criticism in a very intellectual form. I am certain that if you think about what you are about to say, you can find other and perhaps more elegant ways of helping people improve their projects.

Now that we've seen why we share and how to share in a responsible manner, let's discover some of the communities that you may want to join or monitor.

Suggested Communities

There are quite a number of general Windows and Windows 10 websites and online communities that you can visit and join. Most online communities have repositories that you can search for examples, tips, techniques, and complete projects that you can explore. Most also have one or more areas where members can comment, ask questions, or generally communicate with others on the forum. You typically must join to be able to post a reply or ask a question, but viewing is typically permitted by anyone.

The best way to use these resources is to visit them periodically. More specifically, you should read the articles (that interest you) and forum posts regularly. This allows you to keep up to date on current events, new techniques, and even new solutions to challenging problems.

Table 17-1 presents a brief list of online resources that you should consider visiting to keep abreast of the latest news about Windows 10, hardware, and IoT. For each forum or website, I include the general topic, URL, and a brief description. However, you don't have to join any of these communities. And these are not the only forums that you can or should join. In some cases, you may want to simply monitor the site regularly.

Table 17-1.   *Online Resources for Windows 10, IoT, and Hardware*

Topics	URL	Description
Windows 10	`www.windows10 forums.com`	A general forum for Windows 10 organized into a number of categories ranging from hardware to debugging errors to installing software. An excellent source of up-to-the-minute information from the user community.
Windows Technical Documentation	`https://docs. microsoft.com/en-us/`	This site is the quintessential site for all things Windows programming. If you want a one-stop site for learning the ins and outs of Windows programming, there is none its equal. I use this site regularly as a nonmember. Memberships are available for a fee with additional perks.
Windows 10 IoT Development	`https://docs. microsoft.com/en-us/ windows/iot-core/ windows-iot-core`	The home site for Windows 10 IoT Core.
Open Hardware	`www.hackster.io`	General how-to articles, including a number of Windows 10 IoT Core example projects.
General Electronics	`www.adafruit.com`	An online electronics store offering a library of how-to articles covering interesting hardware, including Raspberry Pi, Arduino, and more.
General Electronics	`www.sparkfun.com`	An online electronics store offering a library of how-to articles covering interesting hardware, including Raspberry Pi, Arduino, and more.
Raspberry Pi	`https://thepihut.com/`	An online electronics store offering products for the Raspberry Pi. They also have a number of blogs and tutorials on using the Raspberry Pi in its native operating system.
Raspberry Pi	`www.raspberrypi.org`	An online community devoted to the Raspberry Pi.

Notice that there are websites for Windows 10,[3] Windows 10 IoT, Raspberry Pi, general hardware, and similar resources. Although many do not specifically address or cover Windows 10, most have a growing repository of knowledge that is often surprisingly helpful with Windows 10 IoT Core.

For example, I try to connect to the Raspberry Pi sites to keep tabs on what is going on there. I can often get ideas for IoT projects or simply ideas for features by seeing projects implemented for the Raspberry Pi and its native operating system. I find it is often the case that while the source code can be quite different, most of the hardware (connections, etc.) apply without modification, which makes sense since the hardware is not tied to the operating system running on the device (but the libraries that drive the hardware does).

Tip   Don't discount the power of a keyword search using your favorite online search tool. I have often found obscure gems of information that aren't posted on the more popular sites. In most cases, they are well-written blogs. I often start with a keyword search before I visit the sites listed in Table 17-1. If nothing else, it confirms whether what I am researching is unique or ubiquitous. You should also consider using key phrases from error messages as search terms to get help for specific errors.

There are also a few excellent periodicals that you should consider reading. I exclude the typical Windows periodicals—not because they aren't helpful but because they are far too general for IoT-related research. Rather, the following are periodicals that I've found to be very helpful in my IoT research or any electronics or hobby project.

- *The MagPi magazine*: A monthly magazine devoted to all things Raspberry Pi. Includes many articles on sample projects, news about the Raspberry Pi, peripherals, and general hardware reviews. (www.raspberrypi.org/magpi)

---

[3]Some are even a bit hostile toward Windows users. Should you encounter this absurd bias, simply refrain from replying in kind and ignore the negative remarks. I've found most who post derogatory commentary often know little of which they speak, basing their answers on innuendo, assumptions, and general inaccuracies.

- *Make*: A bimonthly magazine devoted to the broad realm of the maker community presenting sample projects, tutorials, hardware, tools, drones, robots, and more! It truly is a one-stop periodical for all things hacking, tweaking, and general DIY for hobbyists, enthusiasts, and professionals alike. (`http://makezine.com`)

Now that you've seen a number of online communities (but in no way a complete list, as more are added seemingly weekly), as well as a couple of periodicals that you can buy, let's discuss the next step that you can take once you have become a productive contributing member of IoT and IoT-related online communities—becoming a maker.

Become a Maker

The next progression in your growth from novice to enthusiast (and even professional) is to practice your craft regularly and share your knowledge with the world. One excellent outlet for this desire is to become a maker. Makers are widely regarded as experts in their areas of interest. The best part is that a maker's interests can vary greatly from one to another. That is, becoming a maker isn't about learning a specific set of techniques (academic or otherwise). It really is about practicing a craft.

What's a Maker?

Sadly, there is no single definition of what a maker is or should be. This is because a maker is someone with highly creative skills who desires to tinker and work on projects that can range from mechanical sculptures that spit fire to electronic gizmos to new ways to recycle materials to build things on the cheap.

Indeed, a maker is simply an artisan, a craftsman, a hobbyist, or an enthusiast who desires to create things—hence, *maker*. There are many kinds of makers. However, what unites them is the willingness to share their techniques and skills with others. As a result, to truly become a maker, you must participate in the community of makers.

Share Your Ideas

You can become a maker and not contribute at all to the online communities. And while that's perfectly fine, if you really want to help make the community stronger, you should become more involved. The best way to do this is to join one or more of the online forums and start contributing.

That doesn't mean you have to start off by posting some fantastically cool and successful IoT project complete with award-winning documentation and bulletproof code that computer science professors will someday teach young minds to mimic your brilliance. Sure, you may scoff (or even laugh) at that, but I have encountered people who are afraid of posting anything lest they be proven wrong or ridiculed for inaccuracies.

The best way to avoid that is to start out slow by first asking questions of your own. You can also start by throwing out a few positive comments for those ideas and projects you like. As you become more involved with the topics and techniques, your knowledge will expand to the point where you can start answering others' questions.

I encourage you to consider this level of involvement if you want to become more involved with Windows 10 IoT Core. I believe that interacting with the community and gaining a reputation for being helpful and sharing ideas is one of the things that separates a hobbyist from an enthusiast.

Thus, if you want to become a maker, you should share your ideas with others.

Attend an Event

Another way that you can become more involved with the maker community is to attend a Maker Faire (`www.makerfaire.com`). These events are held all over the world. The events allow makers to showcase their creations, teach others, and generally celebrate all things maker. See the Maker Faire website for events in your area.

If you live in or near a larger city, you may find that there are local user groups for Raspberry Pi, MySQL, Arduino, and Windows 10. There are even Maker user groups. Try searching for events and groups in your area to find out where and when they meet. Most groups have an open door policy and invite one and all to attend their meetings. Some, however, do have dues for charter members, but this often comes with additional perks such as access to tools, labs, and discounts on bulk purchases.

Once you have attended an event (even one that has nothing to do with Windows 10 IoT Core), such as a Raspberry Pi meetup, you'll be hooked. You can learn quite a lot from others this way. And who knows, perhaps, one day you will present at an event. Speaking from experience, it can be very satisfying sharing your knowledge with others in an open forum like a conference, user group, meetup, or Maker Faire. That's when you know you've obtained the reputation, skills, and knowledge that a typical maker possesses.

That doesn't mean that by the time you read this chapter you're an expert at all things IoT, or that you even profess to be an expert, but you should be well on your way all the same.

Taking the Next Step—IoT in the Enterprise

Thus far in the book, you've learned a number of new technologies as well as how to write applications to run on Windows 10 IoT Core and how to take the next steps to becoming more proficient in building IoT solutions. During our journey, you have seen a number of example projects that include writing user interfaces, reading sensors, and even how to store the data in MySQL.

These projects have used only a single IoT device, but could easily be expanded to several devices. For example, sensing the temperature and humidity in a large building or across a campus of buildings can be accomplished using multiple devices placed in key locations.

While we have discussed how to store the data in a database, we haven't discussed how to manage a large installation of many devices. Recall from Chapter 1, we learned there are additional Microsoft products designed for the enterprise such as Windows Server IoT 2019 and Windows 10 IoT Enterprise (there are several offerings of IoT Enterprise for purchase as well).

These products are clearly targeted for the enterprise and require a license to use, making them more expensive and therefore not something the beginner would necessarily consider. More specifically, you would have to either purchase a Semi-Annual Channel (SAC) or Long-Term Servicing Channel (LTSC) license from Microsoft to use them. In the case of Windows Server IoT 2019, you should be part of the Microsoft developer network to take advantage of additional resources. That is, Windows Server IoT 2019 is only licensed through the OEM channel under special dedicated use rights.

WHAT ABOUT AZURE IOT?

Aside from these offerings, Microsoft also has a host of IoT products in Microsoft Azure. We had a look at Azure in Chapters 15 and 16. We focused on learning about the basic IoT products in Azure, but as mentioned, there are additional IoT products in Azure that you can leverage should you choose to host your IoT solution in Azure. See the Azure documentation for more details: `https://azure.microsoft.com/en-us/product-categories/iot/`.

In this section, we will take a closer look at the Microsoft IoT products for the enterprise. We won't go into a lot of detail because much of the nuances of setting up and using the operating systems are beyond the scope of a beginner's book. And, as you may have surmised, the hardware for these products differs. We'll take a look at one example of an enterprise IoT board available to run the advanced IoT products.

Let's begin with a deeper look at the Microsoft enterprise offerings.

Enterprise-Grade Systems

In this section, we will take a closer look at the two enterprise offerings from Microsoft as well as compare Windows IoT Core and Windows IoT Enterprise.

Windows 10 IoT Enterprise

Windows 10 IoT Enterprise does not run on your low-cost hardware; rather, it is used to provide manageability and security for enterprise of IoT solutions. It is a full version of Windows 10 and is binary "equivalent to Windows 10 Enterprise," so you can expect the same experience and use all of the same tools as you would on your PC. The difference is in licensing (it is a paid product), and it is intended for use by original equipment manufacturers (OEMs). It is the "next step" in evolving your solution for the enterprise.

Like Windows 10 IoT Core, Windows 10 IoT Enterprise is designed to allow you to run on low-cost computing boards (but not the Raspberry Pi as we will see), allowing you to create Windows applications that can perform IoT operations. Sample solutions include those for the industry, engineering, healthcare, finance, and more.

Tip   See `https://docs.microsoft.com/en-us/windows/iot-core/` `windows-iot-enterprise` for more information about this version.

So, what does it give you over Windows 10 IoT Core? It turns out, quite a lot. Let's look at a few briefly:

- *Full Windows development*: Not only can you develop universal applications, you can also develop using the more classic Windows development libraries and platform.

- *Better security*: You can add better login and user management features as well as device-level security—a must for any application that handles critical data.

- *Device management*: You can manage applications across devices for easier maintenance.

- *More lockdown features*: You can apply more policies to control access on your devices.

In short, if you need to leverage the power of Windows Enterprise, you should migrate to using Windows 10 IoT Enterprise on your devices. However, there are some differences in the hardware requirements. Table 17-2 lists the major differences you should consider when looking to upgrade to Windows 10 IoT Enterprise.

Table 17-2. *Windows 10 IoT Core and Windows 10 IoT Enterprise Comparison*

Feature	Windows 10 IoT Core	Windows 10 IoT Enterprise
Activation/License	Not required	SAC or LTSC required
Target Applications	IoT gateway, devices	Industrial, medical, engineering, point of sale, etc.
Classic Windows Dev	No	Yes
Universal Windows Dev	Yes	Yes
Modern Shell (desktop)	No	Yes
Processor	x86, ARM	x86

Finally, recall that Windows 10 IoT Enterprise is a paid product, and licensing costs will be incurred to use it. Fortunately, Microsoft has several levels you can choose from to keep costs to a minimum. Contact your Microsoft reseller for more details.

But wait, there's more. Developing applications using Windows 10 IoT Enterprise isn't much different than Windows 10 IoT Core save for one very important detail. Most of the enterprise-grade IoT boards will have their own device libraries that must be built into the system image. This process is a bit tedious and involves ultimately creating a special build of the operating system for you to deploy (recall the `flash.ffu` file we used in previous chapters).

Fortunately, Microsoft has outlined the process in a lab experiment in their documentation. See `https://docs.microsoft.com/en-us/windows-hardware/manufacture/desktop/iot-ent-overview` for more details. Or, better, check with your board vendor for their specific set of instructions. You will find the process isn't difficult once you've done it a few times,[4] and you'll see how you can tailor your build to remove or add features including your own applications. Cool!

Windows Server IoT 2019

Recall from Chapter 2, Windows Server IoT 2019 runs on your PC server hardware. It is a special version of the Windows Server 2019 and uses the same tools and has the same experience as the Windows Server 2019. But what does that really mean and what does it give you that you cannot get otherwise?

When IoT solutions grow beyond the limitations of small, dedicated, and independent devices, your demands for more storage, better connectivity, and more processing power will increase. While you can mitigate that somewhat for small solutions, once the solution expands beyond a few hundred (or maybe a few thousand for those skilled developers), you will need something more to manage the devices and applications.

Having the ability to incorporate the powerful features of Windows Server into your solution solves many of those problems. With Windows Server IoT 2019 running on key nodes in your solution, you can handle larger workloads, process more data, and interact with the cloud more easily. Not only that, but with enterprise-grade security and administration, plus high availability features, Windows Server IoT 2019 can fulfill enterprise-grade IoT demands.

---

[4]The process is well beyond the scope of a beginner's book. See the link for more details.

Here are just a few of the possible solutions you can build by incorporating Windows Server IoT 2019:

- Automation of industrial processes and hardware

- Security solutions that integrate video and audio surveillance

- Drone-based data collection and analysis such as real-time traffic and routing

- Healthcare applications for managing medical equipment, processes, and material

- Energy management solutions for a single or many thousands of buildings or municipalities

- Disaster relief efforts

The list is almost endless of what is possible. However, all of these examples have at the core needs for enterprise-grade data storage and processing as well as the need for massive communication to/from devices and, in the case of safety and health (and most), a need for reliability, recovery, and redundancy.

Now I think you can see why Microsoft wants you to join the developer network and become an OEM subscriber. The sophistication needed to implement and properly develop a solution like these is more complex than simply making your devices more powerful. There is a lot to consider and more to incorporate into your designs.

Tip   For more information about this version, see the excellent introductory video at `https://channel9.msdn.com/Shows/Internet-of-Things-Show/Introduction-to-Windows-Server-IoT-2019`.

If you would like more information about Windows Server IoT 2019, see the frequently asked questions at `www.arrow.com/arrow-services/msembedded/wp-content/uploads/sites/3/2020/05/Windows-Server-IoT-2019-FAQ_feb18.pdf` to learn more about the features and some of the challenges to getting started.

Enterprise-Grade Hardware

In Chapter 2, we discovered the hardware available for Windows 10 IoT Core, which included our trusty Raspberry Pi, but it also included a number of enterprise-grade IoT boards. In this section, we will take a look at one enterprise IoT board that hosts a surprising array of features and the power to drive them. These boards include the AAEON[5] UP Squared (compact Intel Atom PC for serious makers), Keith & Koep i-PAN M7/T7 CoverLens (touch panel computers), and NXP i.MX 6/7/8 (multi-core ARM processor boards for advanced graphics, imaging, machine vision, audio, voice, video, and safety-critical applications).

Let's look at one specific board: the AAEON UP Squared board. Figures 17-2 through 17-6 show this board without peripherals from different angles.

Figure 17-2. *AAEON UP Squared IoT Enterprise Board—view 1*

---

[5]An Asus company, so they have a well-defined and respected pedigree.

Figure 17-3. *AAEON UP Squared IoT Enterprise Board—view 2*

Figure 17-4. *AAEON UP Squared IoT Enterprise Board—view 3*

Figure 17-5. *AAEON UP Squared IoT Enterprise Board—view 4*

Figure 17-6. *AAEON UP Squared IoT Enterprise Board—view 5*

Notice in the photos that the board has an impressive array of interfaces as well as a massive, finned heat shield on the bottom of the board. During testing, this board never got warm, so even in the most demanding environments, cooling shouldn't be a problem (you can always direct a fan across the fins).

Diving into the details, we find the board can be configured with several processor options including Intel Celeron, Pentium, and Atom Processors (formerly Apollo Lake).

It comes configured with 2GB, 4GB, or 8GB LPDDR4 and 32GB, 64GB, or 128GB eMMC memory. The expected 40-pin GP-bus provides all you would need to interface with your IoT sensors and actuators. There is even a 60-pin EXHAT for embedded applications. See `https://github.com/up-board/up-community/wiki/Pinout_UP2` for more information about these interfaces.

If that wasn't impressive enough, consider this board also has expansion capabilities that include a native mini-PCI-e slot (with PCIe, USB, and mSATA capabilities), M.2 2230 slot (for Wi-Fi/ Bluetooth modules), and a SATA3 connector for SSD/HDD.

You can find all of these connectors on the board easily (it's only slightly wider than a Raspberry Pi) including two Ethernet connectors, a DVI and HDMI connector for displays, three USB 3.0 ports, one micro-USB 3.0 port, and an internal header for two additional USB 2.0 ports and UART debugging interface, which make the board practical to use with your existing hardware. The following lists the details of the features for the version shown earlier:

- Intel Pentium N4200/Celeron N3350/Intel Atom E3950 Processor SoC

- Onboard LPDDR4 Memory up to 8GB

- Onboard eMMC Storage up to 128GB

- Gigabit LAN x 2, USB 2.0 x 2, USB 3.0 (Type A) x 3, micro-USB 3.0 x 1

- 5V DC-in, 4-6A

- 40-pin GPIO x 1, UART debug x 1

- DP, eDP, and HDMI

- MIPI-CSI (4-Lane and 2-Lane) x 1

- Mini-PCIe /M.2 2230 x 1

- SATA3 x 1

- 60-pin EXHAT

Clearly, this is an advanced board capable of supporting far more than a few sensors. In fact, it is an ideal board for building advanced enterprise solutions for any IoT solution.

Tip   These boards typically cost several times the price of the Raspberry Pi but pack a lot more power and peripheral options.

In using the board, I found it assembled well, the vendor documentation most excellent, and the company a joy to work with. If you need an enterprise IoT board for your solution, don't hesitate to contact them at `www.aaeon.com/en/p/iot-gateway-maker-boards-up-squared`.

You can also find these boards at some online retailers such as Mouser (`www.mouser.com/new/aaeon-up/aaeon-up2-squared-boards/`).

Summary

Taking their IoT skills to the next level is what most people are ultimately inspired to achieve. However, even if you do not want to become a traveling maker, teaching the world, you can learn quite a lot by simply joining the Windows, Windows 10, IoT, and open hardware online communities.

In this chapter, I presented suggestions on how best to interact with online communities, how to join an online community, and how to take your skills to the highest level of enthusiasm and become a maker. We also took a look at the software offerings from Microsoft for building enterprise-grade IoT solutions including an example of an advanced IoT device you can use to build those solutions.

This chapter therefore rounds out our journey down the Windows 10 IoT road. I hope that I have inspired you to continue practicing your skills and to join the community of like-minded enthusiasts.

I sincerely hope that this book has opened many doors for you concerning Windows 10, IoT, and anything open hardware. May your IoT projects all be successful, but when they aren't, may you learn something in the process. Just remember to share your experiences—good or bad—with the world and to give back a little of what you can take from the efforts of others.

Appendix

This appendix presents a list of the hardware required to complete each chapter, along with a consolidated list for the book. The appendix concludes with suggestions regarding acquiring the hardware needed. While these lists are included in each chapter and discussed in greater detail, listing them here helps to see all of the hardware used in the book as a set and helps when planning to purchase the components that you do not already own.

Hardware by Chapter

This section presents a list of the hardware for each chapter that contains a project or sample that you can try on your own. We begin with a look at the base components that apply to all of the chapters with projects.

Base Components for All Chapters

The following lists the hardware needed for all the projects in the book. That is, you should acquire these items to complete the projects in this book:

- Raspberry Pi 2 or 3, MinnowBoard Max–compatible, or DragonBoard 410C

- Wi-Fi dongle (Raspberry compatible) or Ethernet cable

- 5V 2A power supply

- Micro-SD card (minimum of 8GB)

- Breadboard (half or full sized)

- Jumper wires: male-to-female

- Jumper wires: male-to-male

© Charles Bell 2021
C. Bell, *Windows 10 for the Internet of Things*, https://doi.org/10.1007/978-1-4842-6609-0

Some projects that are *headed*, meaning they have a user interface that you can interact with that runs on your device:

- HDMI monitor

- USB keyboard (Raspberry compatible)

- USB mouse (Raspberry compatible)

The following are accessories that you may want to acquire that could make the projects a bit more fun or perhaps enhance your experience:

- Enclosure for your device

- Motorola Lapdock (to provide a monitor, keyboard, and mouse)

- USB hub

Required Hardware by Chapter

This section presents the required hardware list for each chapter that has a project. I include a short synopsis of the main project for each chapter. Those chapters without projects are omitted.

Note   Each chapter is annotated as *headed* (requires a monitor, keyboard, and mouse) or as *headless* (no user interface hardware is needed).

Chapter 3: Headless

This chapter explores the origins of the Raspberry Pi, including a tour of the hardware and a short primer on how to use its native operating system. The chapter demonstrates how easy it is to write programs to control hardware on the Raspberry Pi using a Python script. The project solution is a simple script to turn an LED on and off.

- (1) 560 ohm 5% 1/4W resistor

- (1) 10mm red LED

- (1) breadboard

- (2) jumper wires: male-to-female

Chapter 5: Headed

This chapter provides a crash course in Visual C++ that covers the basics of syntax and constructs of a Visual C++ application, including a walk-through of building a real C++ application that blinks an LED. The project covers XAML, including how to wire events to controls, and even a little about how to use the dispatcher timer.

- (1) 560 ohm 5% 1/4W resistor
- (1) 10mm red LED
- (1) breadboard
- (2) jumper wires: male-to-female

Chapter 6: Headed

This chapter provides a crash course in C# that covers the basics of syntax and constructs of a Visual C# application, including a walk-through of building a real C# application that blinks an LED. The project covers XAML including how to wire events to controls and even a little about how to use the dispatcher timer.

- (1) 560 ohm 5% 1/4W resistor
- (1) 10mm red LED
- (1) breadboard
- (2) jumper wires: male-to-female

Chapter 7: Headless

This chapter provides a crash course in Visual Basic that covers the basics of syntax and constructs of a Visual Basic application, including a walk-through of building a real Visual Basic application that blinks an LED. The project covers how to work with headless applications, including how to manage a startup background application.

- (1) 560 ohm 5% 1/4W resistor
- (1) 10mm red LED
- (1) breadboard
- (2) jumper wires: male-to-female

Chapter 9: Headless

This chapter introduces the Adafruit Microsoft IoT Pack for Raspberry Pi. You look at a project that shows how to read a sensor, thereby making the transition from experiments to actual, usable projects. This chapter shows that the kit is a viable option for getting the right amount of hardware to start building IoT solutions.

- (1) pushbutton (breadboard pin spacing)

- (2) red LEDs

- (2) yellow LEDs (or blue is OK)

- (1) green LED

- (5) 150 ohm resistors (or equivalent to match LEDs)

- (7) jumper wires: male-to-female

Chapter 10: Headless

This chapter presents how to use an ADC, how to connect and set up an SPI device, how to read a potentiometer, and finally how to use the debug feature to write out statements to the output window. While the project itself is rather simplistic, the emphasis and therefore the learning part of the project lie in discovering how to write code for all of these features.

- (1) 10K ohm potentiometer (breadboard pin spacing)

- (2) red LEDs

- (2) yellow LEDs (or blue is OK)

- (1) green LED

- (5) 150 ohm resistors (or appropriate for your LEDs)

- (1) MCP3008 ADC chip

- (19) jumper wires: (8) male-to-male, (11) male-to-female

Chapter 11: Headed

This chapter presents how to use a special library, Microsoft IoT Lightning Providers, to get access to PWM and SPI interfaces. The project introduces how to use the new library to interface with the ADC via an SPI interface, how to read values from an LDR, and how to use PWM to control the brightness of an LED. While the project itself is rather simplistic, the code is another example of more complex programming for IoT solutions.

- (1) LED (any color)
- (1) 10K ohm resistor
- (1) 150 ohm resistor (or appropriate for your LEDs)
- (1) light-dependent resistor (photocell)
- (1) MCP3008 ADC chip
- (13) jumper wires: (5) male-to-male, (8) male-to-female

Chapter 12: Headed

This chapter presents a depth of complexity that demonstrates how to combine a number of advanced tools and techniques from using a code library written by someone else to building a C++ headed application to incorporating a C# and C++ project in the same solution to read weather sensors (a very popular choice for an IoT project). Combining all of these together makes this project the most complex in the book. It provides the best example of the power of Visual Studio and UWP IoT applications.

- (1) Adafruit BMP280 I2C or SPI barometric pressure and altitude sensor
- (4) jumper wires: male-to-female

Chapter 13: Headless

This chapter introduces the concept of storing IoT data in a database. It introduces MySQL, presenting a short tutorial on how to connect your device to MySQL for storing data. The project shows how to add a database component to a C# project as a means to use MySQL to store data and thus write more IoT projects that can persist data for later retrieval and analysis using proven technologies.

- (1) Adafruit BMP280 I2C or SPI barometric pressure and altitude sensor
- (4) jumper wires: male-to-female

Chapter 14: Headless

This chapter takes a minor off-ramp from our regularly scheduled Windows 10 IoT project highway to examine how to control hardware remotely via the Internet. The project for this chapter demonstrates how to build a nifty out-of-office sign with a mechanical flag and LEDs controlled from a web page. This represents the fundamental building blocks for other remote controlled IoT projects.

- (1) SparkFun Servo Trigger (`www.sparkfun.com/products/13118`)
- (1) servo (`www.adafruit.com/products/169` or `www.sparkfun.com/products/9065`)
- (4) red LEDs
- (1) green LED
- (5) 150 ohm resistors
- (1) solder breadboard (optional)
- (11) jumper wires: (10) male-to-female, (1) male-to-male

Chapter 16: Headless

This chapter introduces an optional foray into the world of cloud computing by demonstrating how to incorporate an Azure remote monitoring solution to present data collected from a weather sensor. The chapter shows how easy it is to expand your IoT solution from a simple, local network solution to a worldwide cloud-based solution.

- (1) Adafruit BMP280 I2C or SPI barometric pressure and altitude sensor
- (4) jumper wires: male-to-female

Consolidated Hardware List

This section presents a table that lists the hardware needed to complete the core projects in this book that should be sufficient to satisfy (and challenge) beginners and intermediate readers alike. Chapters 14, 15, and 16 are optional for most readers, due to the added complexity and additional hardware requirements.

Table A-1 present the required hardware for all projects. If the component can be found in a kit, the tables list the source for purchasing the components as a kit. However, most components can be purchased from an online or retail electronics store, such as Adafruit or SparkFun. In some cases, the component is from a specific vendor.

Once again, it is assumed you will have purchased a board (e.g., Raspberry Pi), a power adapter, and so forth, as described in the "Base Components for All Chapters" section.

Table A-1. *Required Hardware (Through Chapter 13)*

Component	Quantity	Sources
Breadboard (half or full)	1	Adafruit Microsoft IoT Pack for Raspberry Pi
Jumper wires: male-to-female	1 set	Adafruit Microsoft IoT Pack for Raspberry Pi
Jumper wires: male-to-male	1 set	Adafruit Microsoft IoT Pack for Raspberry Pi
LEDs (red)	4	Adafruit Microsoft IoT Pack for Raspberry Pi
LEDs (green)	2	Adafruit Microsoft IoT Pack for Raspberry Pi
LEDs (yellow)	2	Adafruit Microsoft IoT Pack for Raspberry Pi
10K ohm resistor	1	Adafruit, SparkFun
560 ohm resistors	1	Adafruit Microsoft IoT Pack for Raspberry Pi
150 ohm resistors	5	Adafruit, SparkFun
Adafruit BMP280 I2C or SPI barometric pressure and altitude sensor	1	Adafruit Microsoft IoT Pack for Raspberry Pi
MCP3008 ADC chip	1	Adafruit Microsoft IoT Pack for Raspberry Pi
Light-dependent resistor (photocell)	1	Adafruit Microsoft IoT Pack for Raspberry Pi

(*continued*)

Table A-1. (*continued*)

Component	Quantity	Sources
10K ohm potentiometer (breadboard pin spacing)	1	Adafruit Microsoft IoT Pack for Raspberry Pi
Pushbutton (breadboard pin spacing)	1	Adafruit Microsoft IoT Pack for Raspberry Pi
SparkFun Servo Trigger	1	`www.sparkfun.com/products/13118`
Micro hobby servo	1	`www.adafruit.com/products/169` or `www.sparkfun.com/products/9065`

Suggestions for Purchasing the Hardware

In Chapter 9, you discover the main source for the components as the Adafruit Microsoft IoT Pack for Raspberry Pi, which contains all of the hardware you will need for the projects through Chapter 13 as well as Chapter 16. Only Chapters 14 and 15 require hardware components not found in the kit. However, there is an alternative, as described in the sidebar in Chapter 9. I repeat the discussion on both options next for reference.

Adafruit Microsoft IoT Pack for Raspberry Pi

The Microsoft IoT Pack for Raspberry Pi 3 comes in two varieties: one with the Raspberry Pi (`www.adafruit.com/products/2733`) and one without the Raspberry Pi (`www.adafruit.com/products/2702`) for those who already own a Raspberry Pi 2 or 3.

The kit with the Raspberry Pi costs about $114.95, and the kit without the Raspberry Pi costs about $75.00. Clearly, if you already have a Raspberry Pi, you can save some money there. In fact, for those who want to use a different low-cost computer board, you can buy the kit without the Raspberry Pi—except for the micro-SD card with Windows 10 and possibly the power supply, all of the other components will work with other boards. The kit comes with a number of handy components, including prototyping tools and a few sensors.

Note You will not need every component included in the kit, but the extra parts will come in handy when developing your own IoT projects.

There are three categories of components: electronic components included in the kit, accessories for the Raspberry Pi, and sensors.

The electronic components provided in the kit include the following:

- (2) breadboard trim potentiometers

- (5) 10K 5% 1/4W resistors

- (5) 560 ohm 5% 1/4W resistors

- (2) diffused 10mm blue LEDs

- (2) diffused 10mm red LEDs

- (2) diffused 10mm green LEDs

- (1) electrolytic capacitor: 1.0uF

- (3) 12mm tactile switches

The list of accessories in the kit is long. The following includes all of the accessories included in the kit. I describe some of these in more detail:

- *Adafruit Raspberry Pi B+ case—smoke base/clear top*: An excellent case to protect your Pi from accidents.

- *Full-sized breadboard*: Plenty of space to spread out your circuits.

- *Premium male-to-male jumper wires, 20 × 6 inches (150mm)*: Jumps from one port to another on the breadboard. They're extra long and come molded in a ribbon so you can peel off only those you need.

- *Premium female-to-male extension jumper wires, 20 × 6 inches*: Jumps from male GPIO pins to the breadboard ports. They also come molded in a ribbon.

- *Miniature Wi-Fi module*: A Raspberry Pi–approved Wi-Fi dongle (not needed for the Raspberry Pi 3).

- *5V 2A switching power supply with a 6-foot micro-USB cable*: Meets the Raspberry Pi requirements for power.

- *MCP3008—8-channel 10-bit ADC with SPI interface*: A breakout board you can use to expand the number of SPI interface channels for larger IoT projects.

- *Ethernet cable, 5-foot*: A nice touch considering the kit has a Wi-Fi dongle. Good to have a backup plan!

- *8GB class 10 SD/micro-SD memory card*: Windows 10 IoT core preloaded!

The sensors included with the kit are an unexpected surprise. They provide what you need to create some interesting IoT solutions. Best of all, they are packaged as breakout boards, making them easy to wire into our circuits. The following lists the sensors included in the kit:

- *(1) photocell*: A simple component to measure light

- *Assembled Adafruit BME280 temperature, pressure, and humidity sensor*: Measures temperature, barometric pressure, and humidity

- *Assembled TCS34725 RGB color sensor*: Measures color. Comes with an infrared filter and white LED

Adafruit Parts Pal

If you are planning to use a board other than the Raspberry Pi, are on a more limited hobby budget, or want only the bare essentials, there are alternatives to the Microsoft IoT Pack from Adafruit. In fact, Adafruit sells another kit that includes almost everything you need for the projects in this book. It doesn't come with sensors, but all of the basic bits and bobs are in there, and you can always buy the sensors separately.

The Adafruit Parts Pal comes packaged in a small plastic case with a host of electronic components (`www.adafruit.com/products/2975`). The kit includes the following components: prototyping tools, LEDs, capacitors, resistors, some basic sensors, and more. In fact, there are more components in this kit than the Windows IoT Pack for the Raspberry Pi 3. Better still, the kit costs only $19.95, which means it's a good deal (and the case is a great bonus). Although you may not need all of the parts in this kit, it offers a great start for building a supply of electronics parts for future projects.

- (1) storage box with latch

- (1) half-sized breadboard

- (20) jumper wires: male-to-male, 3 inches (75mm)

- (10) jumper wires: male-to-male, 6 inches (150mm)

- (5) 5mm diffused green LEDs

- (5) 5mm diffused red LEDs

- (1) 10mm diffused common-anode RGB LED

- (10) 1.0uF ceramic capacitors

- (10) 0.1uF ceramic capacitors

- (10) 0.01uF ceramic capacitors

- (5) 10uF 50V electrolytic capacitors

- (5) 100uF 16V electrolytic capacitors

- (10) 560 ohm 5% axial resistors

- (10) 1K ohm 5% axial resistors

- (10) 10K ohm 5% axial resistors

- (10) 47K ohm 5% axial resistors

- (5) 1N4001 diodcs

- (5) 1N4148 signal diodes

- (5) NPN transistor PN2222 TO-92

- (5) PNP transistor PN2907 TO-92

- (2) 5V 1.5A linear voltage regulator, 7805 TO-220

- (1) 3.3V 800mA linear voltage regulator, LD1117-3.3 TO-220

- (1) TLC555 wide-voltage range, low-power 555 timer

- (1) photocell

- (1) thermistor (breadboard version)

- (1) vibration sensor switch

- (1) 10K breadboard trim potentiometer
- (1) 1K breadboard trim potentiometer
- (1) Piezo buzzer
- (5) 6mm tactile switches
- (3) SPDT slide switches
- (1) 40-pin break-away male header strip
- (1) 40-pin female header strip

The only thing that I feel is missing is the male-to-female jumpers, but you can buy them separately (`www.adafruit.com/product/1954`). For only $1.95 more, they're worth adding to your order!

Index

A

Accelerometers, 341
Accessories, 42, 84, 90, 93, 345, 657
Adafruit, 348
    accessories, 350
    color sensor, 352
    components, 349
    environmental sensor, 351
    MCP3008, 352, 353
Administrative operations, 106
    change ownership, 107
    install/remove software, 108
    run, 106
    shutdown, 109
Alternating current (AC), 308
Analog sensors, 339, 340
Analog-to-digital converter (ADC), 381, 405
Arrow DragonBoard 410c, 34, 35, 37
Audio sensors, 341
az extension add --name azure-iot
        command, 561
az iot hub list command, 561
az login command, 561
az logout command, 561
Azure IoT Hub, 549, 564, 566
Azure IoT solutions, 580
    CLI, 593
    cloud
        code, 593
        data received, 597, 598, 601

        deploy/execute, 597
        Hub class, 594, 595
        Microsoft.Azure.Devices.Client
            NuGet package, 594
    configure, 581
        complete code, 585–587
        components, 581
        hardware, 582
        new project, 583
        sensor class, 584, 585
    create device in Hub, 591, 592
    create Hub, 589, 590
    sensor, deploy/execute, 588

B

Barcode readers, 341
Bar graph LEDs, 382
Begin() method, 445
Biometric sensors, 342
BlinkTimer_Tick() method, 250, 296
BME280 sensor, 351, 436
Breakout boards, 324, 333, 334

C

C++
    arithmetic, 181, 182
    comments, 177, 178
    Curly braces, 175
    definition, 169, 170

661

Printed in the United States
by Baker & Taylor Publisher Services